HAPHAZARD REALITY

THIS BOOK IS PUBLISHED AS PART
OF AN ALFRED P. SLOAN FOUNDATION PROGRAM.

HENDRIK CASIMIR

HAPHAZARD
REALITY

Half a Century of Science

HARPER & ROW, PUBLISHERS, New York
Cambridge, Philadelphia, San Francisco, London,
Mexico City, São Paulo, Sydney

1817

To the memory of my three teachers
Paul Ehrenfest
Niels Bohr
Wolfgang Pauli

HAPHAZARD REALITY. Copyright © 1983 by Hendrik B. G. Casimir. All rights reserved. Printed in the United States of America. No part of this book may be used or reproduced in any manner whatsoever without written permission except in the case of brief quotations embodied in critical articles and reviews. For information address Harper & Row, Publishers, Inc., 10 East 53rd Street, New York, N.Y. 10022. Published simultaneously in Canada by Fitzhenry & Whiteside Limited, Toronto, and simultaneously in Great Britain by Harper & Row Limited, 28 Tavistock Street, London WC2E 7PN, and simultaneously in Australia and New Zealand by Harper & Row (Australasia) Pty. Limited, P.O. Box 226, Artarmon, New South Wales, 2064.

FIRST EDITION

Designer: Sidney Feinberg

Library of Congress Cataloging in Publication Data
Casimir, H. B. G. (Hendrik Brugt Gerhard), 1909–
   Haphazard reality.
   "A Sloan Foundation book."
   Includes bibliographical references and index.
   1. Casimir, H. B. G. (Hendrik Brugt Gerhard),
1909– . 2. Physicists—Netherlands—Biography.
3. Physics—History. 4. Technology—History. I. Title.
QC16.C36A33   1983   530'.092'4 [B]   82–48112
ISBN 0–06–015028–9 (U.S.A. and Canada) 83 84 85 86 87 10 9 8 7 6 5 4 3 2 1
ISBN 0–06–337031–X (except U.S.A. and Canada)
                     83 84 85 86 87 10 9 8 7 6 5 4 3 2 1

Contents

Preface to the Series

The Alfred P. Sloan Foundation has for many years included in its areas of interest the encouragement of a public understanding of science. It is an area in which it is most difficult to spend money effectively. Science in this century has become a complex endeavor. Scientific statements are embedded in a context that may look back over as many as four centuries of cunning experiment and elaborate theory; they are as likely as not to be expressible only in the language of advanced mathematics. The goal of a general public understanding of science, which may have been reasonable a hundred years ago, is perhaps by now chimerical.

Yet an understanding of the scientific enterprise, as distinct from the data and concepts and theories of science itself, is certainly within the grasp of us all. It is, after all, an enterprise conducted by men and women who might be our neighbors, going to and from their workplaces day by day, stimulated by hopes and purposes that are common to all of us, rewarded as most of us are by occasional successes and distressed by occasional setbacks. It is an enterprise with its own rules and customs, but an understanding of that enterprise is accessible to any of us, for it is quintessentially human. And an understanding of the enterprise inevitably brings with it some insights into the nature of its products.

Accordingly, the Sloan Foundation has set out to encourage a representative selection of accomplished and articulate scientists to set down their own accounts of their lives in science. The form those accounts will take has been left in each instance to the author: one may choose an autobiographical approach, another may produce a coherent series of essays, a third may tell the tale of a scientific community of which he was a member. Each author is a man or woman of outstanding accomplishment in his or her field. The word "science" is not construed narrowly: it includes activities in the neighborhood of science such as technology and engineering as well as such disciplines as economics and anthropology as much as it includes physics and chemistry and biology.

The Foundation wishes to express its appreciation of the great and continuing contribution made to the program by its Advisory Committee. The Committee has been chaired since the program's inception by Robert Sinsheimer, Chancellor of the University of California—Santa Cruz. Present members of the Committee are Howard Hiatt, Dean, Harvard School of Public Health; Mark Kac, Professor of Mathematics, University of Southern California; Daniel Kevles, Professor of History, California Institute of Technology; Robert Merton, University Professor Emeritus and Special Service Professor, Columbia University; George A. Miller, Professor of Psychology, Princeton University. Earlier, Daniel McFadden, Professor of Economics, and Philip Morrison, Professor of Physics, both of the Massachusetts Institute of Technology, and Frederick E. Terman, Provost Emeritus, Stanford University, had been members. The Foundation has been represented throughout by Arthur L. Singer, Jr., and Stephen White, and Harper & Row, principal publishers for the program, by Winthrop Knowlton, Simon Michael Bessie and Edward L. Burlingame.

—ALBERT REES
President, Alfred P. Sloan Foundation

Author's Preface

Old men like to tell stories about the past, are critical about the present and worry about the future. I am no exception to that rule, and after my retirement—in 1972—from an active job in industrial research I had plenty of opportunities to indulge these tendencies. I lectured on the development of physics as I had seen it, as well as on various aspects of the relations between science, industry and society, and I took part in meetings and sat on committees dealing with such matters. However, although some of my lectures were published, these activities would not have resulted in a book but for the support and the pressure provided by the Alfred P. Sloan Foundation. I am in particular indebted to Stephen White, vice president, and to Mark Kac, member of the Advisory Committee, for their encouragement.

Other factors helped. From January to May 1979 I held the Erasmus Chair at Harvard, and the lectures I gave there provided the material for several of the chapters of this book. When a year later I gave six lectures at the University of Minnesota at Minneapolis I had already made some headway in writing down my text, so this visit provided an excellent opportunity to present parts of my story to an interested audience. I should like to thank Roger Stuewer and his department for the invitation and for their generous hospitality.

Obviously a book like the present one owes much to informal discussions with colleagues. I trust my friends will forgive me for not having tried to compile a list of acknowledgments: it is in the nature of oral history that its transmitters remain largely anonymous. And if anyone should feel I have plagiarized, let me assure him he is most welcome to incorporate some of my stories and ideas into his own presentation.

I have had the good fortune to have known and even to have worked with some of the greatest figures of a great era of physics, and here a simple word of thanks would be inadequate. I can only hope that my text itself will bring out my feelings of gratitude.

Introduction

This is a personal narrative, but I have tried to write about people I met and knowledge I gained rather than about myself. The story of my own life is of no particular interest: the career of a man who studied under exceptionally favorable circumstances and became a fairly competent physicist, who went into industry and became a moderately satisfactory research manager, and who finally, after retirement, became the somewhat superficial and easygoing amateur historian and philosopher who wrote this book, is hardly an exciting subject for a biographer. But it has been my good fortune to meet, and even to be rather closely associated with, some of the great physicists of our century and to witness the spectacular growth of empirical knowledge and the burgeoning of profound theories, leading to a deeper understanding of nature as well as to surprising technological innovations. That is what I want to tell about: I believe my recollections are worth writing down. My own role in these developments has been a modest one, but a minor actor or supernumerary often has a better opportunity to watch great actors than the great actors themselves. Changing my metaphor, I might say that I am taking my readers on a guided tour through the history of physics and technology. The presence of the guide should give some coherence to what otherwise would become too rambling a walk, but the guide should not impose himself.

I had to write my story largely from memory: I never kept notes, never kept a diary, and I usually destroyed letters as soon as the matter at hand had been dealt with. Moreover, I wanted to write mainly about those things and those events where I have firsthand information, so there would have been no point in starting out with a systematic search in files and archives, something at which I should be bad anyway. However, once I had decided to tell a certain episode, I tried to find corroboration from standard sources and by private correspondence and conversation. In this I was often successful and, I am happy to say, I did not find serious discrepancies. Therefore I feel reasonably certain that my facts are correct even when I found nothing to confirm my recollections. I know this sounds conceited, so let me make it clear that my memory is far from infallible. I am not very good at remembering people and I am quite sure that I have forgotten many things at least as important and interesting as those I remember. My memory is highly selective and tends to recall unessential anecdotes, but what I remember I seem to remember correctly.

Some of my stories, however, are part of orally transmitted "folklore." I have made that clear by introducing them by "it is told" or "I have been told" or some similar expression. One should not despise such stories. Niels Bohr himself quoted on occasion the following aphorism (I am not certain, but its original author may have been Von Kármán, a great name in hydro- and aerodynamics): "When telling a true story one should not be overinfluenced by the haphazard occurrences of reality." (It sounds better in German: "Beim Erzählen einer wahren Geschichte soll man sich nicht zu sehr vom Zufall der Wirklichkeit beeinflussen lassen.") This will shock a professional historian, but if an anecdote sounds true—that is, if it illustrates an important trait of its hero, or makes some of his actions and ideas more understandable—then it may well be worth telling, even if it has no factual basis.

My own recollections, as well as documents supporting them, are expressed in several languages—Dutch, German, Danish, English, French—and I had to face the problem of translation. All translations are my own. A professional translator would have produced more fluent, more natural English, but I think it is no disadvantage that my translations retain some outlandish flavor. Here and elsewhere the reader will have to accept my little essay on broken English (p. 122) as an apology for my particular way of using the English language.

A physicist writing for a general reader has to face a more serious problem of "language." Many of the basic notions of physics are formulated in mathematical terms, and it is almost impossible to convey an adequate idea of the preoccupations of a physicist without recourse to that formalism. I have tried in Chapters 2 and 9 to sketch the development of physics up to 1930 and to give at least a rough outline of physics after World War II, but, as I state in the introduction to Chapter 2, the result may well be trivial to the expert and incomprehensible to the layman. Physicists among my readers will find some further details in the appendixes at the end of the book.

There is another side to my narrative. By going from fundamental research into industrial research management, I exemplified as a person what happens to ideas all the time. Nearly all the innovations of modern technology came into being thanks to preceding fundamental research. On the other hand modern experimental science could not exist without the tools made available by technology. I believe that I have been in a good position to observe this powerful science-technology spiral. It is the subject of my final chapter, one that will possibly be regarded by some readers as extraneous to the rest of the book. I do not see it that way. I am even inclined to take an opposite point of view and to regard the story of my own experiences mainly as an illustration of my ideas concerning the interdependence of fundamentally new knowledge and novel technology.

During my lifetime the attitude of scientists towards their science and the attitude of society towards scientists has changed. Scientists no longer feel safe and smugly virtuous in their ivory towers, and society no longer regards scientists either as ridiculous but innocent, or as admirable and potentially useful: it tends to regard them as potentially dangerous. As to technological progress, it is no longer regarded as an unmixed blessing. Justified ecological concern goes side by side with fanatic anti-intellectualism; "runaway technology" has become a commonplace term, and "alternative" subcultures abound. But all the while the dark clouds of the science-based technology of mass murder are gathering. Although I do not flatter myself that my words of warning will have any effect, I do hope that my analysis of scientific and technological progress may be of some use to people younger and more active than myself.

1

Family Background and Schooldays

Family Background

On 14 May 1864, Brugt Kasimier and Gepke Meersma were married at the town hall in Buitenpost, a village in the east of Friesland. The marriage certificate shows that the bridegroom was then thirty years of age and a farmhand by trade; his parents, Hendrik Brugts Kasimier and Grietje Oosterhof, were both deceased. His bride, a twenty-three-year-old housemaid, was the daughter of Hendrik Meersma and his wife Martje Miedema, carpenters, who had given their consent to the marriage and were present at the ceremony. The document contains the usual legal clauses and is signed by the bride and her parents, by four witnesses, and, of course, by the clerk at the town hall, who adds: "The bridegroom did not sign this deed, the same not having learned to write." That farmhand was to be my grandfather, and from the little I know about him it is clear that he must have been an intelligent and upright man and an exemplary husband and father. In his youth he had been more or less adopted by a family who lived on a farm; later he made a living as a day-laborer, but hated that work. Farmwork was very hard work in those days and he was not a strong man; he was also disgusted by its coarseness and its callous cruelty to animals. Between 1873 and 1875 the family moved to Kollum, the village where my grandmother had been born and where her ancestors had been living for many generations. At Kollum he found a new livelihood: he became a peddler. "Mei it pak rinne"—the Frisian expression means literally "to walk with the pack"—was then a far more respectable profession than it would be today. In those pre-motorcar days with little or no public transport and with farms spread out over the countryside, often a considerable distance from the nearest town, itinerant traders such as my grandfather were an important link in the distribution system. He dealt in textiles, represented a reliable firm in Kollum, and could sell various odds and ends but also fill substantial orders. He had some skill in repairing clocks, and, though he always emphasized that he was no professional clockmaker, this must have made him an even more welcome visitor at isolated farms. I like to imagine him on his long marches, carrying his heavy pack along endless windswept roads, under the wide skies of Friesland. A modest man, and often a tired one; a kindly man and scrupulously honest. He died in the spring of 1878, after a heart attack, when my father, the youngest of six (one more child had died in early infancy) was six months old. One of his last wishes was that, come what may, his sons should not become farmhands. They didn't!

Hendrik, the oldest son, was not quite thirteen years old at the time of his father's death, but he soon became both a competent carpenter and the undisputed head of the family. My father always spoke of him with the greatest respect, possibly mingled with a slight feeling of guilt. He felt that Hendrik had certainly not been less gifted than his two other brothers, one of whom became a prosperous businessman and the other a well-known architect, but that he had sacrificed his own chances in order to provide for the education of his brothers and sisters. Hendrik's own career was honorable but undistinguished. After military service he gave up carpentry and joined the constabulary. Later he became janitor at the court of justice in Groningen. Surprisingly, he spent some time in the United States. When the family doctor diagnosed the beginning of tuberculosis—at that time a frightful scourge, especially among the working classes—Hendrik resolutely set out for America, earned a living as a carpenter in New Mexico, and came back cured, or at least sufficiently cured to live to well into his eighties.

My father liked to say that, in his opinion, Hendrik had in the course of time acquired legal knowledge and juridical insight that would have put to shame many a professional lawyer and that it was no exception that a young lawyer would ask his advice on matters of procedure. Be that as it may, Hendrik's eldest son became a member of the high court of the Netherlands, and a university professor.

It is symptomatic of this age of wavering principles and slack

morals that the word "righteous" has obtained a connotation of prig-
gishness and intolerance, even of hypocrisy. But in its original and
nobler sense it expresses exactly what I want to say: Hendrik, like his
father before him, was a righteous man. I am proud to be named after
him.

My grandmother I remember only slightly—a tall, heavily built
woman of somewhat stern appearance. In her village she was considered to have married below her station: she came from a family of
carpenters and builders. She was a strong-willed person, with a head
for business. She must have encouraged my grandfather in his trade,
and after his death she continued to do some trading herself. "I am
good at 'rithmetic," she used to say. But her interests were much
broader; when she later heard from her grandchildren that one could
study mathematics and physics at a university, she said wistfully,
"that would have been something for me." She might have done well
in medicine too. She often stepped in when no doctor or midwife was
available and later said proudly that there had never been a case of
puerperal fever after the deliveries she had carried out. She then
related how on such occasions she had first cut her nails as short as
possible, had washed her hands thoroughly with ample soap and
rinsed them in boiled water, which must have been more than many
doctors did in those days. It remains a matter of conjecture how the
teachings of Semmelweiss had reached a working-class woman in a
remote Frisian village.

My grandmother was a strict Calvinist, and in those days there
was in the Nederlands Hervormde Kerk (Dutch Reformed Church)
considerable controversy on points of dogma, which in 1886 led to
the secession of the "Gereformeerde Kerk."* The spiritual leader of
this movement, Abraham Kuyper, was also the leading force behind
a new political party, the Anti-Revolutionary Party, which came to
play an important role in Dutch politics. My grandmother followed
these developments with keen interest. She was an avid reader, and
among her readings the books of Abraham Kuyper took an important
place, second only to the Bible. Her energy and her intelligence
made her an exceptional woman, but this type of theological interest

---

*The "Gereformeerde Kerken in Nederland" obtained their definitive form in
1892, when the "dissidents" mentioned above—they called themselves "dolerenden"
(from the Latin *doleo*, "I deplore")—were joined by the majority of the adherents of
a movement that went back to 1834.

was not exceptional. In this respect, there must have been a close resemblance between Frisian and Scottish villagers. In later life my father enjoyed reading the books of an author by now almost completely forgotten, Ian MacLaren (the pen name of the Reverend John Watson), a writer of the so-called Kailyard school, whose well-written though sentimental stories reminded him so much of his native Kollum.

It may seem farfetched, but I think there is also an analogy between the attitude of Dutch Calvinists and the traditions of Jewry. I do not claim to be able to explain fully the reasons for the preponderance of people of Jewish origin in many branches of learning, but I do believe that traditions of religious origin have been an important factor. Even among the poorest of Jews, of any walk of life, the rabbi studying the Talmud was a venerated figure; to study books and to deal with abstractions was not considered a waste of time but a worthwhile occupation, and from Talmud to higher mathematics is perhaps not such a large step. In a biography of David Sarnoff, the creator of RCA, I read that in the village near Minsk where he passed his childhood, it was considered a far worse disgrace to be an ignoramus—an "amhorets"—than to be poor, and he was apprenticed to a rabbi in whose home he almost starved. This attitude must have been prevalent all over the Jewish world. The theological interests of Frisian villagers may have played a similar role; at least they did so in the case of my father. Nobody really taught him to read, but he liked to look over his mother's shoulder while she was reading half aloud from her Bible. And one day—he was not yet five years of age—he surprised her by saying, "Wait a bit, I am not yet at the end of the page." My father forsook the Calvinist faith early in life, but books and reading remained a lifelong interest.

The village school in Kollum was presumably not outstanding, but there was something to be said for this type of institution. Since teachers had to look after several grades, the pupils got tasks they had to cope with on their own, and the really bright pupils could go on at whatever speed they liked. They learned to concentrate on their own work while lessons on an entirely different subject were going on in another corner of the classroom; they also acquired some teaching experience, since they were often called upon to explain things to slower classmates. Some features of the village school have later been reinvented by psychologically trained educationalists. In many

ways my father did all right, but at one time he suffered badly under
an unpleasant and possibly slightly sadistic schoolmaster, who really
ill-treated the children. When he was out of hearing, he was called
"Sjoerd Bokske" (Sjoerd the Little Billy Goat). One day some repair
work was going on at the school and the children had been strictly
forbidden to go near it. However, Hendrik was working there and
my father felt that he had a right to greet his brother. The teacher
spotted him, grasped hold of him, and more or less kicked him away.
My father became very angry and cried, "You ugly Sjoerd Bokske,
who do you think you are?" Whereupon he was sent away from
school immediately. He replied quietly, "That suits me all right; I
intended to leave this school anyway: they teach you nothing here."
The incident was somehow settled; I am afraid my father had to
apologize. My sisters and I loved the story and often asked to be told
about Sjoerd Bokske, and my father was happy to oblige. I believe
this teacher had a profound influence on his later career: his bad
example gave my father an incentive to become a good educator
himself.

A son of Sjoerd Bokske was a well-known notary in Eindhoven.
Came the day that my father was the government-appointed supervisor of school examinations at Eindhoven, and among the candidates there was the son of the notary, grandson of Sjoerd Bokske. My
father came back from the final session and, rubbing his hands, told
us, "Finally I got my revenge." "Did you fail him?" I asked. "Oh no,"
said my father. "Much better, he passed. He was a weak candidate
and I had to use all my skill and experience to let him cut a reasonably
good figure. Without my help he would almost certainly have failed."
"You should not be vindictive," said my mother, "not even in that
way."

In 1888 my father left school and became an apprentice in a
printing shop that published the local weekly. It was the most literate
profession Kollum had to offer. He liked to tell us with some pride
that he had set a record for memorizing the arrangement of the
typecase and that he had done the type-setting of a long letter to the
editor on the burning issue of whether children should or should not
take off their wooden shoes at school. And also of an obituary notice
for J.J.A. Goeverneur (1809–1889), a minor man of letters whose
comic epic about a butterfly collector, *Prikkebeen* (pronounced
"bane")—a free rendering of R. Töpfer's *Histoire de Monsieur Cryp-*

togame—has been the delight of many generations of children and their parents. Some lines, like the warning in the introduction:

> If something might seem strange to you
> It is printed, so it's true.

and his farewell letter to his sister:

> My dear sister Ursula,
> I'm going to America,
> For butterflies the promised land;
> I'm writing this with my own hand
> And, as ever, I remain
> Your devoted Prikkebeen.

have become almost as proverbial in the Dutch language as passages from Edward Lear's *Book of Nonsense* or from *Alice in Wonderland* in English.

Although my father stayed less than three months in that shop—far too short to become a skilled typographer—his apprenticeship did leave its mark. In later life, even when he was writing rapidly, the layout of his pages was admirable. Seen from a distance his manuscripts looked very neat, but from nearby they were disappointing: they were almost illegible. I have copied out many pages. He was also a good proofreader. His apprenticeship came to an end when in 1889 the family moved to Groningen. There he went back to school to complete his elementary education. Once, when the headmaster was ill, the teaching in the two highest grades was temporarily taken over by a young woman, Miss Borkhuis, a brilliant teacher, who at that time was preparing for a higher degree in French. She "discovered" my father and gave him French lessons after hours. I think she also helped him to correct his Frisian accent and somewhat uncouth manners, and she encouraged him to become a schoolmaster. My father was deeply grateful to her. In later years she would stay with our family from time to time and I remember her well. She was then a slight, prim spinster, so enamored of French that she even pronounced Dutch with a guttural Parisian *r*. Sometimes she would still correct my father's accent or manners, and my father, a broad-shouldered giant six feet and four inches tall, would humbly accept such remarks.

In order to become a schoolmaster one had to pass a state-supervised examination; there were, in those days, two ways to prepare for it. Either one could attend a "Kweekschool"—teacher's college—which provided full-time instruction, or one could follow courses, mainly in the evening or late afternoon, at a "Normaalschool," supplemented by practical work in an elementary school under supervision of the headmaster. After a brief period at a very poorly staffed denominational school, my father decided—to the dismay of his Calvinistic mother—in favor of the Rijksnormaal school (state school). There he attended the lessons from 1892 to 1896. The teaching left something to be desired, but it gave him plenty of time to read and study for himself. During those years he must have devoured an enormous number of books. Devoured and absorbed, for he was quick to understand and had a prodigious memory. In Groningen there was a good library organized by the Maatschappij tot Nut van 't Algemeen (the Society for the Common Good), a society founded in 1784 that had been very instrumental in raising the general level of education in the Netherlands.[1]

On 15 April 1896 my father passed the examination, getting 103 points out of a possible 110, perhaps an all-time record. (Marks in the Netherlands run traditionally from 1 to 10, 6 being just passable.) He dropped 3 points on writing, but even for music he managed a 9, although he could not sing in tune and did not play any instrument. The requirements were only theoretical and my father had had no difficulty in memorizing everything that might possibly be asked. This examination made him famous in a small way. My mother, then at Kweekschool, told us that a friend had pointed him out to her, saying, "There goes that fellow with the 103 points," to which she had reacted with, "You would never think he had it in him, would you?" During the next two years my father had several temporary appointments at elementary schools, and obtained a certificate for French in 1897 and the "Hoofdakte" (headmaster's certificate) in 1898. That certificate was required to become headmaster in an elementary school, and the examination was a tough one. Schools did not prepare for it; schoolmasters would usually read for it while fully employed, and for some the examination would be a yearly outing. They would try year after year and fail again and again. There existed various evening courses, and in 1899 my father was appointed as teacher at such a course. That provided him with better opportunities for further studies—he did not need much time for preparing his lessons. It was also there that he met my mother: she was one of his pupils.

To a modern reader these details about the Dutch educational system of those days may seem tedious and irrelevant; yet they were an essential feature of Dutch society. Throughout the nineteenth century and the beginning of the twentieth, almost the only road by which an intellectually gifted working-class boy could gain access to advanced education was by way of the teacher's certificate.* It was a main channel of social mobility. Many would get no further than schoolmaster or headmaster at an elementary school; their children might go to a university. Others obtained additional diplomas and became teachers at secondary schools. And a few, like J.D. van der Waals, who will turn up elsewhere in this book, and like my father, went on to an academic career.

All these schoolmasters had in common that they took learning seriously, that they were devoted to their studies, and that they believed in teaching. In those days some of the most brilliant intellects of the country were found among schoolmasters in elementary schools.

My father was above all a teacher, and his later accomplishments notwithstanding, he may have been at his very best when he taught future schoolmasters. J.B. Priestley in his autobiographical *Margin Released* devotes two moving pages to his father. They are almost literally applicable to *my* father, including the occasional fits of temper. I shall quote only two sentences: "He was no dictatorial and boring pedant: he was lively minded and companionable, always eager to learn himself; but what he knew—and he knew a lot—he had to teach. He believed in Education as few people nowadays believe in anything." You'd better read the remainder of those pages —and the rest of the book if you can find the time—for yourself. I have never been a writer of fan mail, but I should have liked to thank Mr. Priestley for those pages. Now it is too late.†

The next step my father had to take was to pass the appropriate examination qualifying for academic studies. In 1900 he enrolled as

*This was true especially for boys with theoretical and scholarly ambitions. For technicians there existed other possibilities. The teacher's certificate played also an important role in the emancipation of women.

†It would be rash to conclude from this one example that in England the career of schoolmaster played a similar role as in the Netherlands, but I have met several British physicists and engineers who were sons of teachers.

a student of philosophy at the University of Groningen.

Gerard Heymans (1857–1930) held the chair of philosophy at Groningen from 1890 to 1927. He had studied law and philosophy at Leiden University. His doctoral dissertation of 1880 was *Karakter en Methode der Staathuishoudkunde (Nature and Method of Economics)*, but he continued his studies in philosophy in Freiburg im Breisgau with Windelband, a well-known neo-Kantian, whose textbook on the history of philosophy became a classic. At Freiburg, Heymans obtained a second doctor's degree—this time in philosophy. Heymans wrote on many aspects of philosophy and he seems to have been an outstanding teacher, who did not hold up his own ideas as inviolable dogmas. He was also a pioneer in psychology and especially in experimental psychology. My father found in Heymans what is a most important thing for a young scholar or scientist to find: a man he thoroughly admired. My father was certainly interested in, and widely read in, philosophy. He even wrote a concise history of philosophy for the general reader, which went through several editions. I have among my books a copy of the second edition, with an inscription, two lines of verse, dated 1 December 1910 (I was then barely one and a half years old): "To Henk from his father. This book will be old when you, grown up, will read it; but the mind's searching will go on forever." He also said throughout his life that someday he would find the time to sit down and write a magnum opus, in which he would expound a new system of philosophy. But he never got down to it. It became almost a family joke. I am afraid that some of my own projects had a similar fate. Shall I be able to finish this book?

In any case, psychology was closer to my father's interests. For all his scholarly proficiency he was not destined to become a scholar. He wanted primarily to be an educator and an educator he remained. He duly passed his first examination in 1903, but it would be 1909 before he got his master's degree, and he never wrote a doctoral thesis. (The University of Ghent in Belgium later granted him an honorary degree.) For, in the meantime, many things had happened. Notwithstanding his teaching and his studies he had found time for other activities. He had published and lectured on questions of education; he was also a convinced teetotaler and was known to be an eloquent speaker on that subject. In later years he understood that I did not share his objections to alcohol, but he used to say that if I, as a child, had lived next door to a blacksmith who was often drunk when he came home and if I had heard the crying and sobbing of his wife and children when they were beaten up, I would have felt differently; my mother, who before her marriage had been a schoolteacher in a notoriously tough and poor rural area, wholeheartedly concurred.

One of the men he met in the course of his extracurricular activities was Jan Ligthart, and it gives me real pleasure to devote a page of a book that will deal mainly with the progress of science and with prominent scientists to this humble headmaster of an elementary school in a slum quarter of The Hague. For he was a remarkable man, who had a decisive influence on my father's career—and, therefore, indirectly on my own life—and he influenced a whole generation of schoolteachers. And, after all, the influence of primary schools on the culture of a country is at least as important as the influence of higher academic institutions.

Jan Ligthart, born in Amsterdam in 1859, came from an impoverished, rather than from a working-class, family. He went to evening classes, taught in primary schools in Amsterdam, obtained his headmaster's certificate, and in 1885 became headmaster in The Hague, where he stayed until his death in 1916. He was highly intelligent and had a vast knowledge of, and appreciation for, the world's literature, but he was satisfied to stay in his school and to work out in practice his ideas on education. He did not believe in strict discipline, nor in punishment: one should try to understand children and respect their sensitivity. His approach to this understanding was intuitive rather than analytical: "I am too stupid for experiments and too lazy for statistics; I have to rely on my personal experience," he writes in his reminiscences.

On the didactical side he tried to bridge the gap between school and the realities of daily life. Farmers and artisans play an important role in the books that were written by him and his collaborators and that were for many decades the standard reading material in Dutch primary schools. I am afraid they rather underestimated the impact of the industrial revolution, but they are still quite readable. He was no mean writer, and the little volume in which he writes about his recollections, from which I quoted a moment ago, contains many delightful passages. His influence went far beyond the borders of the Netherlands, especially in the Scandinavian countries his writings, first translated into Swedish, later also into Danish, enjoyed considerable popularity. Although the pupils of his school came almost exclusively from working-class families, he was often consulted by well-to-do people about educational problems and even by Queen Wilhelmina about the education of Princess Juliana. The following story may be true; in any case it is characteristic of his approach. One evening a huge, chauffeur-driven limousine stopped in front of Ligthart's modest house; a well-dressed lady stepped out, was admitted, and began to explain the problems she had with a young son. After a few minutes Ligthart cut her short and said, "Get into your car and let the chauffeur drive you home as fast as he can. You will be just in time to tuck him in. Good evening."

Now, a group of parents in The Hague, dissatisfied with the existing secondary schools, had consulted Jan Ligthart and made plans for a new type of school. It was a private initiative and it would have to be financed by private means, which meant that the pupils would mainly come from wealthy families. So it was hardly a democratic enterprise. Yet it was exactly this new type of school—the Lyceum —that later contributed much to the democratization of secondary instruction and became a major road of access to academic studies. Jan Ligthart suggested that my father would be the ideal man for this new venture, so in 1905, one year after their marriage, my parents went to The Hague. There were quite a few obstacles. First of all, my father was not certified to teach in a secondary school and in this respect Dutch legislation was—and still is—adamant. From kindergartens up the required diplomas are rigorously prescribed. Only university professors do not need to have any degree or diplomas at all. Of course most of them do, but there have been some notable exceptions. After two stiff, though unofficial, examinations at the University of Utrecht, a license to teach history and Dutch language and literature was granted by a special instrument. Although my father fully deserved the license, I suppose the governors of the new school pulled some strings at the ministry of education.

A further difficulty arose because under Dutch law there was no provision for a school of the type envisaged, and again some legal ingenuity was required to find a provisional solution. It stayed provisional for more than half a century.

By 1909—the year of my birth—these hurdles had been taken and the school was officially opened. It had already been in operation for three years as an annex to a private elementary school. My father was devoted to his school. The telephone in our home might ring at any time, during meals or late at night and during the weekend. Then he would quote I Samuel 3:9, "Speak, Lord; for thy servant heareth," before lifting the receiver, and he would listen patiently to whatever wish or complaint or problem a parent or a teacher might want to discuss. He made a point of knowing all the boys and girls—the school was coeducational—and spent the first week or two of a term getting thoroughly acquainted with the newcomers. He would never forget them again, and he would remember their performance and behavior. When, for instance, I told him in 1942 that I had met an engineer who had formerly been at his school and mentioned his name, he replied promptly: "Class of 1920, father a professor at Leiden University; somewhat difficult boy to manage, but very bright; average marks well over 8." My father was also very good at picking teachers and, although he could be difficult at times, they were very loyal to him.

As I said before, he never got around to writing his system of philosophy, but he wrote many articles, mainly on education, and was coeditor of several books. In later years he also contributed a weekly column to a newspaper. I remember him, sitting at his desk, writing rapidly, his arm and wrist hardly moving. Sometimes he would put down his pen, put his head on his folded hands, sleep soundly for twenty minutes or so, then quietly take up his pen again and continue where he had left off. I wish I had inherited that ability.

I was curious about many things—history, geography, language, arithmetic—and whatever I asked, I was never told that I was too young to understand the answer. Usually I got it right away. Sometimes he would say, "This will take some time to explain; ask me again tonight after dinner." And very rarely he would confess that he did not know and would have to look it up. There was usually a book in his extensive library that provided the answer, and if there wasn't, well, then of course he had to buy it.

In 1918 my father became a part-time professor of education at Leiden University; this position as "extraordinary professor," as we call it in Dutch, he held until 1950. He retired—for reasons of health —from his school in 1931. He died in 1957.

In order to complete the picture of my family background I should say something about my mother. She was a lovable woman and also a modest one. Whereas she would probably have encouraged me to write about my father, she would not have liked me to write about herself. So I have to be very short. Her father, a son of a hotelkeeper in the Groningen countryside, was a teacher at a school for deaf-mutes; her mother came from a big farm. I have mentioned already that my mother had been a schoolteacher and had attended my father's lectures. She had three sisters and two brothers. The elder brother became a minister in the Dutch Reformed Church—he belonged definitely to the liberal nondogmatic wing of that institution. The younger one, after some unsuccessful attempts at a career in industry and in office work, acquired a considerable reputation (and earned a comfortable income) as a faith healer. Of course, a physicist like myself is not supposed to believe in supernatural powers, so let me only say that my uncle Johan, with his aquiline nose and his piercing blue eyes, was an intriguing person and that I am convinced that he helped a lot of people (and if you want to talk about placebos and psychosomatic disease and the rest it's all right with me, but I don't think it is the whole story). He also painted seascapes.

My mother was, in a way, the opposite of my father. She was rather small, nimble in her movements, good at games, skilful with her hands, but not very strong. She had a good ear for music and played the piano reasonably well although she had had but little training. She always claimed she was stupid and that she only had passed her examinations because the examiners had taken pity on her: she had been so nervous. Though it may be true that she had no particular gift for memorizing formalized data, she had considerable understanding at a deeper level. From my father I may have inherited a reasonably good memory—not half as good as his in his heyday —and some analytical faculty, but if I have occasionally been able to contribute something to physics and mathematics by an intuitive feeling of the direction in which the solution had to be sought, this may well have come from my mother's side. And if my father taught me, my mother educated me. She was no disciplinarian—Jan Ligthart mentions her in one of his books as an admirable mother—but she had strict ethical principles. During the First World War most boys of my age had tin soldiers and would fight big battles with them. My mother did not want me to have such toys and did not like me to take part in such games. War was horrible and detestable. That I should understand and I should not amuse myself making it into a game. One of the sights of The Hague was (and still is) the Gevangen-poort, an old prison where instruments of torture are on exhibition. I was not allowed to go there as a boy and I have never been there later in life. Even today I don't think I could bring myself to enter that building.

She was profoundly religious—and entirely undogmatic. Before her confirmation in the Dutch Reformed Church she had found herself in great difficulty: she felt she could not subscribe to the dogmas of the official profession of faith. But the minister came to her rescue: "Just write down what you *do* believe," he said. After reading her little essay he said, "You are a complete heathen, but a very good girl. I shall be glad to confirm you."

She was also remarkably free of worldly ambitions. She found life in The Hague rather hectic and enjoyed our summer holidays in a cottage in what was then still a very quiet rural area. Sometimes she would look rather wistfully at the school in a nearby village, and she suggested that this might be a desirable future for me: headmaster in a quiet village. Things did not work out quite that way.

The Origin of Our Family Name

I am often asked about the origin of my family name. It is rather obscure, but here is the little I do know.* Kazimierz is a very common Polish name. It was also the name of several kings and princes, of which one, who lived from 1458 to 1484, after an unsuccessful bid for the throne of Hungary, retired from worldly concerns. He was canonized in 1521, became the patron saint of Poland, and his name entered the calendar in the latinized form Casimirus. Since then Casimir has become a fairly common Christian name outside Poland as well, especially in France, but it never really caught on in the Netherlands, although three governors (stadhouders) of Friesland had Casimir as a second Christian name.

How did some form of this name—spellings range from Casimir to Kazemier—become a family name? There was a tradition in my father's family of a Polish origin, but this is not substantiated by the investigations of Mr. Cazemier. He has traced my ancestors back to

*I am indebted to Lucas Cazemier, a schoolteacher in Kollum—and a distant relative, as he found out himself—who informed me about the results of his very extensive genealogical search.

1 January 1688, the day when Stoffer Hindriks, son of Hindrik Stof-
fers, was baptized at the Reformed Church in Tolbert. So the grandfather of Stoffer Hindriks, who must have been born well before
1650, was called Stoffer too. All the members of a well-established
line of descendants were known by patronymics, and all the names
involved are well-known Dutch or Frisian names.

In our line, the surname Casimir (Kazemier, etc.) does not occur
in any of the documents Mr. Cazemier has examined, until my grandfather's grandfather Brugt Hendriks (1763–1852), a carpenter in
Twijzel, adopted the name Kasemir in 1811, when family names—
rather than patronymics—became compulsory under Napoleonic
Law. However, the fact that many relatives adopted the same name
—albeit with some variations in spelling—and that it is found in the
marriage certificate of one relative as early as 1753 suggests that
during the eighteenth century the name must already have had
some currency among the ancestors of the later "Casimirs." Why this
was so I cannot explain.

A few more remarks about the spelling. The diversity that existed
after the official adoption of family names was further increased by
inaccuracies of clerks at town halls and the illiteracy of many of the
men concerned. My grandfather was registered as Kasimier, his two
brothers as Kasimir and Kasemir. Only his second son and his youngest son—my father—were correctly registered as Kasimier; three
brothers (of whom one died in infancy) and one sister were registered as Kazemier, the other sister as Kazimier. My father, believing
that Casimir must be the original form—and possibly attracted by the
more scholarly appearance—began to write his name that way. That
spelling was legalized for himself and his offspring around 1930. In
the meantime, the town hall at The Hague had made one more error:
my first passport was in the name of Kasimir.

I suppose that most people who have heard me expound in this
way on the origins of the Casimirs have thoroughly regretted that
they ever asked the question.

Schooldays

About my early youth and my schooldays I have little to tell. We
—that is, my two sisters (one elder, one younger) and myself—grew
up in a sheltered environment. My father was not exactly rich, but

he earned a good salary and made something extra by his writing. There was always money for things considered important, such as books and music lessons. I was fairly bright, without being in any way an infant prodigy. I could read at five and soon became an avid reader, and I was fairly good at elementary arithmetic by the time I went to school at the age of six. Lessons were therefore sometimes a bit boring, but, on the whole, I liked school. My weakest point was handwriting, a trait I must have inherited from my father. I later managed to cultivate a legible hand, but it doesn't come naturally.

After four years my father and the headmaster decided to let me skip one grade. I wonder whether it was a wise decision. From then on I found myself among boys—and girls—that were on the average a year and a half older than I. At the age of twelve or thirteen that makes an enormous difference. I had no difficulty in holding my own in all scholastic subjects, but it was different with other activities. Had I stayed among boys of my own age I would probably have become a slightly better musician, a slightly better sportsman. But I would not have become a good musician or a champion anyway, and the fact that I got my PhD at the age of twenty-two had some advantages, so I don't really know. Yet my advice to parents is: let kids remain in their own age group; if the school curriculum is too easy to keep them busy, there are other things they can profitably do.

At the age of eleven I entered my father's lyceum. It was a special —and, in those days, still fairly unique—feature of that school that it combined under one roof, and largely with the same teachers, two types of curriculum usually found in two separate institutions: the curriculum of the HBS and that of the "Gymnasium." And again, as a true schoolmaster's son, I feel obliged to give an explanation. The HBS (Hogere Burgerschool: Higher Citizens' School) was instituted in 1863 and was intended to prepare its students for business and for industry, but until 1917 it did not give admission to academic examinations at the universities. (It did give admission to the Technische Hogeschool at Delft.) It taught mathematics, science, and three modern languages, but no Latin or Greek. Only the Gymnasium gave access to universities. Both Greek and Latin were obligatory, but in the two highest grades there were two divisions: *beta,* with science and mathematics and limited Latin and Greek; and *alpha,* with no science, little mathematics, and far more Latin and Greek. There were, however, many students who first went through the HBS and then crammed for a supplementary examination in the classical languages. Many outstanding Dutch scientists—Lorentz, Zeeman, Kramers among them—followed this road. By a law passed in 1917, the situation was changed: the HBS was recognized as sufficient preparation for academic studies in the sciences and in medicine. By the way, these old arrangements, which by now have been largely superseded, were part of our culture to such an extent that in colloquial language we still call a humanist a typical alpha, a natural scientist a beta: in the Netherlands the gap between the "Two Cultures" becomes a gap between alphas and betas. Since sociologists and economists do not fit into either of these categories, they are often referred to as gammas, although there never existed a Gymnasium gamma. Such terms come so naturally to a Dutchman that he is inclined to use them also elsewhere, where they will not be understood, or—even worse, especially for the sociologists—might be misunderstood and be thought to refer to the classification of intelligence in Aldous Huxley's *Brave New World*.

My father's school had the advantage that the choice between HBS and Gymnasium could be postponed for two years, and also that later on a transition was easily achieved. I first went to the HBS, mainly because most of my friends went there, although my father would have preferred the Gymnasium. So we came to an agreement: I got some private tuition in Latin and Greek and after my HBS examination I entered the highest grade of the Gymnasium alpha for one year, where I passed the final examination in 1926.

I mentioned already that my schooldays were not very eventful. I was interested in most subjects, had little difficulty in getting high marks, and was not really fanatic about anything.

Then, why did I decide that I wanted to become a physicist, and more specifically a theoretical physicist? Why didn't I choose German studies? I was much impressed by our teacher of German and if today I speak a passably correct German, that is at least partly due to his influence. Or general linguistics? I have always been interested in words and their history. Or law, or economics? Medicine I felt was out; I was too shy, too uncertain in my contacts with other people, too clumsy with my hands. As to engineering, my inclinations were decidedly towards theory, and I may have vaguely felt that one might always move from basic physics into engineering but that it would be far more difficult to go in the other direction. Also I did not relish the prospect of having to make engineering drawings.

But I must admit that the choice was as much my father's as my own. After all, he was a professional educator and had a lot of experience in advising his pupils. I believe in general his advice was very sound, unless he was thwarted by external circumstances. Once, after he had subjected a boy that was not doing well at school to a lengthy test, he sighed, "That boy would have the makings of a first-class confectioner, but his father is an ambassador, so what can one do?"

My father's choice was influenced by the fact that he was on friendly terms with Ehrenfest, the professor of theoretical physics at Leiden, about whom I shall have a lot to say later on. From time to time Ehrenfest would drop in at our home: he liked to discuss problems of education with my father, although—or perhaps because— they disagreed on many points. My father favored many innovations but he believed in schools and regular programs. Ehrenfest did not. He had been unhappy at school, and felt—at least that was my father's summary—that schools could provide no more than cheap hand-me-down clothes for the mind. Also, my father had a profound admiration for the great theoretical physicist H.A. Lorentz, who had been on the board of his school for several years.

Lorentz took such assignments seriously. "What is worth doing is worth doing well" was one of his maxims. I quote by way of illustration a passage from a letter my father wrote in 1952 to W.J. de Haas, Lorentz's son-in-law:

And this great scientist was attentive to minor details and mindful of the interests of common people. When I proposed to the Board of Governors of the Lyceum to appoint an errand boy to assist the janitor, Lorentz agreed, but on the understanding that the rector should see to it that the boy would learn something. I have been able to take care of that. All my errand boys have later found good jobs, thanks to Lorentz's instructions

Lorentz later became one of the "curators" of the privately endowed chair of education that had been created for my father at Leiden University. In those days many of the professors and especially the humanists did not consider pedagogics a subject worthy of academic attention. The enormous prestige of Lorentz helped my father to gain a foothold.

Another major influence was my physics teacher, Mr. Corver. He was no mathematician; he had studied with Lorentz and later admitted to me that much of the theory of electromagnetic fields had been rather over his head. But he was one of the most devoted and conscientious teachers I have ever met and he made his subject both clear and fascinating. His classroom experiments went far beyond what was then usual in secondary schools. He had a special predeliction for optical experiments, and we suspected that was partly because they gave him the opportunity to smoke. In those more disciplined days it was quite unthinkable that a teacher would smoke during hours. But when he had adjusted his lenses and mirrors or prisms he would with obvious relish light a little cigar and carefully blow out the smoke so that the light rays became visible.

Of course, one had also to take into account the possibilities of finding a job later on. A master's degree in theoretical physics would in those days automatically entitle one to teach mathematics and physics in secondary schools. The possibilities for a career in research, however, were very limited. There were only a few junior positions in the universities, and they were poorly paid. In those days many university professors, even the great Lorentz himself, had started as teachers in secondary schools. I cannot say that I felt a special calling to become a teacher, but I had no objection either. I came from a teaching family; most of my father's friends were teachers; it seemed a normal way of life.

And so I decided to study theoretical physics and expected to become a schoolteacher. My ambitions had moved already one step beyond becoming the headmaster of a village school.

Hendrik Antoon Lorentz

In the preceding pages I mentioned the name of H.A. Lorentz. He was the greatest of Dutch physicists, and even a sketchy biography will show that he was an extraordinary man. Hendrik Antoon Lorentz was born in Arnhem on 18 April 1853. His father was a moderately well-to-do nurseryman—intelligent, hard-working, and willing to support his gifted son. So Lorentz suffered no financial hardships, but his family background did not provide him with any special advantages. In 1870 he went to Leiden; two years later he returned to Arnhem and finished his studies at home. He taught at an evening school, obtained his doctor's degree in 1875, and in 1878 was appointed to the chair of theoretical physics at Leiden. His papers, written in English, German, or French—Lorentz was equally fluent in all three languages—cover a wide range of subjects, and before long he was acknowledged as one of the leading theoreticians. However, until the end of the century he did not seek personal contacts with his colleagues abroad. It seems he did not even send a copy of his thesis to Maxwell; but Maxwell, who had learned to read Dutch in order to study the thesis of Van der Waals, did read Lorentz's thesis too. He refers to it in a contribution to the British Association Report for 1876.

Around the turn of the century this changed, and he was invited to give lectures in Germany and France. Together with Zeeman he received the Nobel Prize in 1902, and in 1906 he went for the first time to the United States. His lectures at Columbia University were published in 1909 as *The Theory of Electrons,* a well-known classic. In 1911 he moved to Haarlem, where a special position had been created for him. This gave him more time for his own work as well as for his other activities, for he had become a central figure in the international world of physics. This is shown by his voluminous correspondence, an annotated edition of which is now being prepared. He was for instance chairman of the Solvay Conferences from their beginning in 1911 until his death. He continued to give his famous Monday-morning lectures at Leiden as an "extraordinarius."

Readers who want to gain a deeper insight into this outwardly uneventful, but scientifically extremely fruitful, life will have little difficulty in finding further material. A beautiful edition of his collected papers in nine volumes was published by Martinus Nyhoff in The Hague. The last volume contains many lectures to general audiences. I recommend them to scientists who want to study the Dutch language without being particularly interested in Dutch literature. The subject matter is fascinating and Lorentz writes excellent prose. If the language strikes us here and there as slightly old-fashioned, it is mainly because most authors today do not write as well as that. Scribner's Dictionary of Scientific Biography contains an authoritative article by Russell McCormach. There exists no full biography, but his daughter, Mrs. G.L. de Haas-Lorentz, edited a charming booklet with her own reminiscences and contributions by others.[2] I wrote the preface to that volume and should like to quote it in part:

Among the physicists who at the turn of the century enlarged the compass of physical science and opened the roads along which the stupendous advance of atomic physics was going to proceed, H.A. Lorentz takes a prominent place. Many of his results have become staple knowledge, his collected works show the width of his interests, but his publications reveal only partly what he meant to his contemporaries. In his own country, little given to hero-worship though it may be, even the man in the street venerated him; in the international world of science he commanded universal respect to a degree attained by few others.

. . . [this book] is not in any way an attempt at a complete biography or at a systematic evaluation of Lorentz's works, but it tries in the form of an anthology to recreate the impression of his personality. To me it seems that these recollections, which portray a charming, modest and unusually gifted man, are of special interest because they deal with a man living in a transition period. Some physicists like Einstein and Bohr would even today decidedly be called "modern," others are now regarded as "classical," but Lorentz was both the one and the other. In the essay that concludes this volume I have tried to analyze this feature of his work in more detail, but it is also in evidence in his personal life; the retired and almost solitary studies of his younger years in contrast with his later international activity. His life began when the industrial revolution had hardly started to make an impression on Dutch provincial towns, it ended in an era of electricity, cars and airplanes. Yet throughout his life Lorentz was not a revolutionary nor was he left behind by the stream of events. In a rapidly changing world he remained a leader without ever losing the serenity of his early years.

Lorentz died on 4 February 1928. He was buried in Haarlem and his funeral became an impressive ceremony. I shall not try to summarize the vivid description given by his daughter[3] but only quote the short and eloquent address given by Einstein.

It is as the representative of the German-speaking academic world and in particular the Prussian Academy of Sciences, but above all as a pupil and affectionate admirer that I stand at the grave of the greatest and noblest man of our times. His genius led the way from Maxwell's work to the achievements of contemporary physics, to which he contributed important building stones and methods.

He shaped his life like an exquisite work of art down to the smallest detail. His never-failing kindness and generosity and his sense of justice, coupled with a sure and intuitive understanding of people and human affairs, made him a leader in any sphere he entered. Everyone followed him gladly, for they felt that he never set out to dominate but only to serve. His work and his example will live on as an inspiration and a blessing to many generations.[4]

I met Lorentz only once. During my first year at Leiden I decided to attend his Monday-morning lectures. When I entered the classroom rather early I discovered—not without some trepidation—that Lorentz was already there: he wished to become acquainted with everyone in the audience. He walked up to me and asked for my name, where I came from, and how far I had advanced in my studies. He then told me he knew my father well and warned me that I might find his lecture difficult to follow, and that was true enough. Although I had studied some electrodynamics, his calculations on the electrodynamics of a spinning electron went over my head. I do remember the beauty of his presentation, though. But what impressed me most, and became more and more important in retrospect, was Lorentz's perfectly natural and simple kindess; it was the simplicity that is the hallmark of true greatness.

2

Development of Physics

Introduction

My first chapter began with a description of the marriage of my grandfather in 1864; it ended with my entering Leiden University in 1926. In the present chapter I shall try to sketch the development of physics during roughly the same period. That is an ambitious undertaking. Heinrich Hertz, the discoverer of radio waves, once said about a lecture he had given to a general audience that it had been trivial to the experts, incomprehensible to the layman, and disgusting to himself. I am afraid that the first two points of such criticism could also be raised against this chapter—and against many other sections of this book. As to the third point, I do not agree with Hertz at all. Whatever the shortcomings of my exposition, I enjoyed writing it. I enjoyed searching for concise yet reasonably accurate formulations and presenting them surrounded by digressions and a parade of hobbyhorses. Perhaps some of my enthusiasm for my subject will get across even to those readers who find it hard to follow my presentation in detail.

A short synopsis may be useful to those who want to skip this chapter, as well as to those who want to read it. There are four sections. The first one deals with classical physics, the physics of the old-fashioned textbook. Its main subjects are mechanics, heat, electricity and magnetism, and optics. I argue that towards the end of the nineteenth century these subjects had reached a state of maturity, and this gives me an opportunity to present my ideas on the nature of physical theories.

The second section deals with nineteenth-century atomistic theory. The notion of atoms and molecules, which played an essential role in the spectacular development of chemistry, could also explain many physical properties of matter and especially those of gases. However, towards the end of the nineteenth century this kinetic theory ran into serious difficulties.

The third section deals with the development of physics during the first decades of the twentieth century. The main line of thought is as follows. At the beginning of the century, atoms were still regarded as hypothetical models, but it was considered self-evident that such models should behave according to the rules of mechanics that had been established for things of larger dimensions. During the first decades of our century it became clear that atoms and molecules, far from being arbitrary models, are very real particles. I discuss a number of experiments that led to this conclusion, and in doing so I touch automatically upon the main topics of the physics of those days. However, at the same time it became clear that these atoms do not follow the laws of classical physics. A new kind of physics, quantum physics, began to evolve. At the beginning of the century, physicists envisaged hypothetical particles but had a firm belief in the laws of classical physics; twenty years later no one could doubt the reality of atoms, but the laws governing their behavior were only imperfectly known. From hypothetical particles and known laws towards real particles and unknown laws, that is my summary of physics during the first decades of our century.

The fourth section is a logical sequel. One finally succeeded in establishing the laws governing the behavior of atomic particles. This discipline is known as quantum mechanics. By the time I started my studies, its mathematical formulation was already in pretty good shape. It is no exaggeration to say that quantum mechanics did for atoms and electrons what Newton's mechanics had done for the solar system. Newton's great treatise is called *Philosophiae Naturalis Principia Mathematica.* Similarly, quantum mechanics is a mathematical theory. Therefore I have to limit myself to a very superficial sketch. Some remarks on the famous discussions between Bohr and Einstein on the interpretation of the theory will be found in Appendix A at the end of the book.

Nineteenth-Century Physics

Pieter Zeeman (1865–1943), famous Dutch physicist, Nobel Prize–winner in 1902, in later life enjoyed telling how, as a young man, he had been warned not to study physics. "Physics is no longer a promising subject," people had said; "it is finished, there is no room for anything really new." That must have been around 1883. Radiowaves had not yet been discovered, neither had X-rays; electrons, radioactivity, superconductivity, all the things that make up today's physics were as yet unknown. As to theory, one had no real understanding of phenomena that *were* known, like ferromagnetism or atomic spectra. How can it be possible that people—and we must assume that Zeeman's advisers were reasonably intelligent and well-educated people—held an opinion about physics that, in retrospect, seems almost unbelievably obtuse? And Zeeman's case was not an isolated one. Reading about the youth of Max Planck (1858–1947), I noticed he had to face similar attitudes.

We can begin to understand if we first limit the field of physics. An elementary textbook of physics, used at my school, defined: physics deals with temporary and reversible changes of the properties of matter; chemistry deals with permanent changes. And our teacher illustrated this by a classroom experiment. Iron powder and sulfur powder can be mixed, using a magnet they can be separated again; that is physics. But if you heat the mixture, a new compound, iron sulfide, is formed; that is chemistry. I must admit that the textbook in question was already out of date when I got my first lessons in physics, but most of my contemporaries and nearly all of my predecessors must have been exposed to this kind of philosophy.

A more picturesque clue is provided by a remarkable stained-glass window above the entrance to the new building for electrical engineering at the Technological University of Delft. It was transferred from an older building, which, at one time, housed the Physics Department. Five allegorical figures symbolize the main notions of physics. A muscular blacksmith resting a sledgehammer on an anvil is *force*, a bearded Helvetian with an alphorn *sound*. A learned man in some kind of academic gown points the index finger of his right hand at a metal sphere suspended on a wire (undoubtedly the weightless, inextensible, and perfectly flexible wire dear to old-fashioned textbook writers) held in his left: *motion*. A lady in a white gown holds in her bare hands two carbon electrodes to which current is fed through nicely spiralized wires; a white glow illuminates her chastely covered bosom: *light*. And in the middle stands *heat*, daringly represented as a buxom female nude. It is a pity that there was apparently no room for magnetism and electricity. An old sailor looking at a compass and Franklin flying a kite in a thunderstorm would not be out of place in this company (and Franklin was also interested in heat).

From such examples we may infer that it was considered the task of physics to observe phenomena and to describe them, but that it was not the task of physics to study the structure of matter. Let me explain this in greater detail by discussing a simple physical phenomenon. Everyone knows Ohm's Law. It has been with us for many years: in June 1976 the physics department of the University of Cologne organized a symposium to celebrate its 150th anniversary. It was a very pleasant meeting, by the way, and the idea of celebrating jubilees of well-known physical laws or equations offers almost unlimited scope for organizing similar festive events. In the lecture I gave on that occasion I quoted an appreciation by Maxwell published, very appropriately, in 1876 in a report to the British Association for the Advancement of Science:

The service rendered to electrical science by Dr. G.S. Ohm can only be rightly estimated when we compare the language of those writers on electricity who were ignorant of Ohm's law with that of those who have understood and adopted it.

Ohm's law states that when a voltage V is applied to a conductor, the current I through the conductor is given by the equation $I = V/R$, where R is to a high degree of approximation independent of V; it is called the resistance of the conductor. For a cylindrical wire of length l and cross-section s this resistance is given by $R = r_{sp} \cdot l/s$, where r_{sp}, the specific resistance, does not depend on the dimensions of the wire. This specific resistance is very different for different materials and also depends on temperature. It is zero (or at least immeasurably small) for so-called superconductors (for instance lead at the temperature of liquid helium); it is almost infinite for a good insulator. Now, could it be that in the way of thinking of Zeeman's advisers, formulating Ohm's Law was physics and that determining the specific resistance as a function of temperature for a wide range of materials was physics, but that they did not consider it a task for the physicist to explain the results of such measurements on the basis of a theory of the structure of matter? Another case in point: when hydrogen is burned, two atoms of hydrogen combine with one atom of oxygen to form one molecule of water, H_2O; that is chemistry. But today we regard it as an important achievement of *physics* that it can explain the nature of the forces that bind the atoms together. This may well have been considered to be outside the scope of physics by those who thought physics a finished subject.

So here we have at least a partial explanation: physics was considered to be complete because one confined its realm to a number of disciplines that together form what we may conveniently call classical—or macroscopic—physics. These disciplines are (the Delft window, though telling, is not an adequate catalogue) first of all mechanics, subdivided into mechanics of particles, of rigid bodies, of liquids and gases (hydro- and aerodynamics), and of deformable solids (theory of elasticity); next, electromagnetism with its various subdivisions, including optics (which can be regarded as a special branch of electromagnetism); and finally heat and thermodynamics. These various disciplines are connected by one unifying principle: the law of conservation of energy.

Now in each of these disciplines there have been spectacular later developments, and an appreciable part of our present competence in these fields is due to twentieth-century work. This is particularly true in the case of hydro- and aerodynamics: one has to be a very supercilious theoretician to brand all the twentieth-century developments in this field as engineering. Yet there was also some truth in the statement that the subject was "finished." To explain this paradox I have to discuss in some detail the nature of physical theory.

My first attempt to formulate some general ideas about the development of physical theories dates from 1939, when I had to give an "inaugural oration" as an "extraordinary professor" in the University of Leiden. (Dutch academic terminology sounds either funny or pompous when translated literally, but what can one do?). Rereading that lecture, which was entitled "Observation, Theory, and Applications in Physics," I find—not a little to my surprise—that I still agree with some of the things I said on that occasion. What follows is partly a free rendering of relevant sections of that 1939 lecture.

In the development of any branch of physics one can distinguish three stages. First comes an exploratory and analytical stage. Observations are made, experiments are designed, results are systematically ordered, and attempts are made to formulate a theory—that is, to find a set of notions and a mathematical formalism* that will account for these results and that will enable us to predict the results for a whole class of new experiments. If all goes well we enter the stage of consolidation. The notions underlying the theory are more clearly defined, the mathematical formalism is perfected, the range of applications is even widened, and entirely new phenomena are predicted. Gradually we enter the third stage, the technical stage. In that stage applications do not aim at a further confirmation of the theory: they are studied for their own sake. Perhaps the term "technical stage" is slightly misleading: the applications I am thinking of are not necessarily of practical or industrial interest; they may relate to other sciences, or even be "art for art's sake." The essential point is that their primary purpose is not a further confirmation of the theory. Mechanics is the outstanding example. The exploratory stage, while going back to times immemorial—David must have had some notions about trajectories of thrown stones when he slew Goliath—culminated in the early seventeenth century, when Kepler epitomized a vast amount of observations on the orbits of planets in three simple laws and when Galileo made quantitative measurements on moving objects, like balls rolling downhill. With Isaac Newton we enter the next stage. The ideas and formulae in his *Philosophiae Naturalis Principia Mathematica* make it possible to derive both Galileo's results and Kepler's laws; his mechanics describes the motions of heavenly bodies as well as those of falling apples, swinging pendulums, and flying bullets. Today Newtonian mechanics is definitely in the technical stage: it is used to calculate the trajectories of a comet as well as of a spacecraft. Mathematical methods for carrying through the calculations are far more refined than in Newton's days, and in recent years large computers assist—and partly replace—the efforts of the mathematician, but the theory remains essentially unchanged.

I expect that, at this point, many readers will object: "Do physicists really dare to pretend that a theory is final, is definitive? Can it not any day be overthrown by some new experiments? Does history not show a coming and going of theories, one revolution after the other, rather than the kind of continuity you suggest? What about

---

*Here, and later on, I use "formalism" in the sense "a set of mathematical equations and the rules for their manipulation irrespective of their physical interpretation". This usage is common among scientists.

relativistic mechanics, what about quantum mechanics, developments that you will certainly have to discuss later on? Did they not supersede Newtonian mechanics, do they not show that Newtonian mechanics is erroneous? You should add a fourth stage: the stage of refutation." Such remarks are not entirely unfounded but I cannot accept this fourth stage. My great compatriot H.A. Lorentz put it clearly—though, being the man he was, very politely—in a discourse pronounced on 27 October 1927 at the centenary of the death of Augustin Fresnel. It was the last of his many eloquent speeches to be published: he died three months later. I translate:

Henri Poincaré once said that theories are fleeting like waves of the sea, one following the other. The comparison is not entirely to the point, for the waves leave no trace whatever, whereas much remains of sound theories. Indeed, much has remained of Fresnel's theories; they are immortal notwithstanding great changes. . . .

Let me put it more strongly. A theory, once it has passed the stage of consolidation and has reached the technical stage, is never completely refuted. It remains valid, or at least its mathematical formalism remains valid, as a satisfactory description of nature for a wide range of phenomena. But this domain of validity has limits. That is not surprising. What is surprising is that a theory originally suggested by a limited number of observations is found to be valid for a much wider domain, that a theory valid for the few planets known to Newton does also hold for the planets that were discovered later on and for all the planetoids; that it holds not only for our moon, but also for a number of interacting satellites, like the moons of Jupiter; that it holds—but I am repeating myself.

Classical mechanics breaks down for very high velocities; there the theory of relativity replaces it. But relativistic mechanics contains classical mechanics as an excellent approximation at low velocities: even at speeds of 3 km/sec—that is, roughly ten times the velocity of sound—relativistic effects amount to less than one part in ten thousand million. Classical mechanics also breaks down for very small dimensions; there quantum mechanics holds sway. But again this contains classical mechanics as a first approximation; and even for quite small bodies, say spheres with a diameter of a thousandth of a millimeter, quantum corrections to its equations of motion are unmeasurably small.

There is no "stage of refutation," but there is all along a process of demarcation and limitation. A theory, once it has reached the technical stage, is not refuted, but the limits to its domain of validity are established. Outside these limits new theories have to be created. They may involve new concepts radically different from those of the older theories, which may therefore be seen in a new light. This is fascinating and exciting, but so long as we regard physical theories as approximate descriptions of a limited section of physical phenomena —which themselves are only a limited section of human experience— this is no real revolution.

However, the gradual evolution of new theories will be regarded as revolutions by those who, believing in the unrestricted validity of a physical theory, make it the backbone of a whole philosophy. It is understandable that many people—physicists as well as philosophers —were so impressed by the logical structure of Newtonian mechanics and by its striking agreement with many experiments that they took this attitude. Physics may even feel flattered by this homage, but it should not be held responsible for the unavoidable disappointments.

Scientific revolutions are not *made* by scientists. They are *declared* post factum, often by philosophers and historians of science rather than by the scientists themselves.

After this lengthy digression I return to the tenet that physics was a finished subject. We can now sharpen the formulation and say: towards the end of the nineteenth century, classical physics had reached the technical stage.

Atomic Physics Before 1900

Nineteenth-century physics, however, was by no means limited to macroscopic, non-atomistic physics. It dealt also with atoms and molecules. The notion that matter is not divisible ad infinitum but consists of small indivisible particles, a notion that already had been upheld by several Greek and Roman philosophers, especially by Democritus, became for the first time of real practical importance towards the end of the eighteenth century. It was an important facet of the rapid development of chemistry that is associated with such names as Priestly, Dalton, Lavoisier, and Berzelius. In those days chemistry developed from an empirical craft into a quantitative science, and at the same time the number of known compounds and reactions grew in a spectacular way. The notion of atoms of elements combining into molecules proved entirely adequate for ordering this rapidly growing wealth of data. Yet it seems that throughout the nineteenth century many chemists, while using chemical formulae as a matter of course—one cannot do much chemistry without writing H_2O—were reluctant to regard atoms as realities. When for instance Van't Hoff, pioneer of stereochemistry, launched the idea that atoms in molecules are arranged in a spatial structure, this must have seemed to many far too concrete an interpretation of the notion of atom. As late as the beginning of the twentieth century Wilhelm Ostwald, a most competent chemist, believed for a while that the whole notion of concrete atoms and molecules was superfluous and should be dropped.

In physics the impact of atomistics came somewhat later. The law of conservation of energy, which was formulated and clearly established around the middle of the nineteenth century, is easily understandable if one assumes that heat corresponds to mechanical energy of the atoms constituting matter. In 1857 Clausius published an article with the title "The Nature of the Motion Which We Call Heat." *Heat Considered as a Mode of Motion* was a well-known book by John Tyndall (1820–1893). Such titles express clearly the main idea. Tyndall, by the way, was born in Ireland, started out in life as a mainly self-taught surveyor and civil engineer, taught for a while at a school in England, obtained a doctor's degree in Germany at Marburg, was appointed to the Royal Institution in 1853, and was superintendent there, as Faraday's successor, from 1867 to 1885. He was one of the most effective popularizers of science of that period. He was also a great mountaineer, one of those Englishmen who conquered the Alps. He was not the first to climb the Matterhorn, but Whymper followed for the first part of his ascent a route already explored by Tyndall. One of my cherished possessions is a copy of his *Hours of Exercise in the Alps* that I once found in Greenwich Village while rummaging in a heap of old books, all of them priced at five cents. Many years later an experienced German mountaineer told me that, in his opinion, Tyndall's *Hours of Exercise* was the most interesting book on mountain climbing he knew. He himself had made practically all the ascents Tyndall described, but he confessed that he would be dead scared to make them with the equipment of those days. Unfortunately, he added, the book is almost impossible to come by. Today I would not sell my five-cent copy for a hundred dollars.

During the second half of the nineteenth century, molecular theories in physics developed far beyond the mere picture of small moving particles. One began to treat the motion of atoms or molecules mathematically, combining Newtonian mechanics with probability considerations and assuming both attractive and—at close range—repulsive forces between molecules. This approach was highly successful in the case of gases. The kinetic theory of gases, developed mainly by Clausius, Maxwell, and Boltzmann, accounted for pressure, for heat conduction, and for internal friction and was able to make several quantitative predictions. The equation of state (that is the relation between pressure, volume, and temperature) derived by J.D. van der Waals even provided guidance in an entirely new field: the liquefaction of gases.

I mentioned already that Van der Waals started out as a schoolmaster. So it took him a few years longer than average to complete his academic studies, but when in 1873 he published his thesis, he became, almost overnight, a man of international repute. Maxwell, in a review of this thesis, writes "there can be no doubt that his name will soon be among the foremost in molecular science" and, on a later occasion—after some critical remarks—"his attack on this difficult question is so able and so brave, that it cannot fail to give a notable impulse to molecular science. It has certainly directed the attention of more than one inquirer to the study of the Low-Dutch language in which it is written." When in 1877 the former Athenaeum in Amsterdam was transformed into a full-fledged university, Van der Waals was appointed to the chair of physics.

Van der Waals had a son and namesake who also became a professor of theoretical physics at the university in Amsterdam. He did creditable work, mainly in the field of Brownian motion, but he was not of the stature of his father. He was known as "young Van der Waals" right up to his death at the age of ninety-two. Once, after attending a meeting of the Academy of Science in Amsterdam, I remarked during dinner "young Van der Waals is getting old these days." My son, then at high school, reacted by saying, "I'm not going to be a physicist." He kept his word and went into economics and computers instead.

Back to our main topic. Applying Newtonian mechanics to the

motion of molecules in gases had been highly successful. This too may have contributed to the curious notion that physics was a finished subject.

Electromagnetism was already briefly mentioned under classical theories. It deserves a special discussion. In the kinetic theory of gases, forces were conceived as actions at a distance, like gravitation in Newton's celestial mechanics. In electromagnetism, however, the development of ideas took a different turn. It was assumed that the electric force acting on a charge is not an action at a distance, but the manifestation of an electric field that is present also when there is no test particle. And that field was considered to be a state prevailing in a ubiquitous medium: the aether. At first it was hoped that electricity and magnetism might be understood by applying the usual laws of mechanics to this hypothetical medium. That did not work. Still, it was held by most physicists that the equations of electromagnetic theory described the properties of a material medium.

The prediction of electromagnetic waves, the subsequent discovery of such waves by Heinrich Hertz, and the unification of optics and electromagnetism were triumphs of the notion of the electromagnetic field. And one was not prepared to think of a "field" as a mathematical abstraction only. Let me quote the final lines of the final section of James Clerk Maxwell's famous *Treatise on Electricity and Magnetism*. Under the heading "A Medium Necessary," Maxwell writes as follows:

In fact, whenever energy is transmitted from one body to another in time, there must be a medium or substance in which the energy exists after it leaves one body and before it reaches the other, for energy, as Torricelli remarked, is a quintessence of so subtle a nature that it cannot be contained in any vessel except the inmost substance of material things! Hence all these theories lead to the conception of a medium in which the propagation takes place, and if we admit this medium as a hypothesis, I think it ought to occupy a prominent place in our investigations and that we ought to endeavour to construct a mental representation of all the details of its action, and this has been my constant aim in this treatise.

Yet, looking at nineteenth-century physics from our present point of view, I cannot help feeling that there was a measure of inconsistency in its approach. Aether was conceived as a continuous medium, transmitting forces and energy, but at the same time continuous media like gases and liquids were regarded as throngs of molecules with interactions that were provisionally described as "at a distance," but which one hoped some day to explain (and eventually succeeded in explaining) as electromagnetic interactions. Therefore, if the aether were a true continuum it would be entirely different in nature from the media we actually know. Otherwise the aether itself would have to consist of molecules, but how should these interact? Through a kind of superaether? And this superaether? As a child I was fascinated by the picture on the tin containing a well-known brand of cocoa. It showed a nurse holding in her hand this same tin showing a nurse, etc. It gave me a kind of presentiment of atomistics—as a matter of fact eight steps take you down to atomic dimensions—and it was also excellent cocoa. But it is not a useful model for theories of matter and forces.

Apparently nineteenth-century physicists were not troubled too much by this situation. Yet the idea of physics being—almost—finished rapidly disappeared as physical knowledge increased.

On 27 April 1900, William Thomson, Baron Kelvin of Largs (1824–1907), one of the leading physicists of the nineteenth century, gave a lecture at the Royal Institution on "Nineteenth Century Clouds over the Dynamical Theory of Heat and Light." It was subsequently published in the *Philosophical Magazine*. His opening phrase is characteristic: "The beauty and clearness of the dynamical theory, which asserts heat and light to be modes of motion, is at present obscured by two clouds." His first cloud has to do with the problem of electromagnetic waves in moving media. He does not find Lorentz's treatment satisfactory and concludes, "I am afraid we must still regard cloud no. 1 as very thick." It was to be Einstein who in 1905 finally dispelled that cloud by his special theory of relativity.

Kelvin is mainly concerned with his second cloud. It was related to the atomic theory of matter. The kinetic theory of gases had led to a systematic discipline of statistical mechanics. Statistical because it does not try to follow in detail the motion of all the millions of millions of millions of particles in a volume of gas, but has found ways to calculate average values. Now that theory made a number of predictions that were in agreement with experiment, but others—notably on the specific heat of gases consisting of poly-atomic molecules—that were not at all. Yet these predictions followed in a simple and straightforward way from a very general and irrefutable theorem of statistical mechanics, the so-called theorem of equipartition. Many physicists felt this was a very fundamental difficulty. Kelvin agreed only to a certain extent: he was not convinced that the conclusions of statistical mechanics were really unavoidable, "did not see validity in the proof," and tried to find counterexamples. His attempts to refute the conclusion were refuted (especially by Rayleigh*), yet I think he remained unconvinced. His final sentences are somewhat ambiguous. After quoting Rayleigh's then-recent remark,

"What would appear to be wanted is some escape from the destructive simplicity of the general conclusion,"

he continues,

The simplest way of arriving at this desired result is to deny the conclusion; and so, in the beginning of the twentieth century to lose sight of a cloud which has obscured the brilliance of the molecular theory of heat and light during the last quarter of the nineteenth century.

Personally, I think this shows that Kelvin really believed one could get rid of this cloud, that it was a passing cloud that owed its existence to a faulty argument. Or do Kelvin's words all the same contain a warning not to forget this cloud? It does not make much difference one way or another. Kelvin was then already an old man who had not kept up with all the recent developments, and his dynamical views were somewhat out of date. However, the clouds, whatever he thought about them himself, were very real. They would soon bring fresh showers to physicists thirsting for exciting new ideas.

As a matter of fact, some of these showers had fallen already at the moment Kelvin made his speech. In 1895 Wilhelm Conrad Röntgen discovered X-rays, called Röntgen rays in many parts of the world. In 1896 Henri Becquerel discovered radioactivity, and soon afterwards Marie and Pierre Curie started their work, which led among other things to the isolation of polonium and radium; by 1902 a hundred milligrams of pure radium had been extracted from uranium-ore. In 1897 J.J. Thomson obtained evidence on the basis of which he concluded that cathode rays consist of electrically charged particles about a thousand times lighter than a hydrogen atom, and in the same year Zeeman discovered the effect that still bears his

*John William Strutt, third Baron Rayleigh, 1842–1919.

name: the splitting of spectral lines in a magnetic field. Together with Lorentz he arrived at the conclusion that the characteristic radiation emitted by atoms must be emitted by electrical particles identical with—or at least very similar to—those that constitute the cathode rays. In 1900 Planck obtained his famous formula for black-body radiation and arrived at the notion of quanta of energy. And then in 1905 Einstein formulated the special theory of relativity and abolished the notion of the aether. In the same year he launched the idea of light quanta.

No wonder that with such a start, the further development of physics during the twentieth century moved along entirely new lines.

The Reality of Atoms

The harder one has worked on the preparation of a lecture, on writing out the text in full in a number of successive approximations, the more difficult it becomes to write later in a different way on the same subject. That was the difficulty I faced when writing this section. I wanted to sketch the main lines of development of physics between 1900 and 1925, but this I had also done in a lecture I gave in Berlin on the first of March 1979, and the special circumstances surrounding that lecture had compelled me to pay more than usual attention, not only to content, but also to form. What follows has become an almost verbatim translation of parts of that lecture—for which I apologize to the reader. The structure of the German language encourages the use of longer and more involved sentences than I would use in English, and although I have here and there broken them up, there probably remains some German ponderosity. But whatever the shortcomings of this English version, it does contain what I have to say, as far as physics is concerned. To other aspects which made giving this lecture one of the important occurrences of my later life, I shall return at the end of Chapter 5.

Instead of trying to give a complete survey, I shall limit myself to one specific feature that seems to me the essence of the fundamental epistemological development of this period. At the turn of the century atoms were still regarded as hypothetical models, and it was considered evident that such models should behave according to the laws of classical mechanics—and possibly of classical electrodynamics. That was, so to say, the very meaning of a mechanical model. During the period I shall consider, it became increasingly clear that the existence of atoms could not be doubted, that atoms were not hypothetical models but real particles. But at the same time it also became clear that these particles do *not* obey the laws of classical mechanics; they behave quite differently. At the beginning of our century: hypothetical particles, well-established laws. Twenty years later: real particles, unknown laws. That is the big revolution of the first decades of our century. The final step, the breakthrough of new quantum mechanics in the years between 1925 and 1928, will be discussed in the next section.

In the preface of his *Vorlesungen über Gastheorie* (1895) Ludwig Boltzmann complains that kinetic theory is no longer fashionable:

Recently the relation between the two branches of the theory of heat [kinetic and thermodynamic] has shifted. The fascinating analogies and differences in the behavior of energy in the various domains of physics led to the creation of the so-called Energetics, a doctrine averse to the idea that heat is molecular motion.

And even Boltzmann does not dare to oppose this development with a clearcut confession of belief in the reality of atoms. All he writes is:

Therefore, clear the road in all directions, do away with all dogmatism, whether atomic or anti-atomistic. Moreover we call the notions of the kinetic theory mechanical analogues. This choice of words shows clearly that we are far removed from the idea that this theory relates in every way to the real nature of the smallest constituents of matter.

In England the idea that heat is molecular motion seems to have been far less called in question than in Germany. Also my great countryman H.A. Lorentz adheres consistently to the molecular picture. So did Van der Waals; in his Nobel Lecture (1910) he remarks: "I was quite convinced of the real existence of molecules . . . I considered them to be the actual bodies. . . ."

Which roads led to the conviction that atoms are real and showed that we are *not* dealing with mechanical analogues but with the real constituents of matter, hence with the same particles chemistry is concerned with? I shall indicate four different approaches:

1. Kinetic theory of heat.
2. Electrons.
3. Lattice structure of crystals and diffraction of X-rays.
4. Nuclear physics and the observation of individual particles.

FIRST APPROACH: KINETIC THEORY

I mentioned Boltzmann's complaint that his views put him into a somewhat isolated position. Yet the kinetic theory was highly successful during the first decades of our century. The equation of state derived as early as 1873 by Van der Waals turned out to be a reliable guide in investigations on the behavior of gases. The liquefaction of helium by Kamerlingh Onnes in the year 1908 may therefore be regarded as a triumph of kinetic theory. It was also the beginning of a new field of research—and, as should soon become clear, a prolific one.

We must further mention the work of the Danish physicist Martin Knudsen on highly rarefied gases. Whenever the free path of molecules, that is the distance a molecule travels between collisions, is no longer small compared with the dimensions of a vessel, then one meets with phenomena that are easily explained by kinetic theory; it would be very difficult indeed to explain them on the basis of a continuum theory. Related to this is the enormous progress in vacuum technology; here the names of Gaede and of Langmuir should be mentioned. It would be hard to believe that these inventive scientists ever doubted the reality of the molecules that they removed so successfully from the vacuum vessel.

And, if all this should not suffice, there was the Brownian motion. Already in 1827 the Scottish botanist Robert Brown observed the irregular, jerky motion of minute particles suspended in water. In 1863 Christian Wiener* showed convincingly that this motion is not due to external influences; it must be caused by inner motion in the liquid. One might say that Brownian motion demonstrates the molecular thermal motion *ad oculos*—to eyes, that is, provided with a microscope.

*Ludwig Christian Wiener (1826–1896) was Professor of Descriptive Geometry at Karlsruhe. It is a curious coincidence that some seventy years later another mathematician of the same name, Norbert Wiener (1894–1964), also made important contributions to the study of Brownian motion.

Between 1908 and 1910 Jean Perrin carried out exact measurements on Brownian motion. On the other hand, Smoluchowski and Einstein, the latter in a very general and pregnant form, had worked out the details of the theory.

It should be mentioned that Perrin was not aware of Einstein's work when he began his observations. On the other hand, Einstein in his 1905 paper ("On the movement of small particles suspended in a stationary liquid demanded by the molecular kinetic theory of heat") "hesitated to identify the two motions. He was not trying to explain an old puzzling phenomenon but rather to deduce a result that could be used to test the atomic hypothesis and to determine the basic scale of atomic dimensions" (Martin Klein in *DSB*). Another way of saying this: Einstein did not *explain* Brownian motion; he *predicted* it, be it post factum.

The excellent agreement between theory and experiment convinced many of the original opponents of atomistic theories; thenceforth they accepted the reality of molecular thermal motion.

SECOND APPROACH: ELECTRONS

It is often said that the electron was discovered shortly before the turn of the century by J.J. Thomson. That is an oversimplification. The idea of an atomic unit of charge was probably formulated for the first time by Faraday in connection with his experiments on electrolysis. But this idea did not easily fit in with the notion of an electromagnetic field[1] which, by the way, also goes back to Faraday.

Hendrik Antoon Lorentz modified and completed Maxwell's theory. In his theory the electric and magnetic properties of matter are interpreted in terms of the motion of charged atomic particles. He arrives at a theory of electromagnetic phenomena in moving bodies —a theory that was later perfected by the theory of relativity. In 1895 he wrote:

I have joined a school of thought that has recently been promoted by several physicists: I have assumed that *all* bodies contain small, electrically charged ponderable particles and that all electrical processes are due to the positioning and movement of these "ions."[2]

Next Thomson showed that cathode rays can be deflected both by electric and by magnetic fields, found always the same ratio between charge and mass, and arrived at the conclusion that cathode rays must consist of particles roughly a thousand times as light as a hydrogen atom. Almost simultaneously Lorentz analyzed the effect discovered by Zeeman—splitting of spectral lines in a magnetic field. He too concluded that lightweight particles existed: it had to be such particles that emitted the characteristic radiation of atoms.

Even before one had accurately determined the charge and mass of these electrons, before the scientific world was fully convinced of their existence, inventors began to play with electrons, began to apply them, probably without worrying about the philosophical question of whether or not they really existed. First came rectifiers, already applied in 1904 by Fleming as detectors for radio waves, then, in 1907 the first triodes (Lee De Forest). That was the beginning of the era of electronic amplification, of transmitting tubes—in short, of electronics. From then on the further development of telecommunication and broadcasting went in parallel with the development of electron tubes and vacuum technology. I shall not go into further details.

A decisive step was taken by Robert Millikan. From 1909 onwards he carried out a series of direct determinations of the atomic unit charge by means of his oil-drop method, reaching an accuracy of about one part in a thousand—at that time an outstanding achievement.

THIRD APPROACH: CRYSTALS AND X-RAYS

The striking symmetry of crystals must have intrigued many scientists, Kepler and Huygens among them. They surmised that a crystal might be constructed of a regular arrangement of atoms, or at least of smaller particles. A serious study concerning possible symmetries of atomic lattices and their correspondence to the well-known symmetry classes of the crystals as a whole was not made before the second half of the nineteenth century.

Let me now quote from Max von Laue's *History of Physics*:[3]

At first, these studies had no effect on physics because no physical phenomenon required the acceptance of the space lattice hypothesis. Among the few physicists who were at all interested in crystallography, some adopted the opposite view that in crystals, as elsewhere in matter, the molecular centers of gravity were distributed irregularly and that only the parallel placing of preferred directions in the molecules produced anisotrophy. Neither was there much discussion of the hypothesis in mineralogy. Paul van Groth (1843–1927) alone upheld the Sohncke tradition in his teaching at Munich. The triumph of this hypothesis came in 1912 through the experiments of W. Friedrich and Paul Knipping who, by means of X-rays, demonstrated the interference phenomena occasioned by the space lattice of the crystal, a finding which verified the prediction of M. von Laue. Because of their short wavelength these waves are able to reveal optically the interatomic distances, whereas these elude radiations of longer wavelengths, such as light. These experiments also furnished the first decisive proof of the wave nature of X-rays.

Never before had the atomic structure of matter had been as manifest as in the X-rays diffraction pictures. From then on the method of X-ray diffraction was continuously improved. It became a powerful tool for studying the structure of molecules as well as of crystals. On the other hand, Von Laue's idea was also the beginning of X-ray spectroscopy, which was of great importance in unravelling the details of the structure of the atom.

FOURTH APPROACH: RADIOACTIVITY
AND NUCLEAR PHYSICS

Again, this branch of science was born shortly before 1900. The discoveries of Henri Becquerel and of Marie and Pierre Curie mark the beginning. In the further development Rutherford played the prominent role. Otto Hahn was one of those who worked with Rutherford: his stay at Montreal and his friendly relations with Rutherford had a profound influence on his style of work. On the other hand Rutherford was forced more than once to recognize the validity of some of Hahn's results he had at first called in question.

Broadly speaking, two aspects can be distinguished in this further development. First, the investigation of "family relations"; in a way this is chemistry rather than physics. They were Hahn's main theme. The surprising difference with customary chemistry was, of course, that in radioactive decay an atom of one chemical element changes into an atom of another element.

In the space of a few short years a whole series of new elements was discovered and transitions between them were ascertained. That is what I referred to when I spoke about family relations. From our present point of view we may be inclined to regard such results as a consolidation and deepening of our conceptions concerning atoms and elements. Traditional chemistry, however, was at first rather skeptical.

The second aspect: the detailed examination of the radiation emitted during radioactive decay is purely physics (although I do not intend to try to define the difference between physics and chemistry). Here, if possible, the atomistic character of the phenomena shows even more clearly. For already early in the game, one had noticed that it is possible to observe the emission of individual particles. There existed even three different methods during the decades considered; today there are several more.

a. *Scintillations.* Put a radioactive, alpha-ray-emitting substance close to a fluorescent substance like zinc sulfide; one observes a glow, which, on close observation by means of a microscope, turns out to consist of numerous short flashes of very small spatial extensions. One was able to conclude that each of such scintillations is caused by one alpha particle (and alpha particles are completely ionized helium atoms). So it became possible to count alpha particles. Since one could also measure total charge, this provided an independent determination of the unit atomic charge. Later the scintillation method was almost forgotten. It came into its own again when one could replace the diligent, observant, and dark-adapted physicist by a modern photoelectric cell.

b. *Geiger counters.* Alpha particles—and, later on with improved techniques, also beta-particles, that is rapid electrons (many of Lise Meitner's papers deal with those)—can also be counted in an entirely different way: they can lead to an electric discharge between electrodes. With a suitable arrangement the discharge will break off again after a short time. These Geiger counters played a very important role in subsequent developments.

c. *Cloud chambers.* Fast particles leave a trail of ions. These ions can act as condensation nuclei for super-saturated vapor. That is the gist of the cloud-chamber method of C.T.R. Wilson.

Kinetic theory and Brownian motion, electrons, X-ray diffraction, and nuclear physics—each one of these confirmed the idea of atomistic structures. It should be noticed that the values—for atomic charge and mass and other atomic properties—derived from such experiments were always in complete agreement with each other. Finally, it should be emphasized that in all these cases I have only sketched the first beginnings of a momentous development that is continuing even today, a development that is essentially based on the notion of atoms and that could hardly be imagined without that notion.

I have arrived at the end of the first part of my reflections. The atoms, conceived by Democritus, sung by Lucretius in a didactical poem, introduced tentatively and treated mathematically by theoretical physicists, have become incontestable reality.

Inadequacy of Classical Theory

I am now coming to the second part of my considerations: the inadequacy of classical theory.

First, a few words about the special theory of relativity formulated by Einstein early in our century. To begin with it was a theory of electromagnetic phenomena in moving bodies; but here I cannot enter into its connection with experimental data—Michelson-Morley, etc.—and with the theories of Lorentz and Poincaré. An essential point is that the aether disappeared. Let us look more closely at this.

Towards the end of the nineteenth century, atoms and world-aether were more or less on the same footing. Atoms were just as hypothetical as that invisible medium filling the whole universe: the aether. As we have seen, more and more proofs of the reality of atoms compellingly presented themselves. The aether, however, remained exclusively the carrier of electromagnetic phenomena—nothing but a mental aid to be able to say that electric and magnetic fields are "something in something." Once more: atoms became real, aether disappeared.

Second, I want to emphasize that the special theory of relativity is neither a macroscopic nor an atomistic theory. It deals in a general way with the space-time reference system of physics.

Third, it taught us that satisfactory theories, in excellent agreement with existing facts, should never be regarded as unshakable truth but only as approximations.

Fourth, it led to the famous relation between energy and mass.

The general theory of relativity—it was formulated during the second decade of our century—is outside the scope of this section. Let me only emphasize that so far, this grandiose construction can only show its real significance in modern astronomy.

For this book quantum effects are more important. We shall have to discuss—briefly—four different aspects:

1. Thermal radiation.
2. The hypothesis of light quanta.
3. Specific heat of solids at low temperatures.
4. Atomic models and atomic spectra.

That is quite a lot, and since I am not writing a textbook on physics I shall proceed more or less in "telegraphese." I hope that all the same I shall not unduly distort the main lines of historical development.

UNDERSTANDING THERMAL RADIATION

In the course of the nineteenth century it became clear that a vacuum in thermal equilibrium with matter at a well-defined temperature—a cavity—contains a definite quantity of radiative energy independent of the nature of that matter: vacuum has, so to say, its own energy. Application of thermodynamics to this situation yielded clearcut results, which were confirmed by experiment.

But attempts to derive by kinetic, statistical theory exact expressions for the distribution of energy as a function of wavelength met with great difficulties. Then in 1900 there came a breakthrough: Planck's stroke of genius. Let me stress, however, that this stroke of genius did not come out of the blue: it was the culmination of many years of work in thermodynamics.

To begin with, Planck obtained a semi-empirical formula that agreed excellently with experiment; next he found that this formula could be interpreted only by assuming that energy exchange between radiation field and matter for radiation with frequency v can only take place in integer multiples—quanta—of hv, where h is a new constant of nature. This was the first example of a radical deviation from "classical" behavior. Of course, Planck's theory was not at once accepted, but his final formula was again and again confirmed by experiment. And an explanation of this formula on the basis of classical theory could not be found.

THE HYPOTHESIS OF LIGHT QUANTA

Next step: light quanta. In 1905 Einstein interpreted photoelectricity—that is, emission of electrons from metals into vacuum under influence of light—in the following way. He assumed that light consists of discrete particles, photons, with energy $h\nu$.

Einstein's hypothesis was primarily connected with the derivation of Planck's formula. However, the theory of the photoelectric effect was the first new result based on it. It explained at once the experimental result of Lenard that the energy of photoelectrons depends only on the wavelength of the light, not on its intensity. It made even precise quantitative predictions that were accurately verified by Millikan, who had set out to *disprove* the notion of light quanta. And in 1923 A.H. Compton discovered the effect that bears his name: dispersion (with frequency change) of X-rays or gamma rays by electrons—an effect that can be simply interpreted as a collision of an electron and a light quantum.

Still, it is obvious that one cannot abandon the electromagnetic wave theory of light. Let me quote the final words of a lecture Lorentz gave in 1923 at the Royal Institution in London:

We cannot help thinking that the solution will be found in some happy combination of extended waves and concentrated quanta, the waves being made responsible for interference and the quanta for photoelectricity.*

One might say that the formalism of new quantum mechanics has indeed achieved such a combination. Whether it would have satisfied Lorentz is somewhat doubtful.†

SPECIFIC HEAT

There exists a well-known law of Dulong and Petit: the specific heat of a solid per gramatom has for all chemical elements the same value, independent of temperature. It follows from the equipartition

*Collected Papers VIII, 17. Of course, similar statements had earlier been made by Einstein.
†It is well-known that it did not satisfy Einstein.

law of statistical mechanics, which asserts that every degree of freedom has the same energy proportional to the absolute temperature. However, at low temperatures—and for diamond and a few other substances also at room temperature—there exist considerable deviations from Dulong and Petit's law, hence from equipartition. Again it was Einstein who solved the riddle: he applied Planck's idea of quantization to atomic motion in a solid.

He oversimplified this motion to vibrations with only one frequency. When Debye in 1911 introduced a better approximation for the frequency spectrum of the atomic vibrations, excellent agreement with experiment was obtained.

In this development Walter Nernst played an important role. It seems however that at first he was not favorably impressed by Debye's work. Debye himself has told me that he once received a letter from Nernst in which he wrote "your latest paper shows once more that you have understood less about quantum theory than my worst pupil." Nernst had a style of his own! In his famous textbook, however, Debye's work finds full recognition.

ATOMIC STRUCTURE AND SPECTRAL LINES

Now for atomic models and atomic spectra, or in German, *Atombau und Spektrallinien*. That is the title of Arnold Sommerfeld's famous book, for me and for many of my contemporaries the first introduction to the fascinating realm of atomic physics. The title expresses exactly what the book is dealing with. From 1913 to 1929 the essential progress of theoretical physics was concentrated there.*
When I started my studies the era of spectra was by no means a thing of the past. Let me try to characterize in a few sentences the evolution in this field.

First the model. Rutherford had arrived at the conclusion that an atom is a kind of "solar system"—a heavy nucleus, the subject of nuclear physics, encircled by electrons. But his model was simply impossible, according to classical theory. Electrons moving in circles should, according to electrodynamics, emit radiation and lose energy until they fall into the nucleus.

Spectral lines: It had been known for quite some time that atoms

*Exception should be made, of course, for the general theory of relativity, but that was a somewhat isolated subject.

emit and absorb light with characteristic wavelengths. These wavelengths are always the same for a given element, even when that element does not find itself on earth but in a distant star. Because in a spectrograph such radiations appear as sharp lines, it is usual to speak about spectral lines. These spectral lines were applied by chemists for identification of elements.

Already Maxwell stressed that the existence of such spectral lines could not, in any way, be understood on the basis of the theory of his day. He even believed that the fact that all hydrogen atoms are exactly equal and always emit exactly the same spectral lines revealed the hand of the creator of all things. To a certain extent he was right. The remarkable stability of atoms is essentially foreign to any classical theory. Only quantum theory accounts for discrete and hence exactly reproducible states.

A decisive step was taken by Niels Bohr in 1913 in his first paper on the hydrogen spectrum. He formulated three postulates. First: The atom can only exist in a denumerable series of discrete states. Second: Light is emitted when an atom jumps from one state to another; the energy difference is emitted as one quantum according to the formula $h\nu = E_2 - E_1$. Third: When radiation is neglected a continuum of states is possible according to classical mechanics. Provisionally the stationary states are selected from this continuum by so-called quantization rules.

This attack at once had great success in accounting for the spectra of hydrogen and ionized helium. And, in the same year, the very idea of stationary states was confirmed by the experiments of Franck and Hertz on gas discharges. It should be mentioned that Franck and Hertz were not yet aware of Bohr's theory when they started their experiments.

The further elaboration of Bohr's theory is connected in particular with the names of Sommerfeld, Kramers, Landé, and later also Pauli and Heisenberg, as well as Uhlenbeck and Goudsmit. Within fourteen years it led to a comprehensive insight into the structure of spectra and the constitution of atoms for the whole of the periodic system.

It also led to an entirely new science of mechanics: quantum mechanics. That development I shall try to sketch in the next section. Here, in this section, I wanted mainly to show that in order to understand the world of atoms it was necessary to renounce the principles of classical theory. This leaves us in the situation: "real particles, unknown laws." Quantum mechanics showed the way out of that situation.

Quantum Mechanics

The *Handbuch der Physik,* edited by H. Geiger (of counter fame) and K. Scheel and published by Julius Springer of Berlin in twenty-four volumes between 1926 and 1929, the so-called Blue Handbook (or even Blue Bible), is a monumental summary of the physics of those days. Volume 23 appeared in 1926 and contains an article by Pauli on the older formulation of quantum theory and on the structure of atomic spectra. It was partly obsolete when it came out, for then the triumphant development of new quantum mechanics had already set in.

Between 1924 and 1928 this development swept physics like an enormous wave, tearing down provisional structures, stripping classical edifices of illegitimate extensions, and clearing a most fertile soil.

A second impression of Volume 24 appeared in 1933. It contained again an article by Pauli, now on the *new* quantum mechanics. He used to say "not quite as good as my first *Handbook* article, but in any case better than any other presentation of quantum mechanics." (I agree with the second part of this statement.) The first paragraphs of this second article relate in a few forcible sentences what had happened.

Quantum theory took a final, decisive turn through de Broglie's discovery of matter waves, Heisenberg's creation of matrix mechanics and Schrödinger's general wave-mechanical equation, which provided a bridge between these two ways of thinking. By Heisenberg's principle of indeterminacy and Bohr's ensuing elucidation of basic principles the foundations of the theory reached—for the time being—a final form.

These foundations are directly related to the twofold nature of both light and matter—as waves and as particles—and provide a solution, long searched for in vain, of the problem of finding a complete and consistent description of the phenomena connected therewith.

The price to be paid for that solution is renouncing the unambiguous objectivation of natural phenomena, renouncing, that is, the classical, causal description of nature in space and time, a description linked in an essential way to the possibility of distinguishing unambiguously between phenomenon and means of observation."

Historians of science are actively studying this period. The number of *dramatis personae* was, to begin with, quite small; to those mentioned by Pauli we should add Born and Jordan, Dirac and Pauli himself. But in their tracks came many others, perfecting the mathematical formalism and applying the theory to a wide range of problems.

Philosophers are still arguing about the meaning of the theory and even about the controversy between Bohr and Einstein.

I am neither a historian nor a philosopher, but I can relate how I reacted to this development, and perhaps throw a little sidelight here and there on some of the great actors in this remarkable drama. Let me therefore slightly elaborate the points mentioned by Pauli.

In 1927 Davisson and Germer in the United States and G.P. Thomson in England published their results on the diffraction of electrons by crystals. Their work was completely independent, but both Davisson and Thomson had been influenced by the Oxford meeting of the British Association, where they also must have met one another. (The earlier suggestion of Elsasser to interpret certain results of Davisson and Kunsman on the reflection of electrons by crystals as diffraction of electron waves, although known to Davisson, does not seem to have influenced him—at least not at the time it was made.)

These experiments showed convincingly that a dualism between waves and particles exists not only for light (or rather for electromagnetic radiation in general) but also for electrons. As in the case of light quanta, theoretical prediction had preceded experimental verification. Some of my colleagues in Leiden were even surprised that de Broglie's and Schrödinger's mathematical abstractions, which they had studied, should be as real as that! And the experiments were not even very difficult ones, if you knew what to look for.

Diffraction of electrons soon became a lecture demonstration in many places. It is characteristic of the difference between an experimental and a theoretical approach that in Pauli's *Handbook* article experiments on diffraction of electrons are not even mentioned. The same is true for several other texts. Probably the authors found it self-evident that such interference effects should exist. But, as I mentioned, among simpler souls they created quite a sensation. However, when in 1929 Stern and collaborators showed that also beams of helium atoms and hydrogen molecules have a wave character (these experiments, by the way, were difficult[4]), this was generally considered a nice, but unnecessary, confirmation of well-established theory. (Further on in my story, on page 144, I shall relate another example where Stern performed an apparently superfluous experiment; in that case the result was quite unexpected.)

Heisenberg's approach, turned into a systematic matrix formalism in collaboration with Max Born and Pascual Jordan, did not admit of such a simple, direct experimental confirmation; also, it was difficult to apply to concrete cases. Calculating the energy levels of the hydrogen atom, for instance, was a hard nut to crack. Pauli solved that problem early in 1926, after having given up some initial resistance to the formalism, but it was quite a tour de force. To attack the helium atom, a nucleus with two electrons, in that way seemed quite hopeless. When, in 1930, Born and Jordan published a book in which they only used matrix calculus,[5] Pauli wrote a somewhat ironic though by no means entirely negative review in which he says, among other things, "Many of the results of quantum theory . . . can in fact in no way be deduced by means of these so-called elementary procedures, others only inconveniently and with indirect methods. (To these latter ones belongs for instance the determination of the Balmer terms, which is carried out by matrix calculus, along the lines of an earlier paper by Pauli on this subject. So one cannot reproach the reviewer with finding grapes sour because they are too high)." Born, who was rather sensitive, resented this review. It seems he was particularly offended by its last sentence: ". . . print and paper are excellent."

Schrödinger's formalism was much more tractable. Energy levels appeared as eigenvalues of differential equations and there existed standard methods of treating them that were already used in classical physics. It was a curious coincidence that a book by Courant and Hilbert on such methods had appeared just before the birth of wave mechanics.[6]

When Schrödinger and, in a more general form, both Jordan and Dirac showed that matrix mechanics and wave mechanics are mathematically identical, that they are just different representations of the same basic theory, quantum mechanics could unfold in its full vigor.

Ernst Pascual Jordan I knew only slightly, and although he was definitely one of the leading figures in early quantum mechanics, he will not occur in the rest of my story. So let me insert some remarks here. He was born in Hamburg in 1902, studied at Göttingen and became one of Born's coworkers. He was appointed to the University of Rostock from 1929 till after the war, when he first moved to Berlin and, in 1951, to Hamburg. He retired in 1971. After his work in quantum mechanics he did some theoretical work in radiation biology and worked together with geneticists. Later the main emphasis of his work was on general relativity and cosmology. One of his, by now almost forgotten, publications which I thoroughly enjoyed as a young man was a note, together with Kronig, on ruminating cows.[7] Cows chew with a rotary movement of the jaw and Jordan and Kronig had observed that every cow always rotates in the same direction, so you have clockwise and counterclockwise cows. During a walk through the pleasant meadows of north Sjælland they had begun to make statistics, and they suggested that it might be interesting to study the heredity of this characteristic.

Jordan's behavior during the Nazi period has often been criticized. On the one hand he was a staunch defender of "modern" physics—quantum mechanics, theory of relativity—and he did not shun freely using the names of Einstein, Bohr, and Born, which in those days required considerable courage. On the other hand, he ostensibly subscribed to the Nazi tenets. I do not want to question his motives, but the Calvinist element in my background objects to the method. Many of us were shocked by the martial prose he contributed to a students' periodical. I recall the sentence: Our future lies "nicht in der Pensionsberechtigung sondern im Trichterfelde des Niemandslandes" (not in pension rights but in the shell-torn ground of no-man's-land). All the same, I suppose he did more for physics and physicists than many who played it safe and just kept aloof.

Now to Pauli's second paragraph and the price to be paid. Since the time of Einstein's 1905 paper on light quanta, one had had to live with a dualism of two pictures. Interaction of electromagnetic waves with material particles, especially with electrons, took place as if electromagnetic energy were delivered by particles with a definite energy and momentum. But the whereabouts of such particles were guided by electromagnetic waves. And now similar considerations had to be applied to electrons, to particles in general. There were differences: here the particles came first; they had an electric charge, they had a rest mass, and their number was conserved. On the other hand there existed macroscopic, classical electromagnetic fields; there did not exist a classical psi-field. Maybe Schrödinger and others at first entertained hopes of a more realistic interpretation of the wave function; they had to abandon all hope when Born showed in 1926 that a consistent interpretation of wave mechanics can be obtained only by interpreting the square of the absolute value of the complex wave function as the probability of finding an electron. This statistical interpretation could then be extended to the whole of the quantum theory. Given initial conditions that can be realized in an experimental situation, more generally given a possible experimental situation, quantum mechanics can make exact predictions on the outcome of measurements. But these predictions are predictions of distribution functions and can in general be verified in detail only by measurements in a large number of identical cases. Heisenberg's uncertainty relations express the limits of the precision with which these distribution functions can define the value of measurable quantities. The fact that because of the quantized nature of interactions, observations perturb a system in a finite way, ensures that no experiment can be designed that defines initial conditions with greater precision than can be described by wave functions, or in other words that defines initial conditions violating Heisenberg's principle.

The foregoing is, of course, a very simplified version of the fundamental notions of the interpretation of quantum mechanics. I believe it is sufficient for a physicist who wants to apply quantum mechanics to concrete problems. At a deeper level the interpretation of quantum mechanics gave rise to the momentous discussions between Bohr and Einstein. Some remarks on that debate will be found in Appendix A. From a pragmatic point of view the role of quantum mechanics can be summarized as follows. Newton created a theory that enabled us to calculate the motion of planets and satellites, of thrown stones, of pendulums, and of rotating tops. Similarly quantum mechanics enabled us to describe the behavior of molecules, atoms, and electrons.

We of the younger generation accepted almost unquestioningly both the formalism of quantum mechanics and its interpretation, and under that banner we went forth into the wonderful world of atomic physics. Those were the days

> When every morning brought a noble chance,
> And every chance brought out a noble knight.[8]

Days when even lesser knights could find rewarding tasks.

3

Early Years at Leiden

"But you were going to be an architect!" Those were Ehrenfest's first words when I went to see him shortly after my final school examination. Since I have no gift for drawing and since my ability to visualize and remember complex structures in space is only average, this was a surprising remark. But, as I told already, Ehrenfest had from time to time visited our home and he had been struck by the concentration with which I built towers and castles with wooden blocks. And since it was one of his tenets that talents manifest themselves early in life, I had to be an architect. (I am afraid he overlooked the fact that if all kids that build towers with wooden blocks were to become architects there would be a surplus of architects.) I explained why I did not think I should make a good architect and he then asked about my work at school and especially what I had done besides the normal curriculum. That was next to nothing, and the fact that I had had no difficulties with any subject impressed him unfavorably if at all. But he was very kind. In any case I had to pass the "candidaatsexamen" before I had to choose between theoretical physics, experimental physics, and mathematics. So there was plenty of time. He gave me an introduction to Uhlenbeck and Goudsmit, two senior students of his who were then preparing their doctoral dissertations (and who were already well known in the scientific world, but that I didn't know) and who were, like myself, living in The Hague. He also advised me to study some calculus during the summer holidays and lent me an elementary, but very useful, textbook by Lorentz. It was obvious, however, that he did not think I would make it as a theoretical physicist.

Sem Goudsmit, in later life, spoke sometimes with bitter irony about the "wonderful days of yore": he remembered also much hypocritical pomposity and petty jealousy. But he was full of enthusiasm when I met him at the home of his parents. Their house, into which they had moved from the apartment near the city center where they had lived during most of Sem's schooldays, had the auspicious house number 137 (auspicious because the number 137 plays an important role in the theory of spectra, Goudsmit's special field). It was pleasantly situated on a street running up a hill, originally a wood-covered dune. Across the street there were no houses and the land sloped down to a canal, on the other side of which the expansion of the city had hardly begun. When Sem returned to The Hague after the Second World War, the house was still there—I think it is still there today—but his parents had been deported and killed by the Nazis.

Such things were far from our thoughts in 1926. I did not understand much of what he told me about physics, about the work of Uhlenbeck and himself on the spin of the electron, and the names of famous people he mentioned—H.A. Kramers was one of them—meant little or nothing to me. But he talked also about the interesting people from many countries you would meet when you worked with Ehrenfest, about traveling and visiting other universities. He advised me to read, as a first introduction to modern physics, Arnold Sommerfeld's *Atombau und Spektrallinien.* One can always find fault with any book and I suppose Sommerfeld's had its shortcomings, but to me it was a revelation and so it has been to many physicists of my generation.

Uhlenbeck was reading modern Dutch poetry when I arrived. His easy manners and his cultured voice were free of any affectation and revealed his family background—the Uhlenbecks were what one might call a patrician family. The year before, he had returned from a three-year stay in Rome where he had been the private tutor of the Dutch ambassador's son, and now he was Ehrenfest's assistant. He was less intense than Goudsmit, but he gave me sound advice on many points of detail.

Studies at Dutch universities were then—and by and large still are—arranged as follows. After a few years (two years of lectures, but most students spent at least one more year reading for the required examinations) one passed the so-called "candidaatsexamen." That made one a "candidate," but that is not equivalent to having a bachelor's degree: a student who stops there is not regarded as a man with an academic degree but as a dropout. There is some reason for this sad state of affairs: the curriculum for the first years does not aim at imparting a more or less rounded-off body of useful knowledge; it is designed as a preparation for the next examination, the "doctoraal." When you have passed that examination you are a "doctorandus," which means literally a man who has to become a doctor. Diehards tried to maintain that doctorandus is not a degree and that only a man with a doctor's degree is a university graduate. But both the practice of society and—of course much later—the law have decided otherwise. Doctorandus, with the funny abbreviation Drs., is now a recognized degree, roughly equivalent to a master's. Amusing detail: the law regards doctorandus as an epicene title, but this did not meet with the approval of all feminists. My mother-in-law, one of the first women in the Netherlands to get a degree in physics and mathematics, was really vexed when, in her mid-eighties, she discovered that after having been a doctoranda for more than half a century she had, overnight, been turned into a doctorandus! The doctoraal is the end of formal education and examinations: to get a doctor's degree one has only to write and to defend a thesis, which has to be approved by a supervisor. The work for the thesis can be done anywhere, and at any time. I remember for instance the case of a man who spent several years writing a thesis—based on his industrial work—after retirement from industry at the age of sixty-two.

Proposals to turn the "candidaats" into a bachelor's degree with a recognized status have so far met with much opposition by conservative older graduates as well as by so-called progressive students.

When I began my studies at Leiden University in September 1926, the number of first-year students was on the increase, but there were no more than a dozen or so who wanted later to take either mathematics, astronomy, or physics as their main subject. It was, in many ways, an ideal situation. The official curriculum was not very heavy. Mathematics was rather old-fashioned and traditional but it was competently taught by two professors and one lecturer. There were no fixed examinations: if you felt you had mastered a lecture subject you asked the professor for a "tentamen," a private examination with no others present. You might have to wait one or two weeks for a tentamen, that was all. Once you had passed the required number of "tentamina," the official candidaats examination was almost a formality. (Usually you were asked about a point you had missed in your tentamen.)

A few years later the number of students had increased so much that the teaching staff felt obliged to institute written examinations at fixed times of the year. This I felt then to be infra dig. That the despicable system of multiple-choice questions should ever gain a foothold within the walls of a university I could not have dreamt in my worst nightmares. Certainly, we sometimes played multiple-choice questions as a parlor game—and, in my opinion, that is the only thing multiple-choice questions are good for, at least in an academic context. It is good fun to invent plausible but entirely wrong definitions for a difficult word (for instance: ionic capital = the total number of ions that is available in a charged battery) and even more fun if somebody walks into the trap, but that is all. I find it saddening that this dehumanizing approach is now being taken seriously by students and professors alike. My worst fear is that multiple-choice questions are taken as seriously as the studies themselves, which can only mean that real studies are no longer being taken seriously.

Physics was rather traditional too, and the teaching left something to be desired. Both professors, De Haas and Keesom, put most of their time and energy in running the low-temperature research laboratories. Ehrenfest and his assistant came to the rescue, although officially they were only concerned with "candidates." The Institute for Theoretical Physics and its reading room (Leeskamer Bosscha) were a pied-à-terre for all physicists. And, largely under Ehrenfest's influence, an organization had been created, De Leidse Fles (The Leyden Jar), that organized special talks, discussions, and so on. Ehrenfest himself often lectured there and Uhlenbeck ran an unofficial seminar where together we read and discussed a recently issued book by Franck and Jordan, *Anregung von Quantensprüngen durch Stösse* (Excitation of Quantum Jumps by Collisions). I had to present the sections on collisions of the second kind. I still have that book, and the relevant pages clearly bear the marks of intensive study. I seem to have done rather well; Uhlenbeck reported to Ehrenfest and, as a consequence, I was invited to attend the famous Wednesday-evening colloquium. About that I shall have more to say later on. It meant that the first hurdle had been taken: Ehrenfest regarded me from then on as a future student of his.

A certain amount of relaxation was provided by "Christiaan Huygens," a debating society with a rather severely restricted membership, where I made many good friends. Real interest in science was an essential criterion for being elected and we took our discussions very seriously. We met every fortnight at the lodgings or home of one of the members. There was also a festive annual meeting when we would stage a little show and poke fun at the professors; I think most of them did not realize that we had appointed an imitator for each of them. Ehrenfest was an exception in that respect also; he was always invited and was always thoroughly amused when we parodied his curious treatment of the Dutch language and some of his other peculiarities.

I worked hard during those first years at the university. I still lived at home and commuted by tram between The Hague and Leiden. No time was lost: I did a lot of reading while traveling. I did very little but physics and mathematics: for the first time in my life I was really completely fascinated by a subject. And I felt that my choice—or should I say my father's choice?—was the right one. At school I had been good at mathematics; now I discovered that I found the so-called higher mathematics—differential and integral calculus—rather easier than complicated (but elementary) plane geometry.

When I began to study the mathematical theory of electromagnetism—Maxwell's theory—I felt that now I began to understand electricity for the first time in my life. During my schooldays I had considered whether I should start playing with radio, like many other boys, and I had bought a simple popular book. Since I could not derive the few formulae it contained, I did not try to build a receiver.

Don't misunderstand me; I have the greatest admiration for people who are able to do things, and do them well, without real basic understanding, or in any case without mathematical understanding, but I am not good at that myself. If I have any special abilities at all they are in combining physical phenomena with mathematical formulations.

I made good progress with my tentamina and I studied a lot of things that went beyond the curriculum, but there was one hurdle: the laboratory work, the "practicum," as it was called. In order to get your candidaats you had to do a number of standard experiments. The equipment was there; there was nothing you had to design or make yourself. At most you had to connect a few wires. The series was well designed and took you through much elementary physics, but the equipment was partly old and worn. De Haas, who was nominally in charge, did not want to spend too much work and money on keeping it up to date, and he always claimed that a beginner learned a lot more from working with poor instruments than with perfect instruments. There may be some truth in that, although I suspect it was not his primary motivation. I hated those experiments; I was bad at doing them, I usually lost my records of the measurements I had done, and I found the idea that I was doing what many students before me had been doing in exactly the same way and with the same equipment very discouraging. I didn't like solving textbook problems either, although it is an indispensable part of learning mathematics. It is much more fun if you have thought of a problem yourself.

In any case I was more or less stuck when Wiersma came to the rescue. Eliza Cornelis Wiersma, born 29 September 1901, was then De Haas's senior assistant and in charge of the practicum. His own work was mainly in the field of magnetism. In Van Vleck's famous book[1] there are already several references to his work, but not yet to his thesis. Wiersma was always so busy with a number of things that he kept postponing writing it. On 23 February 1932 he finally got his doctor's degree; Van Vleck dates his preface June 1931. So when, on page 337, he refers to Dr. Wiersma this is either one of the very few inaccuracies in this remarkably accurate book or a justifiable, though debatable, correction in the proofs. (Not that such hairsplitting is of any importance.) In 1934 Wiersma, together with W.J. de Haas and H.A. Kramers—but it was Wiersma who designed the details of the apparatus and carried out the measurements—began a series of experiments on cooling by adiabatic demagnetization, his most important work. In 1936 he was appointed as professor at the Technische Hogeschool Delft. When the occupying authorities closed Leiden University during the years 1942 to 1945, the Kamerlingh Onnes Laboratories were put under the jurisdiction of the school at Delft and Wiersma was appointed manager. He brought some of his students from Delft to Leiden and supervised the work of the remaining Leiden staff.

Wiersma was a competent all-round physicist and a capable experimenter. I remember his long and skillful hands. But he could be too finicky about details, even about details that were of no real importance for the experiment he set out to do, and this made him lose time and occasionally even made him forget more important features. He could spend days winding a little coil with the thinnest wire he could get—Wiersma winding coils under a microscope, assisted by one or two graduate students and possibly a technician, who could do little else than from time to time hand him a soldering bit or some other implement, became a highly dramatic event. And yet, as I found later when I took over some of his work, a coil roughly wound with somewhat thicker wire would be perfectly adequate for the type of measurements he wanted to do. The Leiden installations and equipment he loved with a love no material and replaceable things deserve, the once outstanding but by then somewhat antiquated Leiden equipment least of all. During the occupation, when a team of Germans headed by Dr. Alfred Boettcher, a German physicist who was then an SS officer, started to requisition equipment of Dutch university labs in order to equip a research unit at Doetinchem, a Dutch town close to the German border, he was utterly upset. Personally I have always regarded this as one of the minor misdeeds of the occupation period. It seems to me that it might even be argued that such requisitioning was permitted by international law. But this I gladly leave to lawyers; the whole idea of a *jus in bello* seems to me a *contradictio in terminis* anyway.

Wiersma died on 31 July 1944. His death was not directly caused by the German occupation, and the Hunger Winter had still to come, but his concern about the plundering of the labs may well have had an influence.

However, Wiersma's curious devotion to apparatus did not mean that he was not interested in people. On the contrary, during the occupation he did a lot of illegal work to help students and took considerable risks while doing so. And, as I said before, without his help I would have been much delayed and might even have got stuck in my studies.

In those days one still worked on Saturdays, and he told me to come every Saturday morning and work under his personal supervision. Each time he would have a look to see how I got on and sometimes he would scold me, for instance when I gave a result to a higher number of decimal places than was warranted by the accuracy of the observations. Under his guidance I made fairly rapid progress, and although I still did not really enjoy doing this kind of experiment, I had soon finished the required number. I should add that I did enjoy writing the laboratory reports; I usually provided them with a short theoretical introduction, and that became my way of learning general physics.

I mentioned W.J. de Haas. I shall have more to say about him later on, but this is the place to tell about my tentamen in general physics. He rapidly asked about one subject after the other; as soon as he saw I knew something about it he moved on to another topic until he hit upon something—the hysteresis curve of ferromagnetic materials—about which I really knew nothing at all. He kept asking me questions about ferromagnetism and I already had a feeling that I had failed when he suddenly smiled and said, "It is all right, but don't imagine that you know everything already." It was a useful lesson.

In June 1928 I passed my candidaats. Now I could start to study theoretical physics in earnest, and that meant a close cooperation with Ehrenfest. I explained already that Lorentz, or rather the image of Lorentz, exerted an important influence on the course of my life. Ehrenfest's influence was far more concrete and direct. Obviously: Ehrenfest was my teacher, later he supervised my doctoral thesis, and I was for some time his assistant. But even if I had studied with Lorentz, his direct influence would have been less. Lorentz was the kindest of men, he was always willing to help people who came to him, but he was reluctant to the point of shyness to intrude into the personal affairs of others. Ehrenfest had no such inhibitions. Some of his students even complained that he exerted—or wanted to exert—too great an influence on their private life. I have never felt that this was so myself, although I don't know how he would have reacted had he thought I was going to marry an unsuitable wife. As things were, he entirely approved of my choice—which showed sound judgment on his part!

Who was Ehrenfest? Readers who are really interested should get hold of Martin Klein's marvelous biography[2]; unfortunately so far only the first volume, dealing with the years up to 1920, has appeared. A short introduction I wrote for the edition of his collected papers[3] should at least convey a superficial impression of his background and character. I quote it here in full:

When in 1912 H.A. Lorentz resigned his Chair of Theoretical Physics at the University of Leiden to take up a research position that had been specially created for him at Haarlem, he recommended Ehrenfest as his successor. Thus Paul Ehrenfest, who had been born in Vienna on January 18, 1880, had studied in Vienna and Göttingen, had obtained his doctor's degree under Boltzmann in 1904 and who since 1907 had been living at St. Petersburg with his Russian-born wife and collaborator, Tatiana Afanassjewa (whom he had married in Vienna in 1904), became a full professor at Leiden. He stayed at Leiden until his death on September 25, 1933.

It is characteristic of Lorentz's width of vision that he should esteem and encourage a young man so entirely different from himself and so utterly at variance with then prevailing ideas about what a professor should be like. Ehrenfest from his side showed towards Lorentz a devoted admiration beautifully expressed in the obituary reprinted on page 559 of the present volume. It is true that both were interested in a broad range of problems and that both were accomplished lecturers to physicists as well as to general audiences, but there all resemblance stops. Lorentz's lectures were masterpieces exposing the subject in a systematic and lucid way in well-balanced and beautifully correct language, whether he spoke his native Dutch, or French, English or German. Ehrenfest's lectures were brilliant too, but in an unconventional way: he emphasized salient points rather than continuity of argumentation; the essential formulae appeared on the blackboard almost as aesthetical entities and not only as links in a chain of deductions. He shunned calculations of any length and numerical constants were often considered irrelevant.

"4π"—mind the quotation marks—could mean almost anything. His language was vivid, full of surprising metaphor, and quite ungrammatical, at least when he spoke Dutch—and more so when he spoke English—but even his German was occasionally spiced with Viennese vernacular. One had very little inclination to go to sleep during Ehrenfest's lectures, but if one ever showed any tendency in that direction one was immediately and ruthlessly called to order.

The difference between Lorentz and Ehrenfest was even more pronounced when it came to seminars and colloquia. Lorentz did not like to speak about a problem before he had arrived at a solution, and he reproved —always in an extremely polite and mild way—those who made remarks without due cogitation. To Ehrenfest, on the other hand, discussions and arguments were an essential part of his scientific activity and the best way to clarify an obscure point. He was never afraid to ask a stupid question and he encouraged others to do the same. How often, during drowsy meetings when an audience is placidly dozing away the hours while incomprehensible speakers are muttering ununderstandable words, projecting irrelevant slides or writing illegible formulae, have I longed for someone with both Ehrenfest's courage to cut a speaker short and Ehrenfest's ability to expose sham and bombast, and also to bring forward the quintessence and to explain the beautiful and important. For Ehrenfest was not only a merciless critic of the stupid and the unclear but perhaps even more a fervid admirer and interpreter of the beautiful and profound, and in this he was entirely unselfish. Few people have shown greater loyalty to their friends and their pupils.

Ehrenfest's own contributions are mainly in the field of statistical mechanics and its relations with quantum mechanics. He has elucidated many basic points in Boltzmann's theories and the article in the *Enzyklopädie der mathematischen Wissenschaften*—written jointly with Mrs. Ehrenfest—is still invaluable for all those who are interested in the fundamentals of statistical theory. His work on quantum statistics led to the formulation of his theorem of adiabatic invariance which played an important role throughout the further development of quantum mechanics.

Now that atoms have become almost tangible realities and the theory of atomic structure can be built up from first principles, it is instructive to realize to what extent the development of quantum theory was initially due to statistical and thermodynamical considerations.

But perhaps Ehrenfest's papers are even more fascinating because of the striking peculiarities of his work. Every single page bears witness to the intensity of his preoccupation with physical science. Most of his papers are concerned with fundamentals—there are hardly any dealing with the application of well-established theory to experimental facts—and many deal with single points rather than with the systematic exposition of a body of theory or the elaboration of mathematical details. He has a great preference for the use of simple models that show the essential traits of a problem—and is a master at inventing them; this is a common feature of his lectures and his writings. Otherwise his written prose is much more restrained. In his papers, one will look in vain for the graphic imagery of his lectures and discussions and only here and there do we gather between the lines of published speeches a glimpse of his exuberant wit. Ehrenfest himself realized that his oral style would not go down well in writing; without his gestures and intonation, without the ambiance that he created around him it would lose its significance. We, Ehrenfest's pupils, shall value these collected papers as a work of reference, as a historical document and as a worthy tribute to the memory of a great physicist, but reading them we shall also wistfully remember a great and inspiring teacher who was for us the central figure in a happy era of physics that will not come again.

Ehrenfest's lectures. There is a lot of talk these days about audiovisual methods. Ehrenfest speaking and Ehrenfest writing on the blackboard were about the best audiovisual presentations I have ever witnessed. All the same, one does regret that no audiovisual apparatus was available at Leiden in those days: it would be nice to have a sound-motion picture or a video tape of a few of his lectures. Now, there remain only a rapidly shrinking group of people who retain a living memory. Klein quotes a letter of recommendation Sommerfeld wrote to Lorentz in 1912:

He lectures like a master. I have hardly ever heard a man speak with such fascination and brilliance. Significant phrases, witty points and dialectic are all at his disposal in an extraordinary manner. His way of handling the blackboard is characteristic. The whole disposition of his lecture is noted down on the board for his audience in the most transparent possible way. He knows how to make the most difficult things concrete and intuitively clear. Mathematical arguments are translated by him into easily comprehensible pictures.[4]

As a matter of fact, I have never seen anyone who could use a blackboard so masterfully as Ehrenfest. We, his students, were instructed in some of the basic principles, such as: "The blackboard begins at the upper lefthand corner" (if you started to write in the middle) or "You may have many good characteristics, but transparency is not one of them" (if you were standing in front of what you had written) and so on. Notwithstanding those useful tips, I have never come anywhere close to his virtuosity. (For clumsy artists like myself, overhead projectors are a great help. You can write the formulae beforehand and uncover them gradually instead of writing.) This does not mean that he despised slides: in popular lectures he would use them very effectively. I have also heard and seen him explain an effect of interference between light waves of different wavelength (the phenomenon of group velocity) by projecting a green and a red comb, one on top of the other, and moving them at different speeds.

He spent considerable time at a cheap cosmetics shop on the Breestraat (Leiden's Broad Street)—a shop that carried the grand and wholly inappropriate name Apollo—rummaging in a box full of pocket combs until he found what he was looking for: a green comb and a red comb, both fairly transparent, and the green one having somewhat smaller spacings than the red one. (Green light has a shorter wavelength than red light.) The salesgirl must have wondered what this man with his close-cropped hair was looking for. The demonstration was very striking. It was part of a lecture—one of a series of five—given during the winter 1931–32 for the "Maatschappij Diligentia," a society arranging lectures and discussions on scientific subjects for a general but well-educated audience. There were, and still are, many such societies in the Netherlands, and "Diligentia" was one of the better known, but I am afraid that the role of such groups in the cultural life of the country has declined. At the request of Ehrenfest I prepared a synopsis that appeared as a little book.[5]

Many of Ehrenfest's expressions became proverbial among his students. "Das ist der springende Punkt" (That is the salient, the essential point) is a common expression. With Ehrenfest it became "Das ist wo der Frosch ins Wasser springt" (That's where the frog jumps into the water). Did he know Matsuo Basho's famous haiku?

> Old pond—
> and a frog-jump-in
> water-sound.[6]

I don't think so, and anyway, Ehrenfest's frog did not make an unexpected splash, interrupting the quiet of an old garden: its jump was the culmination of an actively followed train of reasoning. "Das ist der Patentanspruch" (That is the patent claim) was another favorite. I don't remember that he ever had anything to do wih patents, but he may have discussed them with Einstein, who was a close friend, and who, as is well known, worked for several years at the Bern patent office.

As an introduction to statistical mechanics, he liked to treat the dog-flea model. Two dogs are sitting side by side and have a large number of fleas. On the average about half of the fleas will be on each dog, but there will be fluctuations. If we catch all the fleas and put them on one dog the number of fleas on this dog will exponentially decrease towards the equilibrium state, and so on. In the published version the dogs are urns and the fleas marbles. Somewhat less savory was his remark about someone who had finally disentangled a complicated situation: "Da hat Herr so-und-so schliesslich die ganze Ratte aus der Suppe gezogen" (There Mr. So-and-So has finally pulled the whole rat out of the soup)—a metaphor clearly based on the none-too-sanitary conditions in Viennese soup-kitchens for the destitute.

I give two examples of Ehrenfest's style of discussion. A Dutch doctor's degree is awarded at the end of a session lasting three-quarters of an hour during which the candidate has to defend his thesis. However, since it often happens that apart from the supervisor, with whom the candidate has already discussed his work at length, few, if any, of the faculty members have read the thesis, the candidate has to add a number of propositions—the famous "stellingen"—some of which should not refer to the thesis itself. This ensures that there is always plenty of material for discussion. Now, on 20 October 1925 Van Heel* defended his thesis on optical properties of solids at low temperatures, and his last proposition—number 15— was: "The light one observes with closed eyes is not red, as one might expect, but 'white,' that is, colorless." Ehrenfest told me he had attacked this in the following way. First he asked Van Heel to read his "stelling"; that was the usual thing to do. Then he continued, "Mr. Candidate, close your eyes," produced from under his gown an electric flashlight, switched it on immediately in front of the victim's eyes, and asked, "Mr. Candidate, do you wish to maintain your proposition?" and the candidate humbly admitted defeat. I liked to retell that story, but when Van Heel heard about that he took me to task. "Well," I said, "Ehrenfest himself told me; is it not true?" "Yes," he said, "and at that moment I was so surprised by that sudden red flash that I could not find an answer, but my proposition was, in a way, correct." The point is that the eyelids reduce the intensity of light to such an extent that under most conditions normally encountered the light that reaches the eye through the eyelids will have an intensity in the so-called scotopic range, where color-vision is impossible.

My second example shows Ehrenfest's way of dealing with bombast. At a meeting of the Deutsche Physikalische Gesellschaft, Rupp had given a paper in which he described some experiments with colliding electron beams. He had given it the pretentious title "Ein elektrisches Analogon zum Comptoneffekt." Ehrenfest's comment: "In the same way, when I shoot off a bird's tail I might call it a biological analog to the photoelectric effect." Rupp, by the way, had a remarkable career. In the early days of electron diffraction he did creditable work at Göttingen. Then he joined the AEG laboratories and there, probably overwrought and in a state of nervous strain, he published some alleged experimental results not based on actual observations. This resulted in a psychiatric opinion appearing in the

*Abraham C.S. van Heel, 1899–1966, was later Professor of Optics at the Technische Hogeschool at Delft.

Zeitschrift für Physik. C.P. Snow's *The Affair* is partly inspired by this episode. To the best of my knowledge, Rupp later regained his mental balance and had a satisfactory, though not outstanding, career.

I had better stop here. As I said in the introduction just quoted, Ehrenfest's lecture style does not go down too well in writing. It would take the skill of a greater writer than I am to bring these sayings and examples to life again.

For those students who took theoretical physics as a subsidiary subject for their doctoraal—mathematicians, experimental physicists, astronomers—Ehrenfest did not make life too difficult: they had to follow his regular lectures and to pass a tentamen. The required minimum was treated before Christmas and, if you passed your tentamen before the Christmas holidays were over, you were through with the formal part. For many years Ehrenfest used to reward those that managed to do this with a bar of chocolate. These tentamina were therefore known as "Kwatta-tentamens"—Kwatta being, in those days, a well-known brand of chocolate. Most students continued to follow his lectures also during the rest of the year: they were so enjoyable. And many attended the colloquium.

In order to avoid confusion: I do not call a theoretical physicist a mathematician. To do so was—and to a certain extent still is—common practice in England; this is misleading. A theoretical physicist's primary aim is to describe and explain physical phenomena. To do so he needs mathematics, sometimes even fairly complicated or profound mathematics, but for him mathematics is a tool, not a goal. The mathematician is interested in the abstract structures of mathematics as such. The applied mathematician may work on problems occurring in physics or engineering, but he takes the physical model for granted. For a theoretical physicist, thinking about the model is the essential thing. There have been great scientists who were both mathematicians and theoretical physicists. Isaac Newton, Pierre-Simon Laplace, and Carl Friedrich Gauss were illustrious examples. Also, more recently, professional mathematicians made contributions to theoretical physics—Henri Poincaré, Hermann Weyl, John von Neumann—but theoretical physicists rarely make significant contributions to pure mathematics. It would never occur to me to call Einstein or Niels Bohr mathematicians.

Ehrenfest became far more exacting when you wanted to take

theoretical physics as your main subject. He accepted only a limited number of students. It is questionable whether he was legally entitled actually to refuse a student, but nobody whom Ehrenfest rejected ever tried to force his hand. Although the criteria by which Ehrenfest picked his students sometimes struck us as rather arbitrary, rejection involved no malevolence. Ehrenfest knew that there was little future for a theoretical physicist unless he was rather good. He had himself had considerable difficulty in finding a position: Lorentz's offer came as an unexpected blessing. Also, he wanted to teach his students by almost daily intensive discussions, and he felt he would be unable to do this with someone he disliked personally or whose way of thinking was too different from his own.

Discussions with Ehrenfest were usually about recent publications of one of the leaders in quantum mechanics or the theory of relativity. Together with Ehrenfest we should try to understand them. That was his way of working. If, in the course of such work, one had an idea of one's own, so much the better, but the primary goal should always be to understand, not to write a paper. He did not assign concrete problems to his students to work out.

A certain amount of hero-worship was inculcated upon us from the very beginning. Lorentz, Einstein, and Bohr ranked highest, but men like Pauli and Dirac were respected too. Thomas Carlyle in his *On Heroes, Hero-Worship, and the Heroic in History* writes:

For, as I take it, Universal History, the history of what man has accomplished in this world, is at bottom the History of the Great Men who have worked here. They were the leaders of men, these great ones; the modellers, patterns, and in a wide sense creators, of whatsoever the general mass of men contrived to do or attain; all things that we see standing accomplished in the world are properly the outer material result, the practical realisation and embodiment, of Thoughts that dwelt in the Great Men sent into the world: the soul of the whole world's history, it may justly be considered, were the history of these.

Did Carlyle consider the development of science an important part of Universal History? Apparently not, for then he would almost certainly have included Newton in his cast. Yet I believe that Ehrenfest's ideas on the history of physics came close to Carlyle's on history in general. Personally, I feel that it is a valuable part of education to learn as a young scientist that there are really great and admirable scientists. This does not preclude criticism, and admiration should not turn into idolatry. Ehrenfest in his admiration for Lorentz did not entirely escape that risk.

Ehrenfest's way of teaching also had some disadvantages. One did learn to aim at clarity and logical consistency. One was brought up in a strict discipline of logical thought. But one was not taught in any way the discipline of accurate calculations. I am myself a rather untidy sort of person: my papers are always a mess and I have never been able to do any kind of bookkeeping. It might have done me some good, had I been forced to work out a number of examples down to the last numerical details. But maybe I was hopeless anyway. One thing Ehrenfest did: he forced me to change my handwriting. So now I have two kinds of handwriting: one that comes naturally and that is almost illegible, even to myself, and another that is quite legible, which I can write fairly rapidly, but which I cannot combine with thinking.

Many visitors came to Leiden: to speak at the colloquium, to give seminar lectures, to have a discussion with Ehrenfest; sometimes also for more extended visits.

Dirac* spent some time at Leiden in the spring of 1928, working at his book on quantum mechanics. The book was finished two years later, but he gave a few lectures on the first chapters which, as far as I remember, must have been almost in their final form. Although I had not yet passed my candidaats, I attended those lectures, and was glad to discover that I could understand most of what he told us.

The lectures were beautifully clear, and when Ehrenfest or somebody else asked for a further explanation Dirac would just repeat what he had said: he felt there was no better way of saying it. And, in many cases, it worked. Later, I learned that this was very characteristic of Dirac. For instance, when Ehrenfest, together with his assistant A.J. Rutgers, had had some difficulty in understanding one of Dirac's papers and wrote to him for explanation, the answer was almost exactly the text of the paper itself. But after further study Ehrenfest arrived at the conclusion "Je besser man 's versteht um so besser steht es dort" (The better one understands it, the better it is

---

*P.A.M. Dirac, born at Bristol in 1902, studied electrical engineering before becoming an outstanding theoretical physicist at Cambridge. He was one of the creators of quantum mechanics. In 1933 he received the Nobel Prize, mainly for his relativistic wave-equation that accounts for the existence of positive electrons.

said there). On one occasion, however, there was a question to which he did not at once know the answer: Writing very small he made some rapid calculations on the blackboard, shielding his formulae with his body. Ehrenfest got quite excited: "Children," he said, "now we can see how he really does his work." But no one saw much; Dirac rapidly erased his tentative calculations and proceeded with an elegant exposition in his usual style.

There circulate many anecdotes about Dirac[7] and most of them refer to the mathematical precision of his comments. A few examples. At question time after a lecture, someone in the audience gets up and says: "I did not understand the way in which you" etc. Chairman: "Would Professor Dirac answer that question?" Dirac: "That was not a question, it was a statement".

Someone going for a walk with Dirac apologizes for a rattling noise caused by a tube of aspirin tablets in his pocket. Dirac: "I suppose the noise is loudest when the tube is just half empty."

Someone had made an error of sign (a minus instead of a plus) but asserts that the final answer is correct. "Therefore," he says, "there must be two errors of sign." Dirac: "In any case an even number."

Dirac about sugar: "I think that one piece of sugar is enough for Pauli." Somewhat later: "I think one piece of sugar is enough for anybody." Still later: "I think that the pieces of sugar are made in such a way that one is enough." (The story was told by Von Weizsäcker at Trieste,[8] but I remember that it was told to me by Niels Bohr himself.)

Such stories are not untrue; yet they are misleading. For one thing, they do not reflect Dirac's sense of beauty. Dirac in later life has often emphasized that the beauty of equations had been for him an important guiding principle. And that kind of beauty cannot be formulated in strict mathematical terms. Dirac's perceptivity went beyond mathematical analysis in many other ways. At Copenhagen we once played—I think under Gamow's direction—a rather silly parlor game in which one had to guess a person by asking what day of the week he resembles, what hour of the day, what piece of furniture, what number, what vegetable, and so on. Dirac joined in, and I had expected that such strictly speaking meaningless, yet meaningful, associations would mean little to him. On the contrary, he enjoyed the game and was quite good at it. More seriously: at the Copenhagen conference in September 1933 Dirac was probably the only one who was aware of the extent of Ehrenfest's depression. Mrs. Bohr told me later that he had expressed his worries to her, but there was little one could do. Soon afterwards Ehrenfest committed suicide.

I mentioned that Lorentz was reluctant to intrude into the personal life of his students. Dirac's reserve went even further. He once said to me, "I have never wanted to direct the work of others. I was always afraid to waste their time. I don't mind wasting my own time but I don't want to waste the time of others." Very considerate, of course, but not quite the way to create a school.

Dirac enjoyed puzzles of various kinds. About electrical engineering he liked to say that the winding of armatures presented interesting problems in the theory of numbers. It is told that he once watched a woman for quite some time while she was knitting. Finally, after having understood to his satisfaction the topology of the stitches, he said, "You could also do it a different way," and embarked on a description. The lady cut him short. "Of course you can," she said. "That is purl."

Let me tell how he killed the game of four figures 2. I believe I was introduced to that game in the summer of 1929 when I spent some time at Göttingen. The rules are simple: try to write the integers using four and only four figures 2 and no other numbers (nor letters representing definite numbers like e and π). Well-established mathematical symbols can be used ad libitum. The beginning is easy:

$$1 = \frac{2+2}{2+2}; \quad 2 = \frac{2}{2} + \frac{2}{2}; \quad 3 = 2 \times 2 - \frac{2}{2}; \quad 4 = 2 + 2 + 2 - 2;$$
$$5 = 2 + 2 + \frac{2}{2}; \quad 6 = 2 \times 2 \times 2 - 2$$

and so on, but gradually things get more difficult. Of course discussions arose whether certain notations were acceptable or not. The English had an unfair advantage because of the notation .2 where the Germans had to write 0.2 (which was barred because of the 0). And .2^{-2} brings you at once to 25. So the Germans adopted the British notation. The symbol n !! = n (n − 2) (n − 4) . . . was slightly suspect, but since it occurs in several reputable textbooks it had to be approved, and it was quite useful, for instance 7 = (2 × 2) !! − 2/2. People worked hard and displayed considerable ingenuity trying to beat existing or alleged records.

Somewhat later I learned that Dirac, when confronted with this pastime, had been deep in thought for some time. When asked how far he had got he answered in his usual precise and succinct way, "The number n with three figure 2's." And he gave the following formula,

$$n = -{}^2\log {}^2\log \sqrt{\sqrt{\ldots \sqrt{2}}},$$

where one writes n square-root signs. This was completely within the rules of the game . . . and completely killed it.

A few years ago, during a summer school at Varenna in 1972, I asked Dirac whether he remembered this incident. Yes, he remembered having written down this formula, but he did not remember exactly when and where. It could have been at Göttingen, but he just was not sure. So the authenticity of the formula is well established but place and time of its creation are uncertain, and we will have to leave it at that, unless a reader of this book can supply exact details.

On the same occasion, he told me, "Heisenberg taught me two things, matrix calculus and how to fold a paper cube. But I have forgotten how to fold a cube." Heisenberg had picked up some origami while in Japan, and I suppose his cube must have been what is known as the "water bomb." Just for once I was ahead of Dirac; I could show him how to fold a cube.

For some reason Ehrenfest did not want me to attend his regular lectures. And the closest I came to a tentamen was when one day he asked me to recite and discuss Maxwell's equations—the fundamental equations of electricity and magnetism—while walking up and down a corridor. "Yes, you have understood some of the music," was his final verdict.

During the autumn and winter of 1928–29 we struggled mainly with group theory, a highly developed branch of pure mathematics that had until then not belonged to the usual tools of a physicist. The idea that many qualitative conclusions from quantum mechanics could be reached by formalized considerations about the symmetry of the atom strongly appealed to Ehrenfest. Several specialists came to lecture to us about the subject, Wigner first of all.* That was not

*Eugene P. Wigner, born at Budapest in 1902, studied chemical engineering at Berlin but soon turned to theoretical physics. He was outstanding among those who perfected and applied quantum mechanics and was one of the pioneers of the use of group theory in physics. He received the Nobel Prize in 1963.

entirely successful: he was too polite. Of course, if I were to lecture to professional physicists or mathematicians, and if I should explain in great detail how to solve a simple quadratic equation, the audience would feel insulted. The trouble was that to Wigner many things we found difficult seemed as simple as a quadratic equation, so he did not dare to explain them. After all, according to an apocryphal legend, John von Neumann had taught him group theory during one rainy Sunday afternoon. Heitler had no such qualms, and was therefore a better teacher for beginners. I also remember a brilliant lecture by Van der Waerden. But, for me, the great revelation was Hermann Weyl's *Gruppentheorie und Quantenmechanik*. I think it is a wonderful book. I had—and still have—the feeling that this is the right way to formulate mathematics, that the concepts Weyl formulates have just the right level of abstraction, yet correspond to something "real," whatever that may mean in pure mathematics. If I have later been able to apply group theory with some measure of success to several problems, and if I have even been able to make a small but not entirely insignificant contribution to that field of pure mathematics, this is thanks to Weyl's book. I never met Weyl until after the Second World War at Princeton, when, like Einstein, he had found a position at the Institute of Advanced Studies. Then I was happy to be able to tell him how grateful I was. (Weyl on that occasion paid me a probably wholly undeserved compliment: he introduced me to a colleague as a physicist with the soul of a mathematician.)

Ehrenfest enforced a few strict rules. The reading room was what is known in German as a Präsenzbibliothek, a library where books and journals can be consulted but can never be taken out. The only exception he allowed was that lecturers giving lectures in the same building or in the adjoining physics laboratories could take a book to their lectures, but it had to be brought back immediately afterwards. Once, when during a checkup it was found that a few volumes were missing, Ehrenfest was so exasperated that he closed down the library; it took almost a week to appease him.

Similarly, if you were admitted to the colloquium you were supposed to come regularly. If you were "oscillating," it would be made clear that you were no longer welcome. The colloquium took place on Wednesday evenings, from half past seven until half past nine or ten, with an interval for tea. The boiling kettle on a gas ring remains in my recollections closely associated with the colloquium. An older assistant in the low-temperature lab, Miss van der Horst, took care of the tea. Ehrenfest always respectfully addressed her as Mejuf-frouw. After more than fifteen years in the Netherlands he had not yet discovered that this is a form of address used in formal writing but hardly ever in spoken language. Talks were given by the regular members—mainly on recent publications, more rarely on work of their own—and by guests. To speak at the colloquium was for a young physicist a bit of an ordeal. Ehrenfest might interrupt you at any moment, often with what he called stupid questions, he would im-prove your formulations, and I have even seen it happen that he completely took over. Other participants might interrupt as well and yet discussions never became chaotic: a clear picture finally resulted, or at least a clear statement of the unsolved problems. Guest speakers were treated with greater respect, but they too might have to face a crossfire of questions. There was a certain ceremonial: they had to sign their names with the date on the whitewashed wall. When the Institute of Theoretical Physics moved to other premises and the old building was pulled down, that section of wall was cut out and in-stalled in the new lecture room, where it can still be seen. It shows an impressive list of names. I remember specifically a colloquium given by Schrödinger. Schrödinger was the creator of so-called wave mechanics, which later became one of the formulations of new quan-tum mechanics. He himself was not happy about the way in which his mathematical formulation—which was fully accepted—was later interpreted. It was probably for that reason that Ehrenfest had asked him to speak on a less controversial subject: the theory of colors. Schrödinger had already in 1920 published three papers on the mathematics of three-color theory and in 1926 he contributed an article on the subject to the famous textbook of Müller-Pouillet. It was an interesting lecture on a subject that was new to most of the audience. It ended with Ehrenfest and Schrödinger exchanging recollections about Vienna, and as the evening wore on their lan-guage became more and more Viennese.

I write about a colloquium with Einstein in Appendix A. Another very special occasion was a visit by Max Planck, who had come to Holland to receive the Lorentz medal at the Academy of Science in Amsterdam.[9] Ehrenfest had not asked Planck to speak at the col-loquium, but to attend a lecture by H.A. Kramers on the basic ideas of quantum mechanics. Kramers's opening phrases were characteristic: "Ich fühle mich heute wie ein Kind, das seinem Vater sein Spielzeug zeigt. Ich weiss, er wird schon horchen, eben weil er der Vater ist" (I feel today like a child that is showing its toys to its father. I know he will listen, just because he is the father). I think that it was on that occasion Ehrenfest asked whether Planck did not long for the days of old, when physics worked with clear-cut models and deterministic theories. Planck replied, courageously: "Man soll sich niemals zurücksehnen nach einer Sache von der man eingesehen hat dass sie unrichtig war" (One should never sigh for something which one has come to realize was erroneous).

Ehrenfest's witty remarks and sharp criticisms would sometimes hurt. If the victim was someone he liked, or at least respected, he would apologize in an almost naïve and most disarming way. "Please, don't be angry," or "You aren't angry, are you?" were expressions that came easy to him. But it would be wrong to say—as some biographers have done—that in all cases "his intention was the very best, [that] it was always an attempt to render aid, comfort or to show sympathy."[10]

No, there were cases where Ehrenfest felt a strong antipathy and then he became unreasonably aggressive. Walter Elsasser in his autobiography describes his unfortunate experience at Leiden, where he arrived in October 1927 as a successor to Uhlenbeck.[11] Relations were strained from the very beginning, and in November Elsasser was sent back to Germany (with payment for the rest of the semester). The final crisis came when Ehrenfest one morning detected a whiff of perfume in Elsasser's office: Elsasser had just had a haircut and had not managed to prevent the barber's putting something smelly on his hair. Ehrenfest told Elsasser bluntly to go away. I have no reason to doubt the truth of Elsasser's story; it was well known that Ehrenfest objected to perfumes, just as he objected to smoking and to alcoholic drinks. (It so happened that he had to make exceptions for his two best friends: Einstein was an inveterate smoker and liked a glass of beer; Joffe at Leningrad loved perfumes, though—always according to Ehrenfest—they were perfumes of a most superior kind.) I do remember having heard stories about Ehrenfest kicking somebody out because he smelled of the barber shop, but I didn't know Elsasser was the victim, and perhaps such incidents happened more than once. But Elsasser is in error in his assessment of the prevailing opinion on Ehrenfest. Some students may have been disappointed that Ehrenfest did not accept them, a few may have borne him a grudge, but by far the majority admired him, loved him, took his eccentricities for granted, and were grateful for the many things he did to help them in their careers. Ehrenfest, who had many relations abroad, often acted as an "impresario," not only for his own students, but occasionally also for experimental physicists, mathematicians, and astronomers. Some of the faculty members may have been peeved because they felt Ehrenfest's students did not pay enough attention to their subjects. Once, in 1930, I had to pass a tentamen in differential geometry. I knew enough to get through, but afterwards the professor, who was rather hot-tempered but had been kind to me, rang Ehrenfest and took him to task: I might have done much better had I spent a little more time on the subject—which was true enough. That kind of friction existed; but I don't think it often led to real feuds, and Ehrenfest had many friends among the faculty.

I was too young at the time to spot what really happened. I just remember that Elsasser came, and that he once explained to me that he would like to organize some special classes in connection with Ehrenfest's lectures where, by way of exercise, concrete problems would be put and discussed: that was customary in Germany. Of course, this proposal involved a not entirely unjustified criticism of Ehrenfest's rather cavalier way of treating detailed calculations. Then, one day, Elsasser was gone.

I do not want to exonerate Ehrenfest: I feel Elsasser suffered undeservedly, and I can only guess at the deeper causes underlying the conflict. Where Elsasser goes wrong is when he suggests that the prevailing attitude was something like "Ehrenfest is impossible but we must endure him because he was Lorentz's choice."[12] That was not the real situation at all, and Elsasser's successors did not have comparable difficulties.

In April 1928, when the payments to Elsasser came to an end—he had been appointed for half a year—the position was divided between two Dutchmen, R.L. Krans and A.J. Rutgers. Krans was a solid, sedate, not particularly quick-witted, but extremely reliable fellow. He got his PhD in December 1931, but had left Leiden already in September of that year to take up a position in a secondary school in Arnhem. As a physics teacher he was very successful. He was co-author of a textbook, and later taught didactics at the University of Utrecht and also at Groningen and Amsterdam. He made no real contribution to theoretical physics, but rendered great service to physics teaching at secondary-school level. Rutgers was by far the more brilliant. He had first studied chemistry at Amsterdam, but after his doctoraal, he had come to Leiden to study theoretical physics. His PhD thesis at Leiden (3 June 1930) dealt with the theory of thermoelectricity in crystals. He next worked at Amsterdam with Michels,* and was then appointed to the chair of physical chemistry at Ghent, Belgium, where he did creditable work in electrochemistry. He was an excellent teacher and wrote a very useful textbook. He had a lively sense of humor and was always willing to help younger people. He was also a man of uncompromising honesty. I cannot write more about him: he was and still is one of my best friends.

Let me only mention that in 1930 the two of us accompanied Ehrenfest to the summer school at Ann Arbor. Fermi lectured on the quantum theory of radiation, essentially in the formulation that was later published in the *Review of Modern Physics*. His lectures were marvelous and his English was easy to follow, once you had understood that his phonetics were largely Italian. For instance "finite" was pronounced as "fee*nee*te" with the stress on the second syllable. Someone once put a question containing the word "infinite." Fermi did not understand him, so Ehrenfest came to the rescue and explained it as "not feeneete." We also met Onsager, without realizing his stature. I believe that he had already worked out the main points of his theory of irreversible phenomena and that he discussed them with Ehrenfest. I seem to remember that Ehrenfest's reaction was "He may be on to something, but I am not quite sure." (I shall come back to the Onsager relations in Chapter 8.)

After the summer school, Ehrenfest went on to California, whereas Rutgers and I went home. It was quite an adventure and I can still shudder at the thought of the risks I took as a very inexperienced driver with a very old and uninsured Model T Ford. We sold it in the outskirts of Newark for $12—I had bought it for $35. I am glad that at least I did not try to drive it into New York.

*A.M.J.F. Michels (1889–1969) is known for his precision measurements of isotherms and of dielectric constants of gases, and of electric and magnetic properties of solids, at high pressures. He made important contributions to the technique of high-pressure work, and apparatus he built played an important role in the development of polythene. He was also a very courageous man and is mentioned in the dedication of R.V. Jones, *Most Secret War* (with a slight error in the initials).

It was G.H. Dieke who taught me how to drive. In those days you got a Michigan driver's license after signing a form that you had driven a car for more than five hundred miles. I do not want to swear that I had really covered that distance when Dieke said he was fed up with teaching me and that I should go and sign, but, in any case, I had driven many more miles than that by the time I arrived in Newark. Back in the Netherlands, I discovered that my father was getting a car. The salesman gave me an hour or so of instruction in operating a normal gear shift and then went with me to the office that dealt with driving tests. "Look here," he said, "this young man has an American driving license, and you know how difficult it is to get a license over there, so what about it?" Yes, they knew, and I got my Dutch driving license without any test. So my present driving license is still based on a signature of doubtful validity. But by now I must have driven about half a million miles without major accidents, so I have no great qualms of conscience.

Let me come back to Krans and Rutgers. Although they were very different as physicists and very different persons in general—Krans was a practicing Calvinist, whereas Rutgers was a bit of a libertine—they got on well together and with Ehrenfest.

Nor did I ever have difficulties myself; in any case, not the kind of difficulties Elsasser had. Yet I must admit that for a young man like myself it was disturbing that a professor I admired and respected would from time to time explain that he had lost all confidence in his own abilities, that life was becoming intolerable, and that he considered putting an end to it all. Later, after Ehrenfest's suicide, I often wondered, Could I have done more to help him? But I was young and inexperienced and had put my mind at ease by the well-known but fallacious superstition that people who speak about suicide don't commit it. Moreover, with Ehrenfest you did not always know whether he was talking in earnest or indulging in an intellectual exercise. "Why should I not put an end to my life if I no longer like it?" he once asked. My answer—"You are not allowed to kill anybody, least of all a decent and valuable man, and the fact that you happen to be one and the same person does not make any difference"—seemed to satisfy him, at least for the time being.

An experimental physicist called Van Mierlo was working in the Kamerlingh Onnes Laboratory about that time. I believe he was intelligent but he was somewhat dissipated and had not made much progress in his studies. Finally, he had pulled himself together and felt he could go to Ehrenfest for a tentamen. He was rather nervous though, and although it was only 9:00 A.M., he first fortified himself with a swig of "oude klare" (Dutch gin). That was almost worse than perfume, and after a few minutes Ehrenfest kicked him out and told him that he could not come back until a whole year had past. The boy was furious; he felt that this was unjust and that his knowledge had not been tested at all. So he took his revenge. He went to the colloquium room and there, on the sacred wall, among the names of Einstein and Bohr and others, he wrote in bold characters his name, underlining it with a deep gash in the plaster. When I discovered it, I decided that Ehrenfest would be enraged by this sacrilege and I did not want to get the poor boy into worse difficulties than he was in anyway. So I carefully erased the name—fortunately it was written in pencil—but I did not repair the scratch in the plaster. The young man left Leiden and went to Amsterdam. I later heard that during the war he was active in the Resistance and that he had been caught and executed by the Germans. A few years ago the scratch in the plaster could still be seen, a curious monument to a nameless fighter in the Resistance.

Let me end this section on a lighter note. As I shall relate later, I went with Ehrenfest to Copenhagen in the spring of 1929. Shortly after our arrival I decided to have a haircut. I did not know any Danish at the time, the barber did not know German or English, and when I came out of the shop I noticed with dismay that I was very smelly indeed. I did what I could, but washing facilities at the boarding house where I was staying were not easily accessible during daytime, or provided only cold water; in any case some of the smell remained. I went with some people for a long walk along a windy pier—Langelinie—but in vain. In the afternoon I had to meet Ehrenfest, and I was really afraid. However, someone had told him about my plight and my heroic efforts to remedy the situation. When he saw me he sniffed, smiled, and said: "Du duftest wie eine Frühlingsblume" (You smell like a flower of spring).

Ehrenfest and Pauli

Pauli gave a talk at the colloquium on 26 September 1928.[13] He had arrived in Leiden the day before and his signature on the sacred wall carried the date 25 September 1928. Of course, Ehrenfest and

Pauli had met before: they had both attended the 1927 Solvay Conference,[14] and they must have met at Göttingen, Hamburg, or Copenhagen. In any case Ehrenfest had already coined the name "Geissel Gottes" (Scourge of God)[15], of which Pauli was very proud. But this was Pauli's first visit to Leiden, and it led to a closer friendship between the two.

Wolfgang Pauli was born in Vienna on 25 April 1900. He studied from 1918 to 1921 with Sommerfeld at Munich, and it was there that he wrote his famous comprehensive text on the theory of relativity for the *Enzyklopädie der Mathematischen Wissenschaften,* a text highly praised by Einstein himself. He then turned his attention to quantum theory. He worked for one semester (1921–1922) with Born at Göttingen and from there went to Hamburg. He spent one year at Bohr's Institute at Copenhagen, and returned to Hamburg in 1923. When he came to Leiden he had recently been appointed to the chair of theoretical physics at the ETH at Zürich.*

Ehrenfest and Pauli had several things in common, which brought out how much they differed. They both came from Vienna and went to school there but their family backgrounds were not similar at all. Ehrenfest's father was a rather successful shopkeeper, whereas Wolfgang Pauli senior was a distinguished university professor. Ehrenfest was an almost fanatic teetotaler and was austere in all his habits, while Pauli occasionally had one too many and liked to pose as a man of the world. Not always successfully, by the way: I remember he once handed me a corkscrew to open a bottle of champagne. Both became first known to the world of science by their articles for the *Encyclopedia,* but Ehrenfest was thirty-one when his article was published; Pauli was ten years younger. Both of them were always aiming at an understanding and clarification of basic principles and both were known—and feared—as sharp critics, who did not mince their words. Though Ehrenfest referred to Pauli as "Der Geissel Gottes," he himself could be at least as sharp. Ehrenfest was definitely a better and more lively lecturer than Pauli, who was not at his best in a course of fairly elementary lectures and who hardly ever lectured to general audiences, something Ehrenfest—as Lorentz before him—enjoyed doing. But Pauli's intellectual force and mathematical virtuosity were far superior to Ehrenfest's. Al-

*The "Eidgenössische Technische Hochschule" was then, and still is, an institute of technology of great international reputation.

though Ehrenfest had not been educated as an Orthodox Jew and had forsaken all ideas about a God in heaven at the age of twelve, he had remained in close touch with the Judaism of his relatives and had received some instruction in Hebrew and biblical history. Like Einstein he felt strongly tied to Jewish culture and Jewish tradition. He liked to exalt the merits of Jewish scientists and scholars. He was, at one time, rather disappointed with my performance and said, "If you had one-quarter of Jewish blood, or if it were only one-eighth, you might still grow into something. But you are just too Aryan." A friend of mine found him studying a picture of Lorentz. "It must be one of us," he said, "it must be one of us."

Pauli was half Jewish but had been baptized as a Catholic. I have never noticed that this partly Jewish origin played an important role in his life. All the same, let me relate one little incident that happened at Zürich in the early summer of 1933. Hitler had come to power and many Germans of Jewish origin passed through. There was a young couple, a German Jewish scientist—he was no physicist; I have forgotten both his name and his field—with a pretty, blond, and decidedly non-Jewish wife. We went with Pauli for a swim, and while we were having a snack the conversation turned to racial questions. "The half-Jews look always far more Jewish than the full Jews," Pauli said, and, looking at the young woman, "Your children will look exactly like me, but exactly." She looked at him in some astonishment and was clearly not too pleased with what she saw. And it must be admitted, Pauli in his bathing suit, rather fat and hairy, was not exactly what one would choose as a baby. "Dann muss man sich das doch noch einmal überlegen" (Then one will have to reconsider that), she snapped.

But the main difference between Pauli and Ehrenfest lay elsewhere. "All his life Ehrenfest suffered from feelings of inadequacy and inferiority" (Martin Klein's formulation). Or, as Albert Einstein, his close friend, put it:

... his tragedy lay precisely in an almost morbid lack of self-confidence. He suffered incessantly from the fact that his critical faculties transcended his constructive capacities. In a manner of speaking, his critical sense robbed him of his love for the offspring of his own mind even before they were born.

Pauli, on the other hand, was well aware of his powers. It would be wrong to call him conceited. He could be arrogant, but he knew his limitations. He knew—and perhaps regretted—that he was no Einstein, no Bohr, and even no Heisenberg, although he was at least the equal of all three of them as a problem-solver, as a "craftsman" in that curious craft of theoretical physics.

The 1928 colloquium was, as I said, not at all the first meeting between Pauli and Ehrenfest. Could it be that all the same, just before the colloquium started, Ehrenfest said, "Your papers please me better than you yourself," to which Pauli replied, "With me it's just the other way around" (Ihre Arbeiten gefallen mir besser als Sie selber. Mir geht's gerade umgekehrt)? I am *almost* certain I did overhear these remarks, but they have been so often repeated, have to such an extent become part of traditional legend, that I cannot vouch for it. They may have been made on another occasion.

In 1929, when Ehrenfest and Pauli met again at the conference at Bohr's Institute, about which I have more to say elsewhere, Ehrenfest produced a number of quips at Pauli's expense. For instance: The "Pauli effect [see p. 91] is just a special case of the more general phenomenon that misadventures rarely come singly" (Ein Unglück kommt selten allein). Or "Why is it so irritating when Pauli walks up and down in front of the audience? Answer: Because of the disappointment at the turning points." I heard this remark at Bohr's home and am unable to confirm the standard version that Ehrenfest first made this remark during a lecture, though this may well be.[16]

I think their most amusing—and least known—battle of wits was the one in connection with the awarding of the Lorentz Medal of the Royal Netherlands Academy of Arts and Sciences to Pauli on 31 October 1931. Ehrenfest had written to Pauli that it would be appreciated if he would appear in a black suit. (Although Ehrenfest on the whole was unconventional in dress and behavior, he tried to conform on official occasions. There was even a story, probably invented, that when he first came to Leiden he had misunderstood the regulation that prescribes that the academic gown should be worn over a black suit and, in consequence, went in search of black underwear, which caused considerable astonishment at a haberdasher's.) Pauli replied by postcard, dated Zürich, 26 October 1931.

Dear Ehrenfest,
    Just now I ordered a new black suit at my tailor's. But I shall only put it on at Amsterdam, if you promise me to thank me in public, in your official

allocution at the Academy, for not having saved myself the trouble of going to the tailor. . . .

How did Ehrenfest comply with this condition? I quote from his address:

"At first sight the Pauli principle would seem to concern only those, who occupy themselves with the details of the distribution of electrons in Bohr-atoms, or with their spectra. And the title of your paper seems to confirm this impression. For it is "Ueber den Zusammenhang des Abschlusses der Electronengruppen im Atom mit der Komplex-Struktur der Spectren." But, for all its esoteric appearance, this principle does exert its influence in the midst of our everyday world. It is fascinating to make this clear to ourselves by a simple example.

We take up a piece of metal. Or a stone. A little thought will make us wonder why this quantity of matter does not occupy far less space. Certainly, the molecules are very closely packed. And so are the atoms in the molecule. —All right—But why are the atoms themselves as thick as they are?

Let us look for example at the Bohr model for a lead atom. Why are only very few of the 82 electrons circulating in quantum-orbits close to the nucleus and all others in ever more remote orbits? For the attraction of the 82 positive units of charge of the atomic nucleus is so powerful. Many more of the 82 electrons could come together on the inner orbits before their mutual repulsion gets too large. What then prevents the atom from making itself smaller in that way?! Answer: Only the Pauli principle: "No two electrons in the same quantumstate." Therefore the atoms so unnecessarily thick: Therefore the stone, the piece of metal, etc. so voluminous. You will have to concede, Herr Pauli, by partially waiving your exclusion principle you might free us from many worries of daily life, for instance from the traffic problem in our streets.[17]

So says the printed version, but in his speech Ehrenfest added something like "and you might also considerably reduce the expenditure for a beautiful, new, formal black suit." I attended the ceremony in the spectators' balcony in the Academy, but I do not remember the exact wording. It was just one sentence and it certainly contained the words "schöne, neue, schwarze Festkleidung"; "Und wie man die Auslagen für schöne, neue schwarze Festkleidung herabsetzen kann" should be fairly close. I do remember Pauli's grin and the approving increase of the amplitude of oscillation of his body.

But there was also an interaction in the field of serious science. In 1932 Ehrenfest published a paper entitled "Einige die Quanten-

mechanik betreffende Erkundigungsfragen" (Some Requests for Information Relating to Quantum Mechanics).[18] It is a true *cri de coeur,*
and as usual Ehrenfest's "stupid questions" touch on profound problems. Pauli, who had just finished his article on quantum mechanics
for the *Handbuch der Physik,* answered to the best of his knowledge.[19] It is striking that answering the questions occupied far more
space than putting them!

After Ehrenfest's death in 1933, Pauli wrote an obituary notice.[20]
It deals mainly with Ehrenfest's scientific achievements. But I quote
the beginning and the end where, in a few pregnant sentences, Pauli
expresses his admiration and his deep-felt sympathy:

> On 25 September of this year Paul Ehrenfest carried out his fateful
> decision to free himself of the burden of life that he could not bear anymore;
> under tragic circumstances and to the dismay of his family and his many
> friends and acquaintances. And now we must try to retain the memory of
> his scientific activity and the image of his personality, free from those feel-
> ings of inferiority and those worries that during later years had more and
> more overclouded his mind. It is the image of that man, who like a fountain
> of wit and spirit enters the discussion, with sharp criticism but at the same
> time with a penetrating insight into the foundations of scientific method, and
> who directs the attention to an essential issue that so far has not—or not
> sufficiently—been taken into account.

And Pauli's conclusion:

> If we look once more, in retrospect, at Ehrenfest's scientific activity, then
> it appears as a living testimony of the everlasting truth: scientifically objec-
> tive criticism, however sharp, is always stimulating and inspiring when it is
> consistently pursued to the very end.

But that was in 1933, and now I shall first have to tell something
about the preceding years.

4

Copenhagen

Niels Bohr

April 1929 was the beginning of a new phase in my studies. It was then that Bohr[1] organized the first of what was to become a famous series of conferences. Ehrenfest was of course invited, and he proposed to Bohr that I should accompany him. For me it was a great adventure and I remember many details. Today you fly from Amsterdam to Copenhagen in little more than an hour; in 1929 you could get there in less than twenty-four hours if you took an overnight train; but we traveled even more leisurely.

We first took the day train to Hamburg. Ehrenfest paid keen attention to things and people he saw and had many witty comments; he was, in that respect, anything but an introvert or absent-minded scientist. At the Hamburg railway station he was, for instance, intrigued by the signs indicating where to find the representatives of organizations offering assistance and protection to inexperienced travelers. There was a Roman Catholic organization for young women and one for young men. Similarly there were two Protestant organizations, and one for Jewish girls. I remarked that the Jewish young men seemed to be able to take care of themselves, a remark that thoroughly pleased Ehrenfest. "We certainly can," he said. However, *we* were taken care of by a middle-aged woman, a distant relative of Ehrenfest, who had booked rooms for us at a little hotel near the station. She knew the manager and she told him that the guest she was bringing was a close friend of the great Einstein himself and should be given one of the best rooms. Less than four years later it would be unwise to mention Einstein's name at all.

Next day we continued to Copenhagen. Ehrenfest amused himself with a book of short stories called *Reife Früchte vom Bierbaum*, especially with a rather silly story, "Das Hosentürl," telling about the ruler of one of the many pre-Bismarckian miniature states who had forgotten to button his fly. But somewhere between Hamburg and Warnemünde he suddenly became serious and said: "Jetzt wirst du Niels Bohr kennenlernen und im Leben eines jungen Physikers ist das das wichtigste Ereignis" (Now you are going to know Niels Bohr, and that is the most important thing to happen in the life of a young physicist).

In later years I have increasingly realized how true this was, and I know that this holds true not only for myself but for many others. The scattered recollections and little anecdotes that follow may appear trivial and insignificant in themselves, yet they are important to me because they form part of one great impression that has enriched my life.

Air travel makes one lose a lot of things, often even time. I don't go as far as Ivan Illich, who wants to impose a speed limit of forty kilometers an hour for all traffic, but the good old Warnemünde-Gedser ferry on which we crossed part of the Baltic certainly did not go faster than that. And it gave me the feeling that we were entering a new part of the world. There was still much ice drifting around, for it had been a very cold winter. Leiden people will remember it as the winter the town hall burned down. That town hall was a fine example of a style known as Dutch Renaissance, but I was soon going to see a far more impressive example of that style: Frederiksborg Castle, built between 1600 and 1620 by King Christian IV. Bohr greatly admired that building, and he always took care that visitors should see it.

In the evening we arrived in Copenhagen. The first impressions of a new town and of an unfamiliar language. I was put up in a boarding house where I temporarily shared a room with Gamow. Frøken Have's pension occupied a flat on the fourth floor.* (Here I have to admit a possible error of plus or minus one, but there is in any case no international agreement about the way of numbering

---

*Frøken Have, I soon found out, was a kind-hearted and generous woman, an efficient housekeeper and an excellent cook. By making physicists feel at home she really made a contribution to physics. I shall not go into details but refer to a short but excellent obituary, written by Rosenfeld. *Nucl. Phys.* 31, 689 (1962).

floors.) The window of our room gave onto an inner courtyard. Downstairs there was a restaurant called Café Antiautomat. (There had been a brief early spell of automated snackbars in Copenhagen and that had provoked a reaction. The automatic snackbars had disappeared or in any case were no longer a threat, but there remained more than one Café Antiautomat.)

A rather awful band performed there every evening until midnight. Their favorite number, which they played with some fluency, was a top hit in those days: "Wenn der weisse Flieder wieder blüht" (When the White Lilac Blooms Again), the masterpiece of a certain Doelle, a composer about whom I know nothing at all. (This, by the way, is one case where my memory was at fault. I seemed to remember that the song was by Walter Kollo [1878–1940], a well-known composer of operettas. Hoping it might be an "evergreen" I went to a record shop and asked for records of songs by Walter Kollo. The salesgirl was surprised. *"Walter* Kollo?" she said, "Walter? Do you mean Richard?" And she showed me some records. On the cover appeared the picture of a handsome young man. The text explained his musical background: grandfather Walter was well-known. . . . I felt very old that day. It took me some time to find out about Doelle. Does he too have a singing grandson?) I am not much of a musician, but melodies have a way of getting connected in my memory with certain situations, and I still hear that song and the clumsily heavy "pom-pom-POM, pom-pom-POM" with which the bass underlined the opening bars. "April is the cruellest month, breeding lilacs out of the dead land. . . ." No, I did not feel like that, although at first I felt slightly ill at ease, slightly lonely, perhaps even somewhat homesick. Anyway I did not read *The Waste Land* until years later when Homi Bhabha introduced me to Eliot's poetry.

About the conference itself I shall not say much. Rosenfeld has described it in considerable detail in a booklet sent as a greeting to friends and former collaborators of Bohr's Institute on the occasion of its fiftieth anniversary.[2] I shall quote two passages. Rosenfeld relates:

Ehrenfest came in and went straight to Bohr, followed by a tall, fair-haired, rosy-cheeked youth of rather indolent gait, who did not quite know what to do with his arms. "Ich bringe Dir diesen Knaben" (I am bringing you this boy), he said to Bohr, while he affectionately put his hand on the boy's shoulder. "Er kann schon etwas, aber er braucht noch Prügel" (He has already some ability but he still needs thrashing).

That was me. The second passage relates to an almost unbelievable demonstration of the Pauli effect. Rosenfeld and I compared our recollections and they agreed exactly, so this is definitely an authentic and well-established case. Let me first explain about the Pauli effect. It was alleged that Pauli was so utterly remote from any kind of experiment that his very presence in the vicinity of a research laboratory was sufficient to cause apparatus to break down in the most unaccountable way. Pauli's friends assiduously collected—and possibly invented—examples, and he himself thoroughly enjoyed such stories. He may even have been inclined to believe the effect really existed, but I shall come back to that later on. Of course, because of the legend, experimental physicists tended to get nervous whenever Pauli entered their laboratory, so they may have made errors.[3] But now for Rosenfeld's story.

Pauli, so far as I remember, was rather subdued, except on one spectacular occasion. Heitler, by lecturing on the theory of the homopolar bond, unexpectedly excited his wrath: for, as it turned out, he had a strong dislike to this theory. Hardly had Heitler finished, than Pauli moved to the blackboard in a state of great agitation, pacing to and fro he angrily started to voice his grievance, while Heitler sat down on a chair at the edge of the podium. "At long distance," Pauli explained, "the theory is certainly wrong, since we have there Van der Waals attraction; at short distance, obviously, it is also entirely wrong." At this point he had reached the end of the podium opposite to that where Heitler was sitting. He turned round and was now walking towards him, threateningly pointed in his direction the piece of chalk he was holding in his hand: "Und nun," he exclaimed, "gibt es eine an den guten Glauben der Physiker appellierende Aussage, die behauptet, das diese Näherung, die falsch ist in grossen Abständen und falsch in kleinen Abständen, trotzdem in einem Zwischengebiet qualitativ richtig sein soll!" (And now there is a statement, invoking the credulity of the physicists, that claims that this approximation which is wrong at large distance and is wrong at short distance yet is qualitatively true in an intermediate region.) He was now quite near to Heitler. The latter leaned back suddenly, the back of the chair gave way with a great crash, and poor Heitler tumbled backwards (luckily without hurting himself too much).

Casimir who also remembers the incident notes that Gamow was the first to shout: "Pauli effect." And as an afterthought he adds: "Sometimes I wonder whether Gamow had not done something to the chair beforehand."

I have nothing to add to this story, apart from the remark that unlike Pauli himself, Heitler was definitely a lightweight, which made the demonstration all the more impressive.

For me, the most important consequence of this conference was that Bohr invited me to stay on, and for the next two years I spent more than half of my time in Copenhagen. My father, not too well acquainted with the world of physics, may have had some doubt whether the man with whom I was working was really as famous as I said he was. So he addressed one of the first letters he wrote me: Casimir c/o Niels Bohr, Denmark. Of course, the letter arrived without delay. The Danish Post Office had not even troubled to add an address; they had only scrawled an ø on the envelope. After that I think my parents felt more convinced that I was in good hands; they were even more convinced when they met Mrs. Bohr. I found it so evident that the letter arrived that I did not keep the envelope. A pity; by now it would be a nice collector's item.

During my stay in Denmark I often acted as a kind of private secretary to Bohr, who liked to have someone to talk to about his ideas. Moreover, writing a paper, or rather trying to write a paper, was for him a way of thinking. For that kind of assistance shorthand was not required. Sentences came haltingly, hesitatingly, were often broken off and restarted. But one had to become accustomed to Bohr's voice, which was soft and rather indistinct, whether he spoke German, English, or Danish; and his habit of walking around the room all the time did not help. I put quite a bit of energy into learning Danish (I am still reasonably fluent) and I believe I got fairly good at understanding Bohr.

Working with Bohr in that way was a unique experience. Ehrenfest taught me the importance of clear, crisp formulations and was a master at finding simple examples illustrating the gist of a physical theory. Later Pauli forced me not to shun elaborate mathematical analysis. Bohr was both more profound and closer to reality. As a young man he had done beautiful experiments on surface tension and had built most of the apparatus with his own hands, and his grasp of orders of magnitude went all the way from the atomic nucleus to engineering problems of daily life.

The following anecdote illustrates that point. Close to Bohr's Institute there is a body of water—I hesitate whether to call it a lake or a pond—about three kilometers long and between 150 and 200 meters wide, the Sortedamsø. It is crossed by several bridges. One day Bohr took me on a stroll along that lake and across one of the bridges. "Look," he said, "I'll show you a curious resonance phenomenon." The parapet of that bridge was built in the following way. Stone pillars, about four feet high and ten feet apart, were linked near their tops by stout iron bars (or rather more likely, tubes) let into the stone. Halfway between each two pillars an iron ring was anchored in the stonework of the bridge, and two heavy chains, one on each side, were suspended between shackles welded to the top bar close to the stone pillars and that ring. Bohr grasped one chain near the top bar and set it swinging, and to my surprise the chain at the other end of the top bar began to swing too. "A remarkable example of resonance," Bohr said. I was much impressed, but suddenly Bohr began to laugh. Of course, resonance was quite out of the question; the coupling forces were extremely small and the oscillations were strongly damped. What happened was that Bohr, when moving the chain, was rotating the top bar, which was let into, but not fastened to, the stone pillars, and in that way he had moved the two chains simultaneously. I was crestfallen that I had shown so little practical sense, but Bohr consoled me, saying that Heisenberg had also been taken in; he had even given a whole lecture on resonance!

That bridge, known in Bohr's Institute as the Resonance Bridge, figures also in another story. In the late summer of 1934, I was again in Copenhagen for a conference and my wife had come with me. One evening we were sitting with a small group of people in a café and someone suggested we should walk to and over the Resonance Bridge. This somehow led to a wager between Placzek and myself that I would swim across. The stake, fifty crowns, was high considering our respective financial positions. I later heard that Placzek, himself a bachelor at the time, had based his bet on a theory of matrimony: he had expected that my wife at the last moment would hold me back; little did he know my wife. It was not much of a swim and the water was not cold, but it is a curious feeling to swim so close to the center of a town, accompanied by an escort of surprised ducks, and even more so to walk home to your hotel with the water dripping from your clothes. Next day I had to travel home in the trousers of my dinner suit and in patent leather shoes. A few years ago I made a pilgrimage to that bridge, but it had been widened and the parapet was no longer there, so I could not check the dimensions.

Frisch tells the same story.[4] I must correct him on a few details. The Sortedamsø is not semicircular, it is only slightly curved; the swim was not in order to avoid a detour; that would have been far too rational a reason for swimming; I started my swim no more than 100 or 200 yards from the resonance bridge; the wager was not twenty but fifty crowns; both my wife and I are sure of that and I doubt she would have let me swim for twenty. I did not waste money on a taxi: we walked home; I did not travel home the next day in my complete dinner outfit—only in black trousers and patent leather shoes. This shows how easily legends can arise. But of course the essence is there, even in Frisch's version. I did swim across the Sortedamsø, which is a thing few people—and very few, if any, physicists—have done; I did beat Placzek in a wager, which not so many people have done either; and my wife did not object, which many wives might have done.

Recently I was asked to confirm a sequel to the story according to which I had been met on the other bank by a policeman, who at first took a rather serious view of the matter. I was supposed to have pacified him by explaining that I came from Holland where we have many, many canals and a shortage of bridges, so that we have to swim across all the time. My spokesman was disappointed when I told him there was no factual basis whatever for this version, but he said he would continue to tell the story that way: he liked it too much. I would not be surprised if, in future, he will even add that I personally confirmed it.

My secretarial competence was really put to the test early in 1930. Oskar Klein, for many years Bohr's closest collaborator, had applied for the chair of theoretical physics at Stockholm.* There were three other candidates, all of them good physicists. Bohr had been chosen as one of the experts who had to pronounce judgment and he asked me to help him write the report. I think that Bohr was convinced from the very beginning that the chair should go to Klein. Two candidates, although competent in their respective fields—kinetic theory and hydrodynamics—had not been in touch with modern theory, and for a general chair of theoretical physics competence

*Swedish universities were in those days exceptional in that one did apply for a professorial appointment and that the deliberations on the respective merits of the candidates were made public. Such matters were kept confidential in most European universities.

in quantum mechanics and theory of relativity was an essential requirement. At least so it seemed to me, and I think Bohr agreed. The third candidate, however, was Ivar Waller, and he certainly did meet this requirement. But he was four years younger than Klein, and his work, though excellent, was at that time not as broad in scope as Klein's. (Not long afterwards, in 1934, he became professor of theoretical physics at Uppsala, a position he held until his retirement.)

Such considerations might have been deemed sufficient by many people, but Bohr was far too conscientious to leave it at that. We first went through the publications of the four candidates. I won't pretend that we read every word or that we checked all the formulae, but Bohr analyzed them pretty carefully, and explained to me many of the finer points. He was an expert on kinetic theory, had thoroughly studied both its foundations and its applications in connection with his doctoral thesis on electrons in metals. And he had dealt with hydrodynamics in the work on surface tension I mentioned before. Then came the next phase: writing the report. Bohr wanted to do full justice to the work of each of the candidates. Every sentence, every word was carefully weighed. He tried five or more adjectives to express his admiration for one paper of Klein and rejected them one after the other. Finally he settled for "lødig". I did not know that word, so I asked, in Danish, "What is 'lødig'?" Bohr looked at me in surprise that I should not know so simple a word—by that time he took it for granted that I knew Danish—and said only "gold is."* Of course my knowledge of Danish was far from perfect, but by way of relaxation Bohr and I had often solved Danish crossword puzzles together and I was usually able to supply quite a number of words. The habit of solving crossword puzzles when I feel tired has stayed with me and when, after having been stuck for a while, I hit upon one word that suddenly makes the solution of a whole area of the puzzle evident, I sometimes repeat the phrase Bohr used on such occasions: "Nu kommer det hele" (literally translated: Now comes the whole). Finally we got to the end of the report and Miss Schultz, the extremely competent secretary of the Institute, typed it out. However, that was only the first version, and Bohr started all

---

*The word in its literal meaning is applied to gold or other noble metals and simply means "pure, unadulterated." In a figurative sense it means valuable, rich in content. "Of sterling quality" might then be a good translation. I do not think the word is often used in connection with scientific papers.

over again. The new version was typed out and the whole procedure was repeated. Did the document really gain much after the third or fourth version? I rather doubt it. After all, this was not a case of profound philosophy, it was not a case of formulating the basic concepts of physics. But any suggestion from my side that we might consider the work was done was brushed aside. "Please, do not leave me in the lurch," he said once, when I began to show signs of impatience. And so I kept listening, and writing, and correcting, and cutting up, and pasting together, and—very occasionally—suggesting a slightly different formulation. I believe it was the ninth version —but again I am willing to admit an error of plus or minus one—that was finally sent off. In general, Bohr liked to finish a paper on, or just before, Saturday. "Then it travels on Sunday and so you gain one day" was one of his favorite sayings. I think we made it also this time. In any case on Saturday the first of March the work was done: I have among my books a Danish edition of the Icelandic sagas with the inscription:

Hendrik Casimir

Til Minde om fælles Slid
og med Tak for trofast Hjælp

fra Deres

Niels Bohr

1 March 1930

(In memory of common toil and with thanks for faithful help.) Bohr was a great admirer of these sagas, especially of *Burnt Njal,* and also of the poem Egil Skallagrims wrote after his son had been drowned at sea.[5] It is a sad thought that a few years later he himself would have to face a similar tragedy: he lost his oldest son, Christian, during a sailing trip in the Baltic.

The rough storm has robbed me
Of my best riches. . . .

Discussions with Bohr ranged over a great variety of subjects, both outside and inside physics. I remember for instance how we once discussed the famous one-way optical system proposed by Rayleigh in 1885 and why it does not violate the second law of thermody-namics. Bohr, by the way, greatly admired the work of Rayleigh, especially his work on the resolving power of optical instruments. The device in question works as follows: two nicols* are placed at an angle of 45° and between them there is a substance in a longitudinal magnetic field showing a Faraday rotation of 45°. The light going in one direction will go straight through, light in the opposite direction will be absorbed. In order to explain the geometry, books were placed on tables under appropriate angles, and Bohr, using his foun-tain pen† to indicate the polarization vector, walked to and fro be-tween these "nicols." Later, when I had some concern with mi-crowaves, I had good reason to remember this episode. Hogan, then at Bell Laboratories, did exactly this experiment with microwaves and ferrites and I felt rather ashamed that it had not occurred to me to apply Bohr's object lesson.

While Gamow and Landau were at the Institute, the three of us often went to the movies together, and we had a preference for lowbrow and lurid films. Sometimes we could persuade Bohr to come with us to see a Western or a gangster film we had selected. His comments were always remarkable because he used to introduce some of his ideas on observations and measurements into his criti-cism.

Once, after a thoroughly stupid Tom Mix film, his verdict went as follows: "I did not like that picture; it was too improbable. That the scoundrel runs off with the beautiful girl is logical; it always happens. That the bridge collapses under their carriage is unlikely but I am willing to accept it. That the heroine remains suspended in mid-air over a precipice is even more unlikely, but again I accept it. I am even willing to accept that at that very moment Tom Mix is coming by on his horse. But that at that very moment there should also be a fellow with a motion-picture camera to film the whole business— that is more than I am willing to believe."

Then there was also his famous theory about the advantages of defensive against aggressive shooting, because making a voluntary decision takes more time than reacting in a purely mechanical way. I described this episode in a poem I contributed to the *Journal of*

*This was before "polaroid" had made polarized light a household commodity.
†It was one of the once famous, large, thick, bright orange Parker Duofolds. Why don't they make decently sized fountain pens any more these days?

Jocular Physics of 1935, an unofficial publication on the occasion of Bohr's fiftieth birthday. I wrote it on a camping trip with my wife and a couple of geologists, who were looking for specimens of a special kind of foraminifera on the southern slopes of the Pyrenees; they even found them. For the convenience of the readers who do not understand German I have translated the original German doggerel into English doggerel. I suppose both versions are about equally Dutchy.

> We went to the flicks and Niels Bohr came along,
> And we watched the Black Rider, a man bold and strong,
> In a Western picture, where guns often bark,
> But it's always the hero who first hits his mark.
> At the end of the movie Niels Bohr, deeply moved,
> Set out to explain what the plot really proved.
> "That was a good film," I can still hear him say,
> "There was really a *'pointe,'** it showed in what way
> In a part of the world where all villains are armed
> The innocent men are surviving unharmed.
> In truth, there's no reason for flutter or fear
> If your purpose is pure and your conscience is clear.
> When you're facing a blackguard and he draws his gun
> You quickly draw yours, shoot him down and you've won.
> The scoundrel must make a momentous decision
> And that interferes with his speed and precision,
> But for the defendant there's no such distraction,
> Not a shadow of doubt can retard his reaction.
> So it's easy to shoot in advance of his shot;
> With his gun barely grasped he falls dead, on the spot."
> We, arrogant youngsters, we ventured to doubt
> This thesis of Bohr and we wished to find out
> If really a deep psychological facet
> Of criminal law does make virtue an asset.
> So the three of us went to the center of town
> And there at a gunshop spent many a crown
> On pistols and lead, and now Bohr had to prove
> That in fact the defendant is quickest to move.
> Bohr accepted the challenge without ever a frown;
> He drew when we drew . . . and shot each of us down.

*In Danish, and also in German, the word "pointe" (borrowed from the French but with a change of usage) indicates the gist, the essential point of an argument or a witticism. Bohr used it frequently, even when speaking English, and always pronounced it as "po-angte."

This tale has a moral, but we knew it before:
It's foolish to question the wisdom of Bohr.

Again, this is an insignificant episode, but the moral of the story that one should not doubt the wisdom of Bohr applies to more important things than shooting between gunmen in Westerns.

The story about the gunmen and the advantages of defensive shooting is also told by Dirac.[6] Bohr was really convinced that reactions are much faster than decisions, and our experiments seemed to bear him out. (We bought two identical toy pistols, the type with explosive caps on an automatically transported paper tape; it is easy to hear which shot comes first.) However, I believe this was a special characteristic of Bohr's: he had a sportsman's quick reactions, but he was a slow thinker, or rather, he thought about so many sides of a question that it took him a long time to arrive at a conclusion.

It is interesting to read what his younger brother Harald, the mathematician, had to say about Niels as a soccer player.[7] Harald himself had been a star in the Danish national team, the team that won a silver medal in the Olympic games in London in 1908. Niels did not get further than being a reserve goalkeeper on the first team of a leading Danish club of those days. So he played only in a few first-class matches. And now Harald's comment: "Niels was really quite good but he was too slow in 'running out.' " So here we see the same feature: fast reactions, slow decisions.

Bohr, by the way, kept in pretty good shape, even in later life. He liked to go on long walks, whatever the weather. I have walked with him while a sharp wind was blowing icy rain into our faces and heard him mutter under his breath "Velsigned vejr" (Blessed weather). He liked to ride a bicycle—as a Dutchman I could teach him and his boys one or two simple tricks—did some horseback riding, and went skiing in the winter or spring. At his country home in Tisvilde he would cut down trees, chop wood, and so on.

One of his principles was: if you *see* a tram (streetcar) you can catch it, and we made many a fast sprint together when I could barely hold my own, although he was more than double my age.

Bohr and Genetics

Bohr was always keenly interested in biological problems. He had enjoyed reading the writings of Fabre, that ethologist *avant la lettre,* and was thrilled by the discovery of the curious life history of the eel by E.J. Schmidt. On a deeper level he felt that his ideas on complementarity should be of profound significance for our understanding—and the limits of our understanding—of life.

The spectacular progress of molecular biology has brought with it, so it seems to me, a return to primitive mechanistic ideas, albeit in a slightly more sophisticated form, and also an overrating of physical concepts, like entropy for instance, in a realm where physicists ought to doubt their applicability. I am convinced that sooner or later Bohr's views will prevail.

In later years, however, he became particularly interested in problems of genetics and heredity. He emphasized the importance of a child's early surroundings and experiences; these might well outweigh hereditary factors. I do not think he offered a new point of view. I had often discussed such questions with my father, who regarded this as the fundamental, unsolved, and perhaps unsolvable problem of pedagogics: to what extent are intellectual capabilities, artistic gifts, dexterity, and moral traits predetermined by heredity, and to what extent can they be influenced by education? In recent years there has been a growing realization that experiences in very early childhood, from the moment of birth onward, may have a decisive influence, and also that malnutrition—even malnutrition *in utero*—may cause irreparable damage to the developing brain. There are also many drugs that can have a similar effect.

Bohr tended to the view that hereditary factors are very unimportant indeed, and that every healthy brain should originally have the same potential. He tried to illustrate this with a simple simile. There are many kinds of automobiles, he said, bigger and smaller, faster and slower, but that does not mean that one car should rather go to Paris and the other rather to Berlin. I do not think it was a particularly well-chosen simile; we might perhaps see an analogy between the physical characteristics of a race of man and the specifications of a make of motorcar. I fail to see any parallel to the intellect of man, although I do know that a car suitable for driving through Los Angeles might be inconvenient in the narrow streets of an ancient Italian town. As always, Bohr also looked for a complementary point of view. He had a favorite story about a father of identical twins, who had first given the two boys exactly the same education and the boys were exactly alike. Then a psychologist had convinced him that this was the wrong thing to do, the boys should find their own identity, and so he took a very drastic step: he was going to make them as different as possible, therefore he sent one boy to Harvard and the other to Yale. The one that went to Harvard developed into a real Harvard gentleman, the one that went to Yale became a regular Yale cad. But still there was no difference.

From the point of view of a practical schoolmaster it does not make much difference whether the potentials and limitations of the pupils that are entrusted to him are hereditary or determined by other influences. He can change neither the progenitors nor the early history of his pupils; he has to make the best of it, or rather of them. However, the question is important when it comes to the existence or nonexistence of racial characteristics. Bohr's tendency almost to deny the existence of any genetic factor in intellectual and artistic performance was at least partly a reaction against the mad racism of the Nazis and the supercilious racism of colonial imperialists. And even if we are neither Nazis nor colonizers it is only too convenient to put our conscience at ease by ascribing the poor intellectual performance of underprivileged groups or nations to hereditary factors we cannot change, instead of to material poverty for which we are partly responsible. As a warning against such hypocrisy Bohr's approach was valuable.

It is less gratifying that today in many circles anyone who does not accept as an axiom that all human races are in principle equally gifted in all respects is at once branded as an objectionable racist. That is an unscientific and even an uncharitable point of view. Suppose, for instance, that a serious study would lead to the conclusion that an outstanding gift for music is connected with a special congenital structure of the brain and that this special structure occurs more frequently among Negroes than among white people. And that, similarly, the ability to formalize thought mathematically is more frequent among whites. That would not make one race superior to the other, it would not preclude peaceful and constructive coexistence. Such variations—and I feel pretty sure that what remains after we have eliminated all environmental factors will at most be small variations—would be a pleasant aspect of the international scene. Similarly, it is a good thing if every country is able to feed itself, but if one country has a surplus of good cheese and another of fine fruits, this can further friendly relations.

There exist far less innocent possibilities than slight differences between races. Freeman Dyson in his recent book[8] argues convincingly that a menace to mankind foreshadowed in H.G. Wells's *The Island of Dr. Moreau* and presented with plausible detail by J.B.S. Haldane in *Daedalus, or Science and the Future* (a book in which Aldous Huxley found many of the ideas for his *Brave New World*) is rapidly becoming a very real danger. Modern biological techniques and insight in the structure of the carriers of hereditary characteristics may soon make it possible to interfere with the process of reproduction and the processing of genetic data in reproduction to such an extent that they will "demolish our comfortable world of well-defined species with its impassable barriers separating the human from the nonhuman."[9] Haldane's Daedalus is a sinister figure indeed. He has "superintended the successful hybridization of woman and bull to produce the Minotaur." And Dyson quotes Haldane:

The scientific figure of the future will more and more resemble the lonely figure of Daedalus as he becomes conscious of his ghastly mission and is proud of it.

I am no classical scholar but it seems to me that Haldane's Daedalus figure bears little likeness to its Greek mythological original. That is not an objection. If a writer feels that he can best convey his ideas by a mythological parable and if traditional mythology does not entirely suit his purpose, he is entitled to change traditional mythology. That is exactly what Goethe did when he created his Faust, for there is a fundamental difference between Goethe's Faust, who is ultimately redeemed, and the traditional Faust—and the Faust of Marlowe's tragedy—whose destiny is eternal damnation. Marlowe's closing scene, where Faustus, waiting for the clock to strike midnight, is voicing his utter despair, is deeply moving. Yet Goethe's Faust is more significant.

However, if we stick closer to the original Daedalus story, the moral to be drawn is equally alarming. What is the story?[10] Pasiphaë, Queen of Crete and wife of Minos, fell madly in love with her husband's favorite white bull. It was a curse thrown upon her by Poseidon, who thus took revenge on Minos for not having sacrificed the bull to him. Pasiphaë, in her plight, came to Daedalus, famous architect, builder of the labyrinth, sculptor, and inventor, who had also amused the court with puppets moved by ingenious mechanisms.

And Daedalus found a way to let her satisfy her strange desire. He built a hollow wooden cow, provided with a trapdoor to get in and out; it was covered with cowhide and could roll on wheels hidden in the hoofs. Daedalus showed Pasiphaë how to put herself into this cow in the appropriate position, then wheeled the cow into the meadow where the white bull was grazing, and went discreetly away. The bull, spotting the newcomer, at once mounted her. Pasiphaë was entirely contented; she became pregnant and gave birth to the Minotaur, who grew into a horrible monster.

From this it is clear that it was not at all Daedalus who took the initiative to mate woman with bull; he may not even have given thought to the possibility that the union might be fertile. He only wanted to please the queen, and he may well have found the problem a fascinating one.

This is the lesson I would draw. Whatever mad and perverted ideas may take hold of those in power—even desires far madder than wanting to copulate with a bull, even desires as mad and perverted as wanting to have bigger and more nuclear warheads or neutron bombs—there will always be technicians found who will be willing to oblige, who become fascinated by the technical problem and who do not think about the consequences. To use Oppenheimer's favorite terminology, Daedalus may well have found his cow a sweet animal.

So did the white bull. So do today's bulls. For imitation cows are widely used for collecting bull semen at artificial-insemination centers.

Sir Charles Darwin, prominent physicist, grandson of the writer of the *Origin of Species,* once told me the following story.

"Walking through Cambridge with my two grandsons I met the manager of the university farm. 'Sir Charles,' he said, 'I see you have these two boys staying with you. Why don't you send them one day to the University farm, so that I can take them off your hands for a while? They can ride a pony and drive a tractor, they can see young animals, and, between you and me, sir, it is the best way for boys that age to learn something about the facts of life.' So I sent them to the university farm. They came back dirty like anything, their clothes somewhat torn, but they were in high spirits. It had been a wonderful day. They had been riding a pony and had driven a tractor and had seen a lot of young animals, but what had impressed them most was that they saw a bull performing on a wickerwork cow, the stuff being collected and sealed in bottles to be flown to Russia.

"You know, Casimir," he added pensively, "I'm afraid these boys may have got a slightly distorted view of the facts of life."

So far the follow-up to Daedalus's invention has led to better pedigree cattle, not to a Minotaur. But biological techniques do not stop at artificial insemination. Transplantation of fertilized eggs and fertilization *in vitro*—in a test tube—are further steps. And what is regular practice in cattle breeding will be done, is increasingly being done, for man. And so we get back to the menace discussed by Dyson. In any case, it may well be that the great Charles Darwin's great-great-grandsons got a first glimpse of the facts of life of the future.

Landau

Among the small group of theoretical physicists who began their scientific careers immediately after the breakthrough of quantum mechanics and who explored its ideas to ever greater depth and applied them to an ever widening field of phenomena, Landau takes a prominent place. He was also one of the most colorful figures and one of the most influential teachers. He made significant contributions to many branches of theoretical physics, and the textbooks he wrote together with E.M. Lifshitz are well-known classics. He was the founder of an important school of theoretical physics in the USSR and inspired many experimental physicists. The short article by A.T. Grigorian in the *DSB* contains a highly condensed summary of his scientific achievements but throws little light on his personality.

Far more informative is the biography written by his pupil and coworker E.M. Lifshitz for the *Biographical Memoirs of the Royal Society* (Vol. 15, p. 141). The same author also added a biographical introduction to the third edition of the first volume of the series of textbooks I mentioned. A book by Dorozynski is mainly concerned with the almost miraculous way Landau's life was saved after a severe car accident that occurred on 7 January 1962; the opening chapters contain interesting particulars about his life prior to that accident.[11] Anna Livanova's *Landau* gives only a rather sketchy biography but contains a fascinating chapter on "The School of Landau."[12] It shows Landau at the height of his powers and gives a lively picture of his style and methods. It also contains an extensive and readable account of Landau's theory of superfluidity of liquid helium, one of his major contributions to theoretical physics.

My own reminiscences relate only to a brief period of his life.

Lev Davidovich Landau—Dau to his friends—was born on 22 January 1908 (9 January according to the old calendar, still prevailing in Russia at the time of his birth) at Baku, where his father was a petroleum engineer; his mother was a physician. Landau was an infant prodigy, although he himself denied it. It is told that he read at four and that he had taught himself the high-school curriculum in mathematics by the time he was seven. Legend may have slightly embellished reality, but this much is certain: he finished high school in 1921. At his parents' bidding he attended the Economical Technical College in Baku for one year, which he rather disliked, and in 1922 he entered the university. He graduated in 1924 and then went to Leningrad to continue his studies. By 1929 he had developed into a full-fledged theoretician and he already had several papers to his credit, although, it must be admitted, nothing as thorough and mature as Pauli's *Enzyklopädie* article on the theory of relativity. He next obtained a Russian stipend and a Rockefeller grant that enabled him to go to Western Europe until the spring of 1931. During that period he spent several months at Copenhagen while I was there, and together with Gamow we formed a trio that was most amusing to ourselves, though maybe not always appreciated by others.

Landau's was perhaps the most brilliant and the quickest mind I have ever come across. His knowledge of theoretical physics was both broad and thorough, but he was also willing to start a discussion on any subject even when his opinion was definitely "nicht von Fachkentnissen getrübt" (not obscured by professional knowledge). On the other hand, his physical reactions were slow and he was neither a sportsman nor a handyman. In those respects he was the exact opposite of Bohr. He had no ear for music, but, contrary to Gamow, he was a good linguist and paid attention to details of German grammar and English pronunciation. When he arrived in Copenhagen, where, like Gamow, he lived at Pension Have, he at once began to learn Danish and to build up an appropriate vocabulary. The first two words he needed were the Danish translations of the German words "Minderwertig" (inferior, despicable) and "Spiesser" (philistine), words whereby he expressed his dislike and contempt for what in modern terminology is called the "bourgeois establishment." The translations are "mindreværdig" and "spidsborger." Unfortunately—or perhaps fortunately—the first word is close to another word, "mindeværdig," which means "memorable," so Landau's criticisms were at first occasionally misinterpreted as praise, an error he was quick to redress. Soon he made further progress. One of the books I had been using in my struggles with the Danish language was a *Dänische Konversationsgrammatik* by Karl Wied, a booklet of remarkable German thoroughness that managed to make even the comparatively simple Danish grammar look complicated and that excelled in marvelous insipid conversations, which appealed strongly to Landau's somewhat naïve sense of humor. "Have you sold the horses? No, but I have sold my garden, my house, and my beautiful carriages," was one of his favorites. "Where is the ink? I have poured the ink into a small bottle" sounds innocent enough in English, but it so happens that "blækket," which is Danish for "the ink," sounds almost like the name of the famous English physicist Blackett, and the idea of putting that tall and handsome man into a small bottle struck Landau as very funny.

Next he went on to higher realms of literature. Among Frøken Have's books there was a rather old-fashioned anthology of poetry, formerly used in schools, and Landau began to read and even to learn by heart some of the poems it contained. The one he liked best was "Den glemte Paraply" (The Forgotten Umbrella) written in 1831 by Henrik Hertz (1798–1870)—not to be confused with, though possibly distantly related to, the German physicist Heinrich Hertz, the discoverer of electromagnetic waves. The poet has forgotten his umbrella at the home of friends and asks them to look for it, adding that it is "as good as new." But then he realizes how they might look everywhere and will finally have to report that they have found only a very old umbrella. The poet offers his excuses, but he has not said that it was *almost* new, only that it was *as good as* new. And it is, for it offers exactly the same protection and can be used in exactly the same way (and can be left behind at the home of friends) as a new one. But the Lord knows it is old, "At den er gammel veed vor Herre." That became one of his favorite sentences and he would repeat it again and again, as if he were tasting a delicacy. (Later I noticed that Pauli sometimes would do the same thing.) Even today I can hear him say, "At den er gammel veed vor Herre," and I can see his lean, expressive face and his unkempt shock of long dark hair. He knew quite a bit of poetry but his taste was simple: a strict meter and regular rhymes were an essential requirement. He would not have liked *The Waste Land,* but *Old Possum's Book of Practical Cats* he might have enjoyed. Livanova mentions that he admired Kipling. This is curious, for he must have loathed Kipling's ideology. Apparently in judging poetry he found form more important than content.

Landau often made a point of being aggressive, offensive even. This he rationalized as being part of his contempt for "minderwertige Spiesser," but in reality it was obvious that he was both shy and sensitive. That was understandable. To grow up among boys that are older, stronger, and sexually mature long before yourself, whereas you are absolutely superior in all intellectual pursuits, must leave traces. I believe that I went through enough of that myself to be able to understand, although, of course, both my juniority in years and my superiority in intellect were trifling compared with Landau's.

He liked to talk scathingly about "traditional" virtues: courage, honesty, faithfulness to friends, charity. I remember a conversation in which he claimed that he would gladly kill a man he wanted to get rid of, if he could do so with impunity. I said I considered him a mild kind of person and that I did not think he would do anybody harm. That saddened him and he appealed to Gamow, who was also present: "Johnny, I would kill someone, wouldn't I?" and Gamow, just out of kindness, admitted, "Of course you would." "Not in an open fight, of course." "No, no—'mit einem Brett von hinten' " (with a piece of timber from behind). Did he really think that way? I rather doubt it, but if he did, subsequent events showed how wrong he was. In 1938 he was put in prison in Moscow on the basis of completely unfounded suspicions, and for a year he was slowly pining away. He would have died but for Kapitza's staunch friendship and outstanding courage. Kapitza went straight to Molotov with an ultimatum: he would stop his own work if Landau was not immediately released. Thus goes the story, and I have no reason to doubt that it is essentially true.[13] Hail to the traditional virtues!

However, he did sometimes amuse himself by teasing people in a rather nasty way. A.H. Wilson* spent some time at Copenhagen and for a while he was Landau's chosen victim: Landau regarded him as the personification of British conservatism and did his very best to

*A.H. Wilson, later Sir Allan, born 1906, was a Cambridge theoretician who did important pioneer work on the theory of semiconductors and published a famous textbook on electrons in metals. After the Second World War he went into industry.

irritate him. I remember that I sometimes tried to come to Wilson's defense—although he was quite capable of taking care of himself. I even remember one occasion when Landau wanted me to come along—on a walk, to a movie, or something else, I don't remember —and I wanted to stay at home. "Come along," he said, "or I start teasing Wilson." I went along.

Somewhat exasperating was his habit of rating everything— physicists, papers, girls, motion pictures—with marks from one to five. (One = excellent, two = good, three = passable, four = bad, five = very bad.) I introduced the notion of the inverted five: a motion picture could be so bad that it became amusing again. Later he also had a somewhat different classification for physicists. For some work you needed hard work and perseverance—good trousers, Landau would say—and in most cases a sharp brain was necessary. Physicists could now be classified by simple diagrams:

$\triangle$ sharp and diligent,
$|$ sharp but lazy,
$\square$ diligent blockhead,
$\triangledown$ lazy blockhead.

Sometimes he took an immediate dislike to people. The young French theoretician Jacques Solomon, with his charming wife, a daughter of Langevin, spent some time at Copenhagen. Landau at once and for no obvious reason classified him as a four or five. Mrs. Solomon he rated higher, though not much above three. He may have noticed that I was rather impressed by her. Once she asked me, "Pourquoi êtes vous si timide?" That is a question one does not answer; the correct answer might have been, "Parce que je vous trouve très charmante." Anyway, Landau pulled my leg by saying, "Casimir, wreck that marriage [zerstöre diese Ehe], seduce that woman, 'nur um zu lernen'" (only in order to learn—one of Bohr's ways of beginning one of his penetrating questions). He might as well have ordered me to challenge the heavyweight boxing champion of Denmark.

Solomon was a convinced communist and, I think, what you might call an orthodox Marxist, and this may at least in part explain Landau's dislike. For Landau was a revolutionary, but could hardly be called a Marxist and certainly not a dialectical materialist. I think it would have been equally wrong to call him a Trotskyist; he was just no "ist" at all. He was a revolutionary in the same way as my mother was a Christian. I prefer Christianity.

He believed in doing away with prejudices and with privileges unless they were in recognition of real merit. He felt that the Russian revolution had brought individual freedom to many and wanted to believe that it would do so increasingly.

A memory comes to mind. It is an insignificant incident and I have some difficulty in locating it in time and space. It must have been in Zürich I think, and it may have been in the summer of 1933. But that does not really matter. It may help to illustrate his faith in revolution. We were in a library and on the shelves a beautiful collection of ancient publications of the Académie des Sciences was on display. "Let us have a look," said Landau; "it must be fun to see the nonsense those old fools wrote" (Was für einen Quatsch diese alten Trottel damals geschrieben haben). He took out a volume, opened it, and looked at a paper by Legendre. The next one may have been by Laplace. One classical paper after the other, all of them important contributions to mathematical physics. Landau was silent for a moment, but then his face lit up: "That shows how much the French revolution did for scientific progress," he concluded.

However, the idea that physical theories should be critically assessed on the basis of dialectical materialism struck him as utterly ridiculous. To him it seemed obvious that such a philosophy could not in any way pronounce judgment on, let alone contribute to, physics. Neither could any other philosophy. Landau's ideas on politics and on philosophy came out most clearly on the occasion of a talk he gave at a students' society—Studentersamfund—on 16 March 1931. Prior to that the weekly *Studenten* in its number of 12 March (No. 22) had published an interview with Landau. Since it is revealing—and certainly not easily accessible—I give a full translation (omitting the headlines). A few preliminary remarks are necessary. The Danish original speaks about "Videnskab" and Landau spoke about "Wissenschaft": the interview was in German. Now this word implies *all* branches of learning: the natural sciences, the humanities, the social sciences. In my translation I have simply rendered it as "science." That is a legitimate use of the word: the second edition of Webster's New International Dictionary's second definition is: "Any branch or department of specialized knowledge considered as a distinct field of investigation or object of study." The Oxford English Dictionary gives similar definitions, but, of course, it is no longer the current usage.

A further remark. Whereas Landau's observations about the position of scientists and about humanities, philosophy, and so on sound entirely true to type, I cannot remember that Landau ever spoke in my presence in such detail about labor relations in factories and certainly not with such simple-minded adherence to official doctrine. Landau's actual knowledge about life in industry must have been small, but he was very loyal to his country—although he would deride loyalty as a "bourgeois" virtue. I think that explains his attitude; the possibility cannot be excluded, however, that the interviewer added some touches of his own.

One rarely gets an opportunity to listen to first-hand information on conditions in Soviet Russia, and this is true in particular when the position of intellectuals in the new Russia is concerned. It has been asserted—mainly by emigrants, who have not been in Russia since the revolution—that the vast majority of intellectuals in this mighty empire have been hanged and that the remaining few have to face systematic persecution, which makes it absolutely impossible for them to do any work. Not only, it is said, has the general instruction of the people deteriorated but those in power wage systematic war against the intelligentsia, science wastes away under poor working conditions, intellectual life withers and decays because of lack of sustenance. . . . Most reports by emigrants contain concrete examples showing how ill this or that scientist has fared. Examples which, without any doubt, are based on a cruel reality, but which do not tell us anything about the present conditions for scientists and intellectuals in Soviet Russia.

The Studentersamfund is next Monday organizing a meeting that should help to throw some light on the present circumstances of intellectuals in the Soviet Union, for its officers have been able to induce a young Russian scientist, the physicist Dr. Landau, who at present is visiting Copenhagen, to give a lecture on "The Position of University Graduates in Soviet Russia." Dr. Landau is quite young, only in his mid-twenties [in fact, he was just twenty-two], and after having obtained his degree at Leningrad's university he was appointed by the Russian government to do pure scientific work at an Institute for Theoretical Physics in Leningrad. For the last one and a half years Dr. Landau has been working and studying in Germany and Switzerland, partly on a stipend given by the Russian Commissariat for the People's Instruction, partly on a travel grant from the American Rockefeller Foundation, and now he is taking part, together with a number of Russian and German scientists [as far as I remember there was only one other Russian

—Gamow—but many other nations were represented] in a small, unofficial physics conference with our world-famous compatriot, Professor Niels Bohr, at the Institute of Theoretical Physics of the University. Here we meet with Dr. Landau, whom we asked for an interview; he turns out to be a tall man, somewhat frail of build, with curiously long, black hair [it would hardly be considered curious today] and a pair of dark, intelligent eyes. Our conversation is held in German with a few Danish sentences thrown in, which Dr. Landau has already mastered during his brief stay in Copenhagen.

Has everyone the possibility to become a student in Soviet Russia? we ask Dr. Landau.

Yes, everyone has the possibility, but practical circumstances such as lack of space and the like make it necessary to select by entrance examinations among those pupils that after having finished secondary school apply for admission to universities and institutes of technology. It is expensive to create new institutions of higher learning and of course in this present period most efforts have to be directed towards building up a healthy socialistic production system; taking this into account one might almost be surprised that there have been found means for the many big expansions of universities and scientific institutes that have taken place in recent years and for the considerable sums the state distributes every year as stipends to students.

Is that not due to the fact that there is at present a shortage of personnel with academic training?

Yes, in any case we have no unemployment among graduates, which, I am told, is the case here in Denmark, and especially there is of course at present a great need for capable technicians for building up the new production system and for well-educated teachers for the widespread expansion of elementary schools and for the people's instruction in general. That the intellectuals, because of their better education, receive higher wages than the average worker is a matter of course.

Does that not automatically lead to a class structure of society?

Yes, one might call it that, but it is a structure that is essentially different from the division between the owners of the means of production and the wage-earners in the capitalist world. The directors, the managers and the technicians in a socialist Soviet factory are paid by the state, just as the factory workers, and therefore it cannot happen—which always can happen in the capitalist production system—that the management is interested in overworking the labor force in order to bring the greatest possible profit to the owners of the means of production. The owners are partly the workers themselves, who therefore share the interests of the management. Of course it is not unthinkable that the management might take measures unacceptable to the workers and therefore these are organized in a kind of labor union that negotiates with the managment and has a considerable say in all measures concerning the operations that will influence the workman's situation. In Soviet Russia the many do not work for the few, but every single person works for the well-being of the whole; there is no hostile antagonism between workers and management but solidarity. This is furthered by the fact that most of the younger intellectuals, technicians, schoolmasters (we call them schoolworkers), lawyers, economists, physicians, etc., came from working-class families: other things being equal, they are in most places preferred to people from the old bourgeoisie. Such a comparison takes place for instance among the pupils that after having passed the entrance examination apply for admission to universities and institutes of technology.

Is there truth in the assertion that those in power favor those intellectuals that are communists and persecute those that have other opinions?

It is clear that the state has to resist attempts that clearly oppose the work of social construction; attempts that occasionally have been made by certain emigrants that only came back with the purpose of sabotaging the five-year plan. But it is absolute nonsense that it should be necessary to be a communist to obtain good working conditions, and we certainly don't favor military-drill tendences: *there* you shall work, and *that* you shall believe, nothing like that! I am no communist myself and many of my colleagues among scientists and intellectuals don't take an interest in politics—of course they are permitted to go on peacefully with their scientific work!

Is that also true for the humanities and those theoretical subjects that have no direct and immediate importance for the work of socialist construction?

Yes, of course. Certainly, it should be said that in the years immediately after the revolution one has to a high degree neglected the humanities, favoring technological development instead—that was absolutely necessary —but now one has made up for what one had to neglect. Personally I am even of the opinion that one is now spending too much on pseudo-sciences like history of literature, history of art, philosophy, and so on. What can we do with that; isn't it the main thing that we can rejoice in good literature and art? These literary, art-historical, and metaphysical gewgaws have no value for anybody but the idiots that occupy themselves with them and who believe one can build a science on nothing but a series of words! But, as I said, that is just my personal, subjective opinion, and unfortunately we have in Russia a number of institutes and quite a research staff that practice these "sciences."

Dr. Landau has become quite excited and gesticulates with his narrow, lean hands. When I tell him that we in Denmark have an obligatory course in philosophy with a special examination that is required in order to be admitted to an examination for a degree in any subject, he bursts out laughing and his dark eyes flash cheerfully:

"No, it is not as bad as that in our part of the world!"

During the lecture itself, which became quite an event, Landau did not say anything about industrial relations, at least not as far as I remember, and rereading the reports in the two leading newspapers, *Politiken* and *Berlingske Tidende,* I find my recollections confirmed. *Berlingske Tidende* states that the lecture was hardly a lecture at all but that it consisted of a number of scattered remarks about universities and many other things in Russia, whereas *Politiken* speaks about a "causerie" on conditions in universities. It also mentions specifically, "The lifestyle in Russia is ironic," Dr. Landau further said; "*Pathos* is considered ridiculous and so is *duty;* one wants, on the contrary, to live one's life as merrily as possible. Moscow and Leningrad are the merriest cities in the world." During that evening Landau was not taken to task about that statement but two days later there appeared in *Berlingske Tidende* a cartoon ridiculing that notion. The truth of that sarcastic criticism was going to be clearly shown by later events. At the meeting itself he was asked, in a rather insinuating way, how one did become famous in the Soviet Union. That was an easy one for Landau. "Die Frage wie man berühmt wird ist an sich eine sinnvolle" (The question how to get famous is in itself a meaningful one), he said politely. The distinction between meaningful and meaningless questions that plays such an important role in the interpretation of quantum mechanics always figured prominently in Landau's arguments. Then he went on: "The answer, however, is simple. One has just to do excellent work. If you should ever happen to do some valuable work yourself, even you might become famous."

The next question was a more difficult one. "Wie steht es mit der Lehrfreiheit?" (What about the freedom of teaching?) Landau answered, "One has to distinguish between meaningful and meaningless branches of learning [sinnvolle und sinnlose Wissenschaften]. Meaningful are mathematics, physics, astronomy, chemistry, biology; meaningless are theology, philosophy, especially the philosophy of history, sociology, and so on. Now the situation is simple. With respect to the meaningful disciplines there is complete freedom of teaching and doctrine [vollständige Lehrfreiheit]. As to the meaningless ones I have to admit that there is a preference for a certain way of thinking. But it is after all completely irrelevant whether one prefers one kind of nonsense to the other [ob man den einen oder den anderen Quatsch bevorzugt]." Poor Landau. He got away with it then, although he must have known that in Russia the theory of relativity and quantum mechanics had been under attack. A few years later the Lysenko story would put an end to all illusions about the freedom of the "meaningful" disciplines. To defend Soviet science policy by labeling all philosophy, all social sciences, and much of the humanities as nonsense was a courageous attempt, but even that gesture of extreme arrogance would soon turn out to be insufficient.

The last weeks before his return to Russia Landau went shopping. He had saved all the money he could to buy presents for friends and relatives at home (What about traditional virtues?). One of them wanted a camera, the classical type of those days. That is: 9 × 12 cm, Zeiss Tessar $f{:}4.5$ or, better still $f{:}3.5$, Compur shutter, to be used with plates or film pack. I helped him to find a suitable second-hand one and tried it out for him. Gamow rose at once to the occasion and started to arrange various groups, which I was then ordered to photograph. The result was a number of somewhat crazy pictures. One of them, featuring Teller on skis, Gamow on a motorbike, and Landau on a toy tricycle, with Aage and Ernest Bohr in between and the Institute in the background, has been reproduced in more than one place.[14]

It shows at least that the camera worked all right. We only had to spend some time on making it look a bit more second-hand without really damaging it. We also treated new tennis balls with a bit of coal dust.

On the afternoon of his departure we found Landau in his room in a state of utter despair: he just couldn't get all his stuff packed. One of his acquisitions was a beautiful woolen blanket. He had put it loosely in his suitcase, not leaving much room for anything else. I don't claim that I was then—or ever later—one of the world's great packing experts, but I had at least some camping experience and knew how to roll a blanket really tight. So we began by making a neat roll—I think Frøken Have provided us with some material to wrap it up—and that had to be tied to his suitcase. After that the rest was fairly easy. Dorozynski writes about Landau's moving to Moscow shortly after his marriage: "and for the first time in his life his suitcase was neatly packed."[15] I feel justified in taking exception to that statement.

In the evening we saw him off. I think he boarded a ship near Langelinie.

I did see him again in 1933. He came to Bohr's conference, which that year took place in September. During an evening reception at Bohr's home some Brahms was performed. Landau—who had, as I have mentioned, no ear for music—was pulling grimaces all the time and was making rather a nuisance of himself. Afterwards, Dirac got hold of him and said, "If you don't like the music, why don't you leave the room?" Landau, as usual, had a ready repartee. "It isn't my fault, it was Mrs. Casimir's fault [I had recently married, and the Copenhagen conference was the end of our honeymoon]. She isn't much interested in music either, so I said, 'Let us go out of the room together.' Why didn't she come with me?" To which Dirac replied in his usual quiet and precise manner: "I suppose she preferred listening to the music to going out of the room with you." Landau, for once, had no answer.

At the beginning of this section I emphasized that Landau was a brilliant physicist. The anecdotes I related may throw a sidelight onto his character, but they should not make us forget that physics was his main concern. In 1930 he published his paper on diamagnetism of free electrons, a most ingenious and elegant piece of work that later turned out to have far-reaching consequences. Also in 1930 he published, together with Peierls, a paper on fundamental aspects of relativistic quantum mechanics. This paper, which met with Bohr's disapproval, was certainly a contribution to the discussion of the basic ideas of physics, but it left fewer permanent traces. Also in later years Landau's special genius lay in his ability to find elegant and satisfactory approximate solutions for problems that would at first sight seem to be absolutely insoluble. His theory of phase transitions, his theory of the intermediate state of superconductors, his theory of superfluid helium are striking examples. His contributions to basic principles are less striking. Could it be that here his contempt for any kind of philosophy restricted his scope?

I mentioned already that Landau was an extremely quick thinker. The discussions I had with him at Copenhagen offered no opportunity to find out to what extent he was familiar with the more advanced branches of modern mathematics. In any case he considered mathematical rigor incompatible with sound physics. But the mathematics he needed he always had at his fingertips. He was certainly not afraid of complicated calculations, but he tried to steer around them. "Das würde der Herr Gott nicht zulassen" (Our Lord would not permit such a thing), he would say when a formula got too complicated. I sometimes tried to oppose him by saying, "Your theology is all wrong; to Our Lord a Bessel function of complex order is just as simple as a sine or cosine is to you," but I didn't stand firm. Yet, aiming at simple formulae can be risky for someone who doesn't have Landau's acuity of judgment. That I found out to my disadvantage. I once published a short note "On the Internal Conversion of Gamma Rays" (*Nature*, 20 December 1930) in which I had managed to obtain a simple and elegant formula—which even met with Landau's approval and is quoted in Gamow's book—by courageous and unjustifiable simplifications. It led to completely erroneous conclusions, which were fortunately later corrected by others, to whom, I am happy to say, my first attempt was not entirely useless.

According to the Copenhagen files Landau came back to Copenhagen once more in 1934, but I do not remember having met him on that occasion. After that he did not get permission to leave the country for many years. After Stalin's death in 1953 conditions in that respect slowly improved. He would certainly have been allowed to go to Stockholm to receive the Nobel Prize, which was awarded him in the autumn of 1962. But he had not sufficiently recovered—and as a matter of fact he never completely recovered—from his motorcar accident. And I did not visit Russia myself until well after his death on 1 April 1968.

Gamow

Besides acting as a private secretary to Bohr I sometimes also acted as a scribe to Gamow.* He was a man with many abilities, but speaking and writing foreign languages correctly was not one of them. Probably he just didn't care. Also, he was by far the most

*George Gamow, born Odessa, 4 March 1904; died Colorado, 20 August 1968. Left USSR 1931; professor, George Washington University 1934–1956; University of Colorado 1956–1968.

playful of all the guests at Bohr's Institute during my time at Copenhagen, and sometimes his jokes involved writing a letter; that was where I came in. For instance, he noticed in an issue of *Science Abstracts* an abstract of a paper by two women scientists he had known at Cambridge, and he told me at once to help him write an indignant letter to the editor. It came out as follows:

Sir,

Abstract . . . in a recent issue of your journal . . . contains a most disturbing mistake. The authors are there referred to as Miss . . . and Miss. . . . Both persons having been married a number of years the prefix Mrs. would have been more suitable. If however you want to quote the authors under their maiden name—a practice against which I would have no objection—I can inform you that these names are . . . and. . . ."

There came a prompt reply with an apology. It was addressed to Mrs. G. Gamow and began with Dear Madam. I like to think the editor knew who Gamow was, in which case he deserves credit for an excellent answer.

A somewhat more serious letter was written to the Oxford University Press. Gamow was writing part of his book on the nucleus[16] while he was at Copenhagen and the press had delegated a young physicist, Miss Bertha Swirles (in 1940 she married Sir Harold Jeffreys, a well-known astronomer and applied mathematician) to translate Gamow-English into English. She was working in the library with an almost permanent smile on her face because of the grammatical and idiomatical eccentricities of Gamow's manuscript. ("But there is an occasional correct sentence," she admitted.) Now one of the points Gamow stressed in his book was that the notion of electrons in a nucleus is fraught with difficulties and that it was as yet impossible to make a reasonable theory of beta-decay. (The first successful theory of beta-decay was formulated by Fermi in 1934; it involves both neutrons and neutrinos.) So Gamow had a rubber stamp made with a skull and crossbones with which he marked the beginning and end of all passages dealing with electrons. The Oxford University Press suggested that this should be replaced by a less gloomy sign and asked for Gamow's consent. Gamow—and I—then wrote a letter in which it was said that he agreed and that "it has never been my intention to scare the poor readers more than the text itself will undoubtedly do." I have written elsewhere that thereupon the press replaced the skull and crossbones by asterisks.[17] I am sorry to have to admit that there I made a "most disturbing mistake" myself. They were replaced by a kind of tilde in heavy print.

Gamow could be the soul of any party and was a great organizer of parlor games. He was happiest when he had the whole company sitting on the floor and, for instance, playing a kind of volleyball with toy balloons. On such occasions he liked to tell about a Russian game called "Cuckoo," in which people would sit on the floor of a big room, completely in the dark and with loaded rifles. Each of the participants had to call "Cuckoo" in turn, and then the others would shoot in the direction from which the sound came. I don't remember whether the last survivor or the first man killed was considered to be the winner. Such stories acquired a special flavor because of Gamow's very fluent but highly individual German or English. In my study on "Broken English" Gamow was certainly one of my models.

Like many theoretical physicists he kept late hours and did not like to get up early. Above his bed he had a picture with a lowing cow, bleating sheep, a crowing rooster and a sleeping shepherd with, underneath, the poem:

> When the morning rises red
> It is best to lie in bed.
> When the morning rises grey
> Sleep is still the better way.
> Beasts may rise betimes, but then
> They are beasts and we are men.

(I apologize for possible inaccuracies; I have not tried to find a copy of that remarkable work of art.)

Gamow loved humorous scientific or pseudoscientific publications—he admired Jordan and Kronig's cow paper—and on one or two occasions he pulled the leg of the editor of a scientific journal. Unlike most practical jokers, however, he did not, in general, try to make other people look ridiculous.

Although he was somewhat sloppy on details and had no great mastery of mathematical technique, he had a keen intuition, and even in later years he managed to be at the forefront of science, contributing original ideas to nuclear physics, cosmology, and molecular biology. But he rarely presented results in a definitive form.

His playful spirit and his talents as a draftsman he turned to good account in a number of popular books, especially in his Mr. Tompkins stories, based on a fantastic exaggeration and distortion of the fundamental notions of modern physics.

I think it was also Gamow who started the tradition that, during a conference at Bohr's Institute, one afternoon or evening should be devoted to some light entertainment.

In any case, in 1931 he produced a remarkable shadow-pantomine, *Den Stjaalne Bakterie* (The Stolen Bacteria), a spy story with a professor, a beautiful daughter, a lurid gang of spies, threatening bacterial warfare—in short, all the major elements of the bad films we enjoyed seeing. A distinguishing feature was that the actors were seen only as black shadows, but they acted in colorful surroundings provided by slides, drawn by Gamow himself or by that versatile talent, Piet Hein. For lack of something better I provided the music by scratching a kind of "fiddler on the roof" performance on the violin.

Piet Hein was a remarkable young man. His ancestors had been living in Denmark for many generations, but he was a descendant of Pieter Pieterszoon Heyn (Piet Hein for short), a Dutch admiral who lived from 1577 to 1629, who may have been a great naval organizer, tactician, and commander, but whose memory is particularly cherished by my money-loving compatriots because, in 1628, he intercepted an extremely rich Spanish silver transport. Young Piet had studied both mathematics and art, and I believe the stirring drinking songs that were sung at the students' debating society "Parentesen" were at least partly his work. He moved, so to speak, on the fringes of Bohr's Institute. Later he would become famous for his "Gruks," short poems containing some witty point, or sharp observation, or even some by no means trivial wisdom, illustrated by elegant little drawings. He wrote quite a number in English, too, and then they are called "Grooks." I should also mention his association with the "superellipse," a special curve, which he applied in interior decoration, in town planning, and in many other ways. And here I feel sorely tempted to embark upon a lecture on the superellipse and to explain why I consider it an unsuitable shape for a busy road, but I shall suppress that urge.

For some reason I did not attend the 1932 conference when the famous Faust parody was enacted. The idea itself was not new: innumerable Faust parodies must have been enacted by schoolboys and students: the words lend themselves beautifully to corruption and misinterpretation. However, I think the Copenhagen Faust—as I said, I did not see it and did not participate in its creation; I only read the text—was in a special class. Bohr as our Lord in Heaven, Pauli as Mephisto, Ehrenfest as Faust—it was more than a childish joke.[18]

Later efforts were not quite so ambitious, but there was always a good show. Max Delbrück, with an imperturbable face and a top hat, was a magnificent master of ceremonies, Vicky Weisskopf provided both poetry and song, and Weizsäcker was an acknowledged master of the "Schüttelrhyme," an art form not much practiced—and rather difficult to practice—in English. Of course, Spoonerisms are well known (you have hissed my mystery lessons, you have tasted two worms, etc.), but Schüttelrhyme is meaningful poetry where the rhymes are Spoonerisms.

> Butterfly,
> Flutter by!

is the shortest I can think of. In Germany the Schüttelrhyme has about the same social position as the limerick in England and America (and it has the same innate tendency to become indecent).

> Pauli sich beim Lesen wiegt
> Weil das in seinem Wesen liegt.

is a very noble example (Pauli rocks when he reads, because that is part of his being).

On one occasion Bohr had just come back from a long voyage. He himself told us that during that trip—I believe in Hawaii or thereabouts—when, somewhat absentmindedly, he had almost missed his ship, an official had said, "Sir, you are the foggiest person I have ever come across." To us this suggested a parallel with that other great traveler, Phileas Fogg, and our performance that year pictured a new voyage around the world. We wanted to give Aage Bohr a part, so we invented an excursion across the Himalayas into Tibet, where Weisskopf played the Dalai Lama. I am still rather proud of my Himalayas, a double stepladder entirely covered with white paper. Into that "ice" Aage cut steps with a real pickaxe, and with Rosenfeld on a rope he crossed. Now there had recently been built an extension to Bohr's Institute that had to house a cyclotron, but when the magnet arrived it turned out that the door was too small and part of the wall had to be broken down. In our performance the same had happened in Tibet, but here, Weisskopf explained, another solution was chosen. After all, according to quantum mechanics there was a finite though very small probability that the magnet would go in by itself. They did not know whether the higher powers were actually in favor of the experiments they envisaged, so they would just sit and wait, confident that these higher powers would make use of the freedom left by quantum mechanics if they were favorably disposed.

There was a year when we arranged a series of contests. One game of my invention—but I suppose it must have been invented before—was a competition in "shower production." It had been found in work on cosmic rays that one particle of very high energy may give rise to a whole "shower" of secondary particles. Heisenberg had dealt with the theory of this phenomenon and Paul Ehrenfest, Jr., who was working in Paris, had done some experimental work.* Now they had to show what they could do with real showers. They had to sit down on chairs about four meters apart, and above their heads two tins full of water were suspended in such a way that they could easily be tipped. Each player was provided with a basket with half a dozen old tennis balls and the game could start: each had to try to hit the tin above his opponent's head. During the first round both missed all shots, but in the second or third round Paul finally placed a hit. Heisenberg got rather wet and looked sad. I said to Weizsäcker that I thought Heisenberg did not mind getting wet, but that he did not like to lose. "You seem to know him well," was his answer. But, in retrospect, we should have disqualified young Ehrenfest—he rose from his chair while throwing, which was against the rules; or we should have awarded Heisenberg one or two penalties. Not long ago I mentioned this story to Heisenberg's widow. She had never heard it before and was thoroughly amused; she agreed that Heisenberg didn't like losing, not even in a silly game, and that he would certainly have been irritated to lose by foul play.

When we had the first conference after the Second World War, Weisskopf and I hoped that now younger people would rise to the

*Owing to his father's ideas on education, Paul junior had never followed a regular curriculum but he was a promising experimenter all the same. A few years later he was killed by an avalanche when skiing in the French alps.

occasion and that we would be the elderly and critical spectators. Nothing of the kind happened, and I had to assume the role of master of ceremonies. My main contribution was a pseudo-linguistic analysis of Broken English. I later wrote it down for Volume III of the *Journal of Jocular Physics* published on the occasion of Niels Bohr's seventieth birthday. (I have already referred to my contribution to Volume I.) It was reprinted in the *Scientific American* and elsewhere.[19]

I am afraid that of all my publications this one is the most frequently quoted. Several years ago I met the great Russian theoretician Bogoljubov. "Casimir," he said, "are you the man that wrote about Broken English? We often quote your paper, for that is the language we speak at my Institute." And he went on to explain that he always had visitors from many different countries and that this was the language they had in common. "But don't you speak broken Russian?" I asked. He thought this over for a while and then made the remarkable statement: "No, Russian does not break well." Here follows the complete text.

BROKEN ENGLISH

There exists today a universal language that is spoken and understood almost everywhere: it is Broken English. I am not referring to Pidgin English—a highly formalized and restricted branch of B.E.—but to the much more general language that is used by waiters in Hawaii, prostitutes in Paris and ambassadors in Washington, by businessmen from Buenos Aires, by scientists at international meetings and by dirty-postcard peddlers in Greece—in short, by honorable people like myself all over the world.

One way of regarding Broken English is to consider it as a more or less successful attempt to speak correct English, but that is a pedantic, schoolmasterish point of view that moreover threatens to stultify the speaker of B.E. and to deprive his language of much of its primeval vigor. Of course a corresponding point of view would be entirely justified in the hypothetical case of an Englishman trying to speak Dutch (or any other language for that matter) because he will necessarily be a lone figure, but the number of speakers of Broken English is so overwhelming and there are so many for whom B.E. is almost the only way of expressing themselves—at least in certain spheres of activity—that it is about time that Broken English be regarded as a language in its own right. It is then found that B.E. is a language of inexhaustible resources—rich, flexible and with an almost unlimited freedom. In the following I shall try to establish some of the fundamental principles of B.E. in the hope that others, more qualified than myself, will take up the subject and help to secure for it a prominent place in linguistics to which it is justly entitled.

Phonetics. The immense richness of B.E. becomes at once evident if we try to represent its sounds. Two short lines of keywords (44 in all) at the bottom of a page of a 25-cent Merriam-Webster are a sufficient clue to the pronunciation of standard American. And the famous English pronouncing dictionary of Jones has only 35 keywords. Compare these pedestrian figures with the wealth of sounds current in B.E. The whole international phonetic alphabet is hardly sufficient to meet the case. Take one simple letter like *r*. It may sound like an Italian *r* beautifully rolling on the tip of the tongue, like a guttural Parisian *r* or like no *r* at all. In this last case the speaker usually suffers from the illusion that he speaks pure Oxford English. Similarly *th* may sound as a more or less aspirated *d* or *t* or as a simple *z* and sometimes (especially in the case of Greeks) almost like *th*. Then there are elements entirely foreign to English, like the Swedish musical accent and the Danish glottal stop (some people pretend that the glottal stop is hard to pronounce, but that is nonsense; it is very easy in itself and gets difficult only if you try to put it into a word).

But even more important is the principle of free choice. It is well known that the combination *ough* can represent at least five different sounds. The educated speaker of B.E. is well aware of this fact; but, whereas the speaker of standard English can only use one pronunciation in one word, the speaker of B.E. is at complete liberty. Some speakers make their choice once for all: they decid that they are going to pronounce *doughnut* as *duffnut* and stick to it. Others may use their freedom in a more subtle way and say *doffnut* or *dunut* depending on the hour of the day or the weather. Still others create distinctions and say *dunut* when referring to pastry but *downut* (like in *plow*) when referring to a circular discharge tube used in modern physical apparatus. The pronunciation *dupnut* (like in *hiccoughs*) is rarely heard, although it is certainly correct B.E. It is dubious however whether *donut* (like in *go*) is acceptable.

Then there is the *accent.* In standard English this is a queer business. During the development of the English language the accent has had a tendency to move to the front of the word, but it has not gone all the way and it has shown a curious inclination to linger on the most irrelevant and meaningless syllable of a word. Words like *barometer* and *turbidity* will illustrate the point. Whether this is one more example of the traditional British sympathy for the underdog I do not know, but the result is baffling, and to the convinced speaker of B.E. the realization that he has nothing to do with these weird intricacies comes as a great relief. The dogmatics will use their freedom by putting the accent always on the first syllable, whereas the rationalists will stress what seems to them the most important syllable. The quixotics try to imitate standard English. This is obviously impossible, but the result has sometimes a certain slightly pathetic charm.

Grammar. Much of what is said about phonetics applies to grammar too. Again a great richness, again a principle of free choice. The gain in power of expression that can be derived from, for example, a judicious use of the article is impressive. If a man invites you to a party it may very well turn out to be a dull show, but if he says, "Today we will have party and shall drink the cocktail," you can almost be certain that you are in for a lively time. Changing the sequence of words gives new flavor to old sayings—"this is the moment when the frog into the water jumps," one of my teachers used to say at the critical spot in a mathematical proof. Although past tenses and third persons are entirely superfluous, it should be emphasized that occasionally brilliant effects can be obtained by borrowing a correctly inflected verb from standard English.

Vocabulary. Also here there is a great freedom. Of course complete Humpty-Dumptyism is impossible, but B.E. is the closest approach compatible with a measure of understandability. (To explain the term Humpty-Dumptyism: "The question is", said Alice, "whether you can make words mean different things." "The question is", said Humpty Dumpty, "which is to be master, that is all.") It is characteristic of the genius of Lewis Carroll that he, who was by birth and breeding excluded from obtaining a mastery of Broken English, came by sheer artistic intuition to one of its basic principles.

Notwithstanding the great liberty in the use of words, there is one case where all speakers of B.E. seem to agree: in Broken English Broken English is called just "English."

Idiom. It is remarkable how old and trite sayings in any ordinary language may acquire new glamour when translated in B.E. The only danger is that one may unintentionally come to use an existing English proverb. "Who burns his buttocks must sit on the blisters" sounds all right to me, but heaven knows whether a similar saying does not exist in Standard English. As a matter of fact this is always a grave peril also with respect to phonetics and grammar: one may lapse unawares into trivially correct standard English.

I am afraid that this very short survey will have to do for the present. But I have still two important remarks to make. First: in view of the stupendous wealth of B.E. it will at once be evident how completely ridiculous, ludicrous, preposterous and ill-advised are the attempts to introduce for use by foreigners a so-called Basic English, a language not richer but even poorer than standard English. Second: it is often stated that at the age of sixteen or so one loses the faculty to learn English correctly. This again is entirely wrong: nothing of any importance is lost; what is gained is the faculty to create one's own brand of Broken English.

Physics and Play

I am afraid that—occasional references to serious work notwithstanding—readers may gain the impression that we, young physicists at Copenhagen, spent far too much time on rather childish jokes. Otto Frisch was well aware of the possibility of such criticism. Let me quote what he writes about this:

Why is it that scientists are liable to waste their time with such childish pranks? These were all grown-up men, men in their late twenties,* with a considerable reputation for scientific achievement. Then why this schoolboy behavior? Well, I think scientists have one thing in common with children: curiosity. To be a good scientist you must have kept this trait of childhood, and perhaps it is not easy to retain just one trait. A scientist *has* to be curious like a child; perhaps one can understand that there are other childish features he hasn't grown out of."[20]

That may be a partial explanation but I believe there is another side to the question. Scientific work, and especially theoretical work, requires intense concentration and therefore—at least temporarily—a conscious avoidance of other serious thought. Yet the concentration cannot be maintained over very long periods and so one sought refuge in these childish pranks that brought relaxation without interfering with the real work. In order to underline the necessity of this escape I could point to a few sad cases where indulgence in alcohol gradually replaced the childish amusements, with disastrous effects.

Even Bohr, who concentrated more intensely and had more staying power than any of us, looked for relaxation in crossword puzzles, in sports, and in facetious discussions.

Bohr, by the way, was no great lover of detective stories, another favorite outlet. When Conan Doyle in later life became a firm be-

*Actually I was barely twenty-five at the time of my swim and not yet twenty when I arrived for the first time in Copenhagen.

liever in occult phenomena, Bohr's commentary was: "Of course he is more easily fooled than any of us. *We* know that we don't know anything about the detection of crime and the unmasking of impostors. Neither does Conan Doyle, but he thinks he does, which puts him at a great disadvantage."

Yet another aspect. We youngsters that had been growing up with new quantum mechanics felt that we had at our disposal entirely new methods for tackling the problems of atomic physics and that we looked at natural philosophy in an entirely new way. I remember Kramers in a colloquium criticizing a speaker treating chemistry with rather crude electrostatic models, the speaker reluctantly admitting that there might come a day when one would be able to go beyond that, and Kramers exclaiming, "But dawn is there, already!" Kramers belonged to a slightly older generation, but on that occasion he expressed exactly what we youngsters felt. We had, of course, the greatest respect for the creators of this new discipline, but we were inclined to look down on all those who had not absorbed this new lore, or worse still, objected to it. Our nonconventional behavior was just one more way of showing that disdain.

I won't deny that we did waste some of our time and that the childish amusements may sometimes have become an aim in themselves. Connected with this may have been the fact that we gave little thought to the possible influence of our physics on the world in general. I shall have to come back to that later on. But though we sometimes acted as immature schoolboys, we were serious enough where physics was concerned. The nonprofessional reader will have to take my word for it that this was so. Those interested in the details of physics should turn to Appendix B.

5

Berlin, Zürich, and Back to Leiden

Berlin

After getting my doctor's degree—on 3 November 1931—I stayed at Leiden as Ehrenfest's assistant. It was not a very productive period in my life as a physicist. Things were happening in solid-state physics, but I followed them with no more than a lukewarm interest. I did not see a problem I wanted to tackle myself, and Ehrenfest, although dissatisfied with my lack of activity, could not really help me. In those days he himself needed help. I remember distinctly the evening of 23 February 1932. Wiersma had got his doctor's degree that day, we had been celebrating and were enjoying a final brandy at Wiersma's home under the disapproving eye of Ehrenfest, when he started to complain that even giving his normal lectures was becoming difficult for him. Of course, we had all manner of suggestions how we might help him, but they all came to nothing. On the other hand, during that same period he gave the brilliant Diligentia lectures at The Hague I mentioned before. Writing the booklet based on those lectures was at least one productive thing I did as an assistant.

C.J. Gorter got his degree the week after and his parents organized an enormous dinner party in The Hague.[1] The debating club "Christiaan Huygens," of which he was a prominent member, was well represented. It was decided to put on a little review describing and parodying Gorter's career, and a young lady, Josina Jonker, and myself were commissioned to write a text. We produced an impressive epic and decided to continue our collaboration on a more permanent basis—we were married in August 1933. Insignificant pastimes can have serious consequences! Some people may think my ideas on marriage old-fashioned, but to have found as a young man someone who was going to be my companion throughout my further life is something for which I am deeply grateful.

It was in early 1932—27 February was the exact date of the first publication—that the great upsurge of nuclear physics that was going to continue throughout the thirties set in with the discovery of the neutron. That was not just the discovery of a new particle, although that in itself was remarkable enough. If you believe you need only two elementary particles to explain the constitution of all matter, then it is quite a shock to find yourself confronted with a "fâcheux troisième." But soon there came an even greater surprise: it was not "fâcheux" at all. On the contrary it was most helpful. It gave an entirely new view of the constitution of nuclei: they consist of protons and neutrons. The skull-cum-crossbones—or the symbols that replaced them—in Gamow's book could henceforth be omitted. That does not mean that all problems were solved overnight, but at least one saw a road towards a solution, especially after it had become clear that one needed both a neutron and a neutrino.

High-voltage generators were being built in many places and Gamow had explained to me that it should be possible to obtain nuclear reactions with protons of fairly low energy and that he expected the Cambridge people would soon have results. Here might be a new field of activity for a young theoretician, and Ehrenfest suggested that I should work for a while in Berlin at the Institute of Lise Meitner. So, on the first of June, I set out for Berlin and stayed there until the middle of July.[2]

It was a hot summer; in the streets of the crowded city center the heat was oppressive. And it was a hot summer also in a figurative sense. Brawls between Nazis and communists were the order of the day; one had often to make a detour of a few blocks to avoid being caught up in a street fight, and university buildings were sometimes closed because of riots. I had, on occasion, taken part in political discussions—with Landau and with others—but they were discussions of an academic type; I had kept away from the realities of political conflict. Here in Berlin I understood for the first time the meaning of political tension. As yet the fighters were only small groups, seeking an escape from the dull despair of a people hit by a grievous economic depression, but that would soon change. Perhaps

I should be blamed for having so far led an idyllic existence in the green pastures of academic science. There were many people out of work in the Netherlands too; there was poverty and humiliation and I had hardly paid attention to that. Here in Berlin one could no longer escape the cruel reality. The image of a man, collapsed at the side of the road, exhausted after walking all day in search of a job he did not get—just a routine example of thousands of similar occurrences, I suppose—stayed with me as an epitome of this human tragedy.

Dahlem, the suburb where the Kaiser Wilhelm Institutes were located, was different. There was more fresh air there, even though it was warm, there was no street fighting, and there was no manifest poverty. Still, it was clear that the once well-to-do inhabitants of the beautiful villas of that quarter had difficulty in making ends meet. I found a small room in the big house of a high government official, a Herr Geheimrat. His wife had to take in lodgers in order to keep going, but she did it gracefully, with an air of bestowing a great favor. Some of the graduate students at Lise Meitner's Institute were obviously hard up, some may have been undernourished, but still they were among the privileged students.

On 24 June I attended a lecture by Schrödinger organized by some students'-aid society (one had to pay an entrance fee). Before introducing Schrödinger, the president, an eloquent young woman, sketched the plight of many students. It was a grim picture indeed. Schrödinger, by the way, spoke on "Die Naturwissenschaft im Kulturmilieu" (Natural Science in the Cultural Environment). He tried to show that there existed striking parallels between the main tendencies in modern science and the characteristics of contemporary society. It was a witty lecture, but he did not convince me. Neither did he convince Lise Meitner. When I remarked afterwards that I thought that it was at least "ein witziger Vortrag"—a witty lecture—she answered somewhat reluctantly, "Ganz witzig"—Quite witty. (*Ganz* in German suffered the same fate as *quite* in English. As the later editions of Fowler put it, *quite* with a special intonation means "not quite."[3])

Economic depression and political threats notwithstanding, Berlin was in those days still an important cultural center, and Berlin Dahlem was a good place to live and to work. There were many guests from abroad, and the Harnack House, where I usually had my meals—it might be compared to an American faculty club—provided interesting contacts. Among the things we discussed were Aldous Huxley's *Brave New World,* which had recently appeared, and the motion picture based on Brecht's *Dreigroschenoper.* Two Italians, Rasetti and Bernardini, were working in Lise Meitner's Institute. We often went out together, and Bernardini became a lifelong friend.

Experimental techniques in the Meitner laboratories were rather simple although experiments were done with great skill and care. Lise Meitner herself once related that she had had to do an experiment involving a short-living radioactive substance. She had practiced all the manipulations required "like a difficult sonata" before doing the experiment with the real preparation. As I remember it, equipment consisted mainly of gold-leaf electrometers, string electrometers, ionization chambers, and Geiger counters. Some of the younger physicists were keen to try their luck with electronic apparatus, but for the time being Lise Meitner was rather reluctant to let them have their way. Today's physicists may have difficulty in imagining the way in which particles were counted in those days. Geiger-Müller tubes have gradually been replaced by other counting devices but for many years they played an important role. They are thin-walled cylindrical tubes filled with a gas at reduced pressure in which an electrical discharge of short duration between the wall and a wire along the axis of the cylinder is triggered by every fast electron that passes through. Each discharge gives rise to a voltage pulse across a resistance, and these pulses can be fed into an electronic counting device. But, in the Meitner laboratory, they were at that time observed by means of a string electrometer: a discharge produced a jerk of the string and these jerks were ticked off by hand. Work for a thesis might easily involve hundreds of hours of counting, that is of peering through a microscope and ticking off.

I don't think I was able to make myself very useful although I may have given some help in understanding quantum theory. I proposed one silly experiment. Someone had suggested that the weak radioactivity of potassium might be connected with an emission of neutrons. His arguments were not very convincing but I thought it might be worthwhile to look into the matter. So Rasetti designed an experiment and within one day he was able to show that no neutrons were emitted.

I attended the theoretical seminar that was run by Schrödinger and Richard Becker. Becker's assistant was F. Sauter, later professor at Cologne. We had corresponded about the photoelectric effect: he had made very thorough calculations for low energies of the absorbed light quantum and I rather flighty ones for very high energies. He introduced me to other physicists; I remember in particular a visit to the laboratory of G. Hertz, who was then working on the separation of isotopes by diffusion.* He had an installation with twenty-four mercury pumps—for those days very spectacular. The diffusion plants for separating the uranium isotopes are based on this work.

I gave a talk at the seminar on the theoretical treatment of the interaction between electrons recently published by Chr. Møller in Copenhagen. It met with Becker's approval; Schrödinger was also kind but somewhat noncommittal. There was a rumor—probably unfounded, but characteristic of his somewhat passive attitude at these events—that he was in the habit of plugging his ears with wax whenever he attended a seminar. On another occasion there was a lecture on superconductivity by Schachemeier. I wrote to my fiancée that it was nonsense. I have not the slightest recollection of what he said, but given the state of knowledge about superconductivity in those days I believe it is fairly safe to assume that my verdict was right.

I also attended some sessions of the main colloquium. Kurt Mendelssohn has described this institution in enthusiastic terms as the place where the most prominent physicists of the day pronounced judgment on the most recent developments.[4] It did not strike me that way at all; compared to Ehrenfest's colloquium, discussions were both formal and perfunctory. But it was an experience to listen to Walter Nernst.† In a fairly soft, yet penetrating, rather high-pitched voice he would proclaim that he had already said some of the things presented at the colloquium in his book, and complain that people apparently did not recognize that as a publication. He struck me at

*Gustave Hertz, 1887–1975, Nobel Prize (with Franck) in 1925, was professor at the Technische Hochschule, Berlin, from 1928 to 1935.

†Walter Hermann Nernst, 1864–1941. Professor at Göttingen 1891–1905, at Berlin 1905–1933; Nobel Prize in Chemistry 1920. Made important contributions to the theory of electrochemistry. Formulated the so-called third law of thermodynamics, which played an important role in the early developments of quantum theory and is an essential principle in low-temperature physics.

the time as a ridiculous figure, and that impression was strengthened by the many stories that circulated about him and by his being a "Bonze." (This term, literally a Buddhist priest, was used in Germany for a VIP in a pejorative sense.) Later, I realized that some of his remarks had contained a rather subtle point. In 1964 the centenary of his birth was celebrated at Göttingen and I was invited to give the main talk.[5] On that occasion, I studied his published work more closely and was impressed. True, there were some irritating mannerisms and his mathematics was shaky, but his work shows throughout a remarkably clear and often prophetic vision. And so I had an opportunity to atone in public for an error of judgment I had never voiced.

A meeting that I remember quite clearly took place on the seventeenth of June in the Deutsche Physikalische Gesellschaft; it dealt with the first experiments in electron microscopy. Of course, electron beams in vacuum had been studied before, they had been focused and deflected, but here the idea of electric or magnetic lenses, the idea of electron optics, was for the first time brought home to me, and I felt at once convinced that this was the beginning of a new and promising development.

From a theoretical point of view the history of the subject has an amusing twist. William R. Hamilton (1805–1865) had already shown the mathematical equivalence of mechanics and geometrical optics. Felix Klein (1849–1925) took up that question again in 1891 and 1901, and although he complains somewhere that his considerations had not attracted the general attention he had envisaged for them, the main idea was staple knowledge among theoretical physicists. But it was regarded as a purely formal analogy: nobody made much of it until in the hands of Schrödinger it became an important heuristic principle that led to the formulation of wave mechanics. The reasoning—in a highly condensed form—went as follows. We know mechanics and geometrical optics are mathematically equivalent: the rays of light obey similar laws as the trajectories of particles. We also know that geometrical optics is an approximation to the real, physical wave optics. Could it not be that also behind mechanics there is a wave mechanics of which classical mechanics is an approximation? And so Schrödinger, of course partially inspired by the work of De Broglie, set out to find such a wave mechanics and the attempt proved most successful. Only then did experimental physicists begin to take the analogy between geometrical optics and mechanics seriously: they went back from Schrödinger to Klein and to Hamilton; that became quite clear during this meeting. However, that retrograde way of looking at things had its advantages. Since the wavelength of electrons is much smaller than that of visible light, it followed at once that the resolving power of electron lenses could be much better than of optical lenses. Modern electron microscopes, developed since the war years, make good use of that feature.

Fritz Houtermans was in those days working at the Technische Hochschule. He was living with his wife and a few-months-old baby in a minute appartment. He was a colorful person—so colorful, indeed, that one is liable to forget that he was also a good physicist. He was a friend of Gamow, had published together with Atkinson some speculations on the origin of the chemical elements, and had published also a survey of nuclear physics in the *Ergebnisse der exacten Naturwissenschaften*. So I must have discussed with him the big news from Cambridge: Cockcroft and Walton were disintegrating nuclei by bombardment with protons accelerated by a voltage of half a million volts. Even newspapers carried this news. I believe but have been unable to confirm that at least one of them added that further important progress had been made in Berlin by Brasch and Lange. As a matter of fact Brasch and Lange had already been working with pulsed high-voltage sources for quite some time.* (Before that they had tried to harness atmospheric electricity, but they had given up the attempt after a fatal accident.)

I met many other physicists—Wigner, Szilard, Kallman, F., London—but for my own work my meeting with Schüler was even more important. Hermann Schüler—he later went to Göttingen—was already in those days one of the foremost experts on the so-called hyperfine structure of spectral lines. I visited him at his laboratory at the Einsteintower in Potsdam. (There I also met the astronomer Freundlich, who explained to me that his measurements of the deviation of light by the gravitational field of the sun were in his opinion more accurate than those of Eddington and coworkers, and revealed a systematic deviation from Einstein's formulae. I did not know what to say about that, and I still don't know.) Schüler had just found some curious effects in the hyperfine structure of mercury; could I explain

---

*Brasch later went to the United States. His attempts to apply pulsed high voltage equipment to the sterilization of foods and of bandages were only a limited commercial success.

them? Now, as I explain in Appendix B, hyperfine structure was an old and somewhat unhappy love of mine. I tackled the problem right away and although I could only explain part (albeit the major part) of the discrepancies found by Schüler, this time I published my results right away, and my contacts with Schüler were most useful later on.

So all in all, my stay in Berlin was a profitable one, but I still have to mention the most important happening of all. While in Berlin I received a letter from Pauli, asking me to come to Zürich as his assistant in September. That was a great opportunity and I was happy to accept.

I never returned to Berlin during the Nazi era, and after the war, during my Philips years, I paid only one or two cursory visits. It was in 1979 that I really revisited Berlin. It was the year of the centenary of the birth of Albert Einstein. All over the world there were celebrations: at Princeton, where he had spent the last twenty-two years of his life; in Bern, where for seven years he worked at the patent office, to begin with as "Experte dritter Klasse," doing some of his greatest work in his spare time (and possibly during some of his office hours, although he did not at all neglect his official duties); in Jerusalem, for he was an early supporter of Zionism and had even once been asked to become President of Israel; at the Academy of Science in East Berlin, the continuation of the Prussian Academy with which Einstein was connected from 1914 to 1933 and in whose proceedings he published many important papers. This is certainly not a complete list of events, and at many universities there were series of lectures, seminars, and so on. And then there was the meeting in West Berlin, to which I had been invited, and that had been organized by the Max Planck Gesellschaft, the Bunsengesellschaft, and the Deutsche Physikalische Gesellschaft. It differed from other meetings, in that— to put it irreverently—it was a package deal. Albert Einstein, Otto Hahn, and Max von Laue were all three born in 1879, and one had added Lise Meitner, born on 7 November 1878, "who had always been a precocious girl."

West Berlin today is certainly a place full of industrial, cultural, and scientific activity, but it is a somewhat artificial activity maintained under precarious circumstances. The meeting must have brought back to many of the visitors memories of past grandeur. A few of them—like Ewald—remembered the Berlin of the pre–World

War I era, when it was the thriving capital of a proud—alas too proud
—empire. Many had lived and worked there in the twenties and
early thirties; as I mentioned before, it was in those days still a remarkable center of creativity, economic depression notwithstanding.
Spree-Athen (Athens on the River Spree) it was sometimes facetiously called. I have referred already to Kurt Mendelssohn, a German physicist who emigrated to England, and his book, *The World
of Walter Nernst.* It is not a very thorough biography of this remarkable physico-chemist, but it does give a lively picture of Germany
and Berlin before the Hitler regime. For physicists who, like Mendelssohn, had had to leave Germany, coming to Berlin on this occasion—he himself was not there—must have been a kind of pilgrimage, and even for those who had remained in Western Germany this
element cannot have been entirely absent.

As to the meeting, there were lectures on each of the four centenarians. A lecture on Von Laue by P.P. Ewald, who a month earlier
had celebrated his ninety-first birthday, was particularly impressive.
There was also an interesting exhibition of books, papers, letters,
photographs, and instruments. I remembered that during the weeks
I was in Berlin in 1932 Von Laue, who had the reputation of being
a rather intrepid driver, had once driven me into town to attend a
colloquium. I seemed to remember his car was an open Steyr, but
after almost half a century I was not entirely sure. It *was* an open
Steyr: there was a photograph in the exhibition.

At the opening session, which was attended by over a thousand
people, the Bundespräsident Walter Scheel gave a highly relevant
and incisive address on physicists and politics. I had to follow—not
without trepidation—with a lecture on physics during the first
decades of our century. (It was that lecture I mentioned in Chapter
2 and from which I took most of the material for the fourth section
there.)

The four centenarians were not only great scientists, they were
also persons of great integrity and courage, perhaps Von Laue most
of all. I tried to express this in my peroration:

Ladies and gentlemen, the four scientists we honor today lived and
worked here in Berlin. They were—each in his own way—representatives
of things valuable and excellent in German-speaking culture. It is meaningful that we celebrate their hundredth anniversary here in Berlin. But they

were also prominent members of the international community of scientists. In the preface to the English translation of his *History of Physics,* Von Laue writes:

> The first draft of this book was written in the summer of 1943 and is accordingly a war product. Precisely because of this circumstance, the thought of the culture that is the common property of all nations, and which was then so despicably mistreated, was ever present throughout the period of composition.

Perhaps the same thought has induced the organizers of this meeting to invite a foreigner as a main speaker. In any case I am grateful for this opportunity to express also on behalf of the international world of physics my sincere admiration.

Zürich

In September 1932 I arrived in Zürich. The first thing I had to do was to find a place to live. Armed with a map and the advertising pages from a newspaper, I went in search of a room. It did not take me long; the first house on my list pleased me at once because of its location. It was on the Susenbergstrasse, in those days the highest street on the slope of Zürichberg, and it was only a short distance— and a climb of two hundred meters or so—from the physics laboratories. To the woman who answered the bell I explained in my best German that I had read her advertisement. Was the room still available and could I see it? She eyed me rather suspiciously and then said: "I don't really know. Are you German?" When I told her I was a Dutchman her face lit up and we soon came to terms. From then on she insisted on addressing me in Swiss-German, which I soon learned to understand although I never tried to speak it. When I told her that I was assistant to Professor Pauli, she repeated the name and I saw the ghost of a smile on her face. I would understand the reason for that the first evening after I had moved in.

My room had a balcony that commanded a magnificent view of part of the city and of the Uetliberg, the range on the other side of the lake. The street was very quiet at night, but one could hear the clocks of many churches striking the hour. Church clocks in Switzerland ran, of course, fairly accurately, but they were not exactly synchronized, so around the hour there would be half a minute or so of striking clocks. Rilke has said it beautifully:

Die Uhren rufen sich schlagend an,
Und man sieht der Zeit auf den Grund.[6]

(Clocks, striking, are calling one another
And one sees to the very depths of time.)

I had gone to bed early, but I did not yet sleep. The doors of my balcony were wide open and I was listening to the clocks and to the silence that followed. Then I suddenly heard an enticing female voice calling, "Pauli, Pauli, come, please do come." I could not help myself; I got out of bed to have a look. A young woman was standing below; she was watching the bushes on the other side of the street and called again and again, "Pauli, komm doch, komm doch." Finally, out of the bushes there appeared a little dog, and the couple went home together. That scene repeated itself almost every evening. When I told Pauli about it, he laughed and said he might hide himself in the bushes and suddenly jump out saying, "Here I am," but he was not the man to play such a practical joke. Gamow or Houtermans in his place might have done it.

During the year I was at Zürich, Pauli had no graduate students working for a doctoral thesis and only one or two who took theoretical physics as a main subject for their "diplom" (master's degree). He gave a course of fairly elementary lectures on theoretical physics, but he never asked me to assist him in the preparation of these, or with correcting examination papers. These lectures did not enjoy a very good reputation among the students. Pauli could be a first-class lecturer if he prepared himself carefully, but this he rarely did. And he was not one of those lecturers who at a moment's notice can give a clear, well-structured, and beautifully presented lecture on any subject. It took a good student to look through the superficial imperfections—mediocre blackboard technique, errors in numerical factors or plus and minus signs, soliloquies muttered under his breath when he was not satisfied with a deduction—and to appreciate the power of his intellect and the depth of his understanding. I attended only the special course he gave on relativity, where such shortcomings were of far less importance than in, say, a course on thermodynamics for future applied physicists or engineers. Yet I remember that once when he was speaking about the so-called red shift (that is the phenomenon that the frequency of a spectral line emitted by a light source in a strong gravitational field is shifted towards lower frequencies), he obtained an expression with the wrong sign, which meant a shift towards the violet instead of towards the red. He then began to walk up and down in front of the blackboard, mumbling to himself, wiping out a plus sign and replacing it by a minus sign, changing it back into a plus sign, and so on. This went on for quite some time until he finally turned to the audience again and said: "I hope that all of you have now clearly seen that it is indeed a red shift."

One of the regular attendants of the lectures on relativity was Stodola,* professor emeritus of turbine fame, who was interested in the philosophical aspects of physics in general and in the relativistic aspects of rotating frames of reference in particular. I believe that he in some ways disapproved of me and my probably none-too-well-veiled lack of respect for age and established reputation as such—it was reported to me that he once complained about "that Casimir, so young and already such an *Eigenbrötler*."† But he invited me to have coffee with him once a week after lunch and to discuss modern physics. In a way these discussions were quite interesting: he was a man of stature. He did not impress me though when he told me about his life and said about one episode: "And then I became what I since then have always been—*eine Persönlichkeit*" (an important personage). He lacked the simplicity of real greatness, the simplicity of Lorentz or Bohr.

At first there were no other theoretical physicists working with Pauli, but that did not mean that I was lonely. Pauli had a few rooms at his disposal in the physics building, where Scherrer was in command, and the atmosphere there was pleasant and cosmopolitan. There was Egon Bretscher (1901–1974), who later went to Cambridge, was a member of the British mission to Los Alamos, and after the war became divisional head at the British Atomic Energy Establishment at Harwell. He and his young wife were a most charming and most hospitable couple. On the other side of the spectrum there was the German Bühl. He was a pupil of Lenard's and was to become a fervent Nazi.[7] During his Zürich days he was a pleasant enough fellow. He also struck me as a sound though somewhat narrow-

---

*Aurel Stodola (1859–1942), professor at Zürich 1892–1929, is generally regarded as the foremost authority on thermal engines of his day. His book on *Dampf- und Gasturbinen*, which went through six editions, was the bible of all engineers in this field. For further details cf. *DSB*.

†The word is hard to translate. It describes a person that goes his own way heedless of others, which may—but not necessarily does—make him an eccentric.

minded physicist. Alexopoulos from Greece later played a prominent role in the academic life of his country. There were also several Hungarians; one of them, A. Nemet, afterwards settled down in England, where I met him from time to time; he did quite well for himself in X-ray instrumentation.[8] In November Homi Bhabha arrived. He came from Cambridge (England) with a recommendation from R.H. Fowler—that is to say a recommendation of a kind. Fowler realized that Bhabha was very gifted, but he also thought him opinionated and unruly, so he felt Bhabha needed a strong hand. "You can be as brutal as you like," he wrote. This Pauli enjoyed immensely; he showed me the letter and repeated over and over again, "I can be as brutal as I like." I wonder whether Fowler was subtle enough to understand that this letter was the best way to make Pauli well-disposed towards Bhabha. But it did work out that way and they became good friends, although it must be admitted that Bhabha sometimes turned to Wentzel* rather than to Pauli when he wanted to discuss an entirely new idea. Bhabha, who came from a wealthy Parsee family in Bombay, was a man of many parts. He was not only a gifted physicist but also a painter; he was widely read and knew much about music. Also he had the knack of becoming befriended by interesting people (and with very pretty girls). My wife liked to refer to him as the fairy-tale prince.

Elsasser passed through Zürich after he had had to leave Germany. He refers to Bhabha's painting as being the darkest kind of painting he had ever come across.[9] That impression must be partly connected with the frame of mind he himself was in. I do not think it was as dark as that; I remember specifically a painting he called "theme with variations," with rather colorful abstract shapes against an admittedly dark background.

Bhabha and I became good friends and often had our meals together. On such occasions we might first go to a restaurant of the Zürich Women's Association for Alcohol-free Restaurants, which provided plenty of quite good food for little money, after which we would adjourn to a less reputable pub and wash away the priggish aftertaste with a glass of beer.

Bhabha later became leader of the Tata Institute in Bombay and rendered invaluable service to his country by building up both basic

*Gregor Wentzel, 1898–1978, was then professor of theoretical physics at the University of Zürich.

and applied research. Also he was president of the first Geneva conference on the peaceful applications of atomic energy. He died in an air crash on Mont Blanc in 1966. I still have a beautifully carved Indian picture frame he sent me on the occasion of our marriage "to hold the picture of Archimedes Casimir." (He always insisted I should call my first son Archimedes: it struck him as an impressive combination. My son is grateful that I did not follow his advice.) In the spring David Inglis—I had first met him at Ann Arbor—also stayed for some time in Zürich.

The first task Pauli assigned to me took a curious and, for me, rather fortunate turn. Jacques Solomon, the same young French theoretician who had, without much reason, incurred Landau's wrath at Copenhagen, had been in Zürich the winter before and he had written a paper in which he had worked out some further details of the mathematical structure of Einstein's most recent five-dimensional relativistic field theory. Now he was sending Pauli the second proof of the paper, which was to appear in the French *Journal de Physique,* and he suggested that it should be published as a joint paper by Pauli and himself: he felt that Pauli's share had been so great that it would not be honest if he were to publish it under his own name only; also he would be very proud to have a joint publication with Pauli. Pauli thought this over, and told me that what Solomon wrote was true: he himself had not only suggested the problem but also explained exactly how to tackle it, but if the paper was going to bear his name I had better check the formulae. In those days I was reasonably good at juggling with tensors (the type of mathematics involved), and anyway to check the calculations—not the reasoning —of somebody else is just a routine job. To my dismay I found that most formulae, at least the essential ones, were wrong. I told Pauli, who at once cabled Paris that the paper would have to be rewritten. Soon afterwards there came a letter from Jean Langevin, the editor of the *Journal de Physique* and Solomon's brother-in-law, with apologies: the second proof had not come back in time, there had been hardly any misprints in the first proof, and so the issue of the *Journal* containing the paper (with all the errors and with Pauli's name as co-author) had gone to print and was already sent out to the subscribers. Pauli was less angry than he might have been, but he resolutely tackled the problem. He redid the calculations himself—and, to my satisfaction, confirmed my results—and wrote a new version of the article with a masterly introduction. He sent it to Solomon, who translated it into French, but who was told not to change one single formula. And then a curious thing happened. Pauli came into my room and looked worried, almost embarrassed. "Look here," he said, "you found those mistakes; perhaps you should be a co-author, or at least I should put in an acknowledgment. But this is a rather awkward situation and after all, I should not like to harm Solomon's future. How do you feel about it?" I answered that I did not think I deserved an acknowledgment. It might have been different if I had spotted the errors on my own initiative, but I had just been told to check the formulae and that I had done "like a good schoolboy" (wie ein braver Schüler). The great man looked really relieved. I am relating this incident because it shows that Pauli, his sharp tongue notwithstanding, was very conscientious and even generous in dealing with the work of others. It also helped to establish friendly relations between master and assistant. I feel quite happy that I have made at least one contribution to general relativity, albeit an entirely anonymous one.

Pauli's concern for Solomon's scientific career turned out to be of no avail. Solomon became more and more interested in politics and during the Second World War he was caught and executed by the Germans; his wife survived a concentration camp and later remarried—I never saw her again.

The two papers, Pauli & Solomon I and II, are easily accessible. One need not look for them in old issues of *Journal de Physique*: they are reproduced in Pauli's collected works. But I do not think that anyone is seriously looking at the erroneous formulae of paper I or the correct formulae of paper II these days. I even wonder whether anyone ever did, and I am pretty sure nobody ever applied them to a real physical problem. May they rest in peace.

Pauli in those days was not a very happy man. His first marriage had broken up after a few months and he had not yet found his second wife, who would be such a wonderful companion until the end of his days. Also, he had put a lot of work into his article on quantum mechanics for the *Handbuch der Physik,* and it almost seemed as if he did not quite know where to go from there. He did not like solid-state physics and some of the other straightforward applications of quantum mechanics. And I do not think he really believed in Einstein's approach to unified field theories. One evening we walked home together after having been drinking wine at the home of one of Pauli's colleagues. "We are living in a curious time," Pauli said, "in a cultureless time. Christianity has lost its grip. There must come something else. I think I know what is coming. I know it exactly [Ich weiss es ganz genau]. But I don't tell it to others. They may think I am mad. So I am rather doing five-dimensional theory of relativity although I don't really believe in it. But I know what is coming. Perhaps I will tell you some other time." He never did, and I never did find out what he had in mind, but I am convinced it was something fairly definite. The wine he had been drinking may have made him more communicative, but it did not interfere with the clarity of his thinking.

Pauli, for all his outstanding competence in exact physical theory, was far from being a downright materialistic thinker. He once told me he thought there might be some truth in astrology, that is to say, that there might be a non-causal correlation between someone's fate and the time of his birth, but that astrology used a curious and unnecessarily complicated way of measuring time. He did not a priori deny the possibility of extrasensory perception and he emphasized the importance of archetypical concepts (that is concepts not derived from logical discussion of experimental evidence) in the structure of theoretical physics. He later published a study on the role of "Archetypes" in Kepler's work. It appeared in the same volume as Jung's theory of synchronicity,[10] and that can only mean that he took Jung's ideas seriously.

Jung is convinced that there can exist meaningful correlations between causally unrelated events. For instance: a man is dying and at the moment of his death a flight of birds alights in his garden. The man is not dying because the crows are there: his disease is taking its inexorable course. The birds are not there because the man is dying: their migration follows its customary path. Yet the coincidence of the two events is meaningful. I am afraid Jung would not have liked the following prosaic example, but I believe it helps to understand the gist of "synchronicity." I want to boil an egg. I put the egg into boiling water and I set an alarm for five minutes. Five minutes later the alarm rings and the egg is done. Now the alarm clock has been running according to the laws of classical mechanics uninfluenced by what happened to the egg. And the egg is coagulating according to laws of physical chemistry and is uninfluenced by the running of the clock. Yet the coincidence of these two unrelated causal happenings is meaningful, because I, the great chef, imposed a structure on my kitchen.

I have not made up my mind about this idea of synchronicity. Yet, for me, there is a clear lesson to be learned. If a physicist like Pauli thought along such lines, no physicist needs to feel ashamed if he looks in his own way for structures and relations beyond the realm of physical theory.

One of Pauli's consolations was his car. He had had some difficulty in passing the driving test and he was not a good driver, but he managed, and as far as I know, his accidents were limited to some dents and scratches incurred while leaving or entering his garage. For mechanical advice he relied on Charles Mongan, an American experimental physicist then working in Scherrer's laboratory, who was a superb mechanic.

Scherrer* sometimes complained that as soon as the apparatus Mongan had constructed worked well he would take it to pieces and start to improve it, instead of carrying out measurements. In 1979 I met Mongan again at Cambridge, Massachusetts. He was still doing high-precision instrumentation, at the Draper Laboratories.

To my knowledge Pauli himself never took a real interest in technical problems, not even in the mathematical problems arising in technology. In that respect he was quite different from his teacher, Sommerfeld, who did important work on the hydrodynamics of lubricants, on the propagation of radio waves, and on the theory of spinning tops and their applications. And H.A. Lorentz devoted several of his later years mainly to the calculations of the influence of the big dam across the Zuider Zee on the tides along the adjoining coasts.

Once Pauli had to act as chairman of the general colloquium—the professors chaired by turns—when a lecture was given on novel types of radio tubes. After the first few sentences he turned to me and whispered, "That is amusing; I don't understand a word." And as the lecture proceeded he laughed more and more. "But that is amusing; I don't understand one single word. Do you understand anything at all? Das ist aber wirklich lustig." At the end he thanked the speaker politely and expressed the hope that the lecture had

*P. Scherrer, 1890–1969, became professor of experimental physics at the ETH in 1916. He was well known for his work in X-ray diffraction, and also for his magnificent lecture demonstrations.

satisfied those members of the audience that happened to be interested in that kind of thing.

More important was a colloquium where Stern,* an old friend of Pauli and a regular visitor to Zürich, reported on his measurements of the magnetic moment of the proton, the hydrogen nucleus. It was a fantastically refined experiment and the result was most surprising: the moment was about two and a half times larger than was expected by most theoreticians, including Pauli, who had at first even been inclined to consider the experiment useless! Pauli was also partly responsible for an erroneous estimate by one of his students of the magnetic moment associated with the electronic charge distribution in a rotating hydrogen molecule. Stern and Pauli liked to engage in verbal skirmishes in which Pauli often scored, so Stern did not pass up this opportunity to get some of his own back: in his lecture he spoke with obvious glee about his regrettable experiences with theoreticians.

Sometimes Pauli would take me out in the evening and we would have a quiet meal in a country inn. His car kept him then from having too many, which was just as well.

To the last-mentioned point, however, there occurred one notable exception. That was on the occasion of the spring meeting of the Swiss Physical Society in Lucerne. Pauli had driven us—that is, David Inglis, Elsasser, Felix Bloch, and myself—from Zürich to Lucerne in the morning, and apart from Pauli's slightly disconcerting habit of saying from time to time, "Ich fahre ziemlich gut" (I'm driving rather well)—a statement he underlined by turning around to his passengers and by releasing his hold on the wheel—nothing untoward happened. In the evening Pauli was drinking fruit juice with a wry face. Suddenly he changed his mind and ordered a whisky and soda. That was still all right, but when he had ordered a second one and showed no signs of wanting to stop there, we became really worried. So we held a council of war and decided to offer him more drinks and then Inglis could drive us home. The first part of the operation went according to plan but, when we suggested that Inglis should drive, Pauli bluntly refused. He was going to drive and we could either

*Otto Stern, 1888–1969, Nobel Prize 1943, was from 1923 to 1933 professor at Hamburg, where Pauli worked from 1923 to 1928. In 1933 he emigrated to the U.S.A. and found a position at the Carnegie Institute. Estermann in *DSB* gives an adequate survey of his work; Frisch (loc. cit.) relates some personal touches.

come along or stay in Lucerne. By then the last train to Zürich had left, and anyway we did not want to let Pauli go all by himself. So we went, with Inglis sitting beside Pauli, ready to grasp the wheel in an emergency. We even had one more passenger, Bhabha, who had missed the last train and was sitting on the floor of the car. Pauli sounded his horn several times, hit one curb, swerved to the other side of the street, where he hit the other curb, then managed to find his bearings and got going. It was a memorable trip.

Pauli would still say from time to time, "Ich fahre ziemlich gut," but when the car went screeching around curves Inglis would say sternly, "Dass heisst nich gut fahren" (that is not called good driving), which had a somewhat sobering effect. Once a rising moon came just over the crest of a hill and Pauli started swearing at the driver who did not dim his headlights. Once Pauli said, "Here I know a shortcut," and suddenly turned into an unpaved track. It came to an end at the wagon shed of a farm, and after some angry comments about people that put wagon sheds across his shortcut, Pauli turned round and went back to the main road. But that was all; we came safely home.

Maybe I have slightly embellished the story: a letter I wrote shortly afterwards to my future wife is more sober, but probably I did not want to scare her. So much is certain: when fifteen years later I met Inglis again and I asked him, "Dave, do you remember that drive from Lucerne to Zürich?" his answer came promptly: "Will I ever forget it?" In a way he had felt even more worried than the rest of us: he felt he had a certain responsibility, but he also felt he was almost powerless.

Pauli certainly did not mince his words when he thought you were a fool, but I found it far easier to work with him than with Ehrenfest, whose sarcasms were only a thin cloak for his own vulnerability. And he really put me to work, suggesting profitable problems and discussing the ways in which I might tackle them. I am by nature inclined to be lazy, but Pauli gave me little opportunity to indulge that tendency. "Er hat geächzt" (he has groaned), he later said with obvious pride to a common friend—I believe it was H.A. Kramers.

During the whole year I was in Zürich we had only one serious altercation. It began at a colloquium with a difference of opinion about multipole radiation[11] and went on afterwards in Pauli's office. I felt that just for once I was right and Pauli was wrong, so I did not want to give in. Our quarrel had a happy ending, but in order to explain that, I have first to relate an entirely different matter.

I have mentioned Walter Nernst. He must have been a remarkable man. He was an authoritarian "Herr Geheimrat," he was a shrewd businessman, and of course he was a great physicist (although chemists might claim him as one of theirs). He is also a striking confirmation of my earlier remark that it is wrong to call a theoretical physicist a mathematician. Nernst's third law of thermodynamics is an important contribution to theoretical physics, but the mathematics he uses is always extremely simple and sometimes he presents the most elementary mathematical deductions with ridiculous pomposity. Notwithstanding this, I believe that even Pauli was somewhat awed by Nernst. In any case one of his beloved stories—I heard him tell it more than once—was the story about his famous conversation with Nernst ("mein berühmtes Gespräch mit Nernst"). When telling it, Pauli used to give exact data about time and place, but they have slipped my memory. Here is what I remember. Shortly after his appointment at Zürich, Pauli had occasion to visit Nernst and the following conversation developed.

"Ah, Herr Pauli, I recently heard a lecture you gave; was not bad. Perhaps still a bit *schülerhaft* [somewhat like a schoolboy] but otherwise quite good. And where are you now?"

"I am at Zürich, Herr Geheimrat."

"Do you work there with Herr Meyer?"

Now Edgar Meyer was the well-intentioned but none-too-bright professor of experimental physics at the University of Zürich. He was a member of the little wine-drinking circle I mentioned before and I think Pauli rather liked him. However, the idea that Pauli should work under Edgar Meyer was farcical.

"No, Herr Geheimrat, I am at the ETH."

"Then you are with Herr Scherrer?"

"Well, Herr Geheimrat, as a theoretician I am really independent, but of course I am in the same building as Herr Scherrer and I meet him regularly."

By then it began to dawn upon Nernst that he had misjudged the situation, and he felt he should do something to put things right. So he went on:

"So you are Ordinarius [full professor]." On Pauli's affirmative answer, he continued:

"But tell me, *lieber Herr Kollege* [dear colleague], can you live on that [Können Sie denn davon leben]?"

And now back to my quarrel with Pauli. We finally reached an agreement and then Pauli said, "There, we have really been barking at each other," and he continued rather surprisingly, "Here, I'll show you what Springer is paying me for my *Handbuch* article."

The amount was not staggering but I was happy to accept this generous, though implicit, apology.

I should have liked to stay one more year at Zürich and Pauli was quite willing to keep me there. Sometimes I still dream about the problems I might have tackled and the results I might have obtained had I stayed. But around Eastertime there came an urgent message from Ehrenfest that he insisted on my coming back to Leiden in September 1933. Both Pauli and I felt that we had to comply with that wish. Only later did it become evident to me that already at that time Ehrenfest had almost decided to put an end to his life. And he also knew that it would take some time before his successor would come to Leiden. Therefore, he wanted to be sure that there was at least an assistant who could look after the reading room, keep the colloquium running, and so on. The final sentence of this message was: "Ach Caasje, setze deine breiten Schultern unter den Karren der Leidener Physik." (Put your broad shoulders under the wagon of Leiden physics.) In retrospect it is clearly a farewell, and yet he must have hesitated to the very last. We still saw him at Copenhagen in September. Then, on 25 September 1933 he carried out his fateful decision.

Back to Leiden

Duino, on the Bay of Trieste, is a lovely place. It must have been even more lovely at the beginning of this century, before modern expansion began to close in upon the fishing village and the ancient castle of the Princes del Torro e Tasso. It was to Duino that Ludwig Boltzmann went for a holiday in September 1905, and there on the fifth of September he took his life. The rejection of atomistic theories by leading scientists such as Ostwald and Mach had aggravated his melancholic despair; it may even have been its main cause.[12] That makes the event still more tragic, for the victorious advance of atomistics that I have sketched in Chapter 2 had already set in. Had he paid more attention to what was happening in the Anglo-Saxon world and had he realized that H.A. Lorentz had to be taken far more seriously as a theoretical physicist than Mach, he might have found the courage to endure the silly criticisms and might have lived to witness an almost universal acceptance of his ideas.

The event must have made a profound impression on Ehrenfest. He owed much to Boltzmann, whose brilliant lectures had been a source of inspiration and a decisive factor in making him a theoretical physicist. They must also have influenced the development of Ehrenfest's own particular style of lecturing. In a way he amply paid back what he owed: for many years his main efforts were directed towards the elucidation and vindication of Boltzmann's ideas. However, as Martin Klein points out, neither in the obituary notice of 1906 nor in his later writings is there one single personal remark, and I do not remember that in our many discussions, including those in which he spoke about suicide, he ever talked about Boltzmann the man, although he very frequently talked about Boltzmann's physics. Had their personal relations been difficult, as Klein suggests? Did Ehrenfest somehow feel guilty? Had the shock been too deep, or did he even resent Boltzmann's action? Did he avoid the subject of Boltzmann's death because he vaguely felt it might be a foreboding of what might happen to himself? We shall never know. We do know he defended Boltzmann's theories with an almost quixotic loyalty, which even led him to grossly underrate the superior power of Gibbs's formulation.

The remembrance of Boltzmann's suicide may have haunted him all his life, but the circumstances of his own death were quite different. Ehrenfest did not have to contend with adverse criticism from others; it was only his self-criticism that finally destroyed him.

His youngest son was born in 1918. It was one of those sad cases of mongolism that occur rather frequently when the mother is over forty—and Mrs. Ehrenfest was almost forty-two at the time of his birth. In the afternoon of the 25 September 1933, Ehrenfest went to the institution that took care of this child, drew a revolver, and shot first the child and then himself. It was a gruesome deed, carried out with uncanny resoluteness and precision.

I have already mentioned Ehrenfest's lack of self-confidence, his apprehension of not being able to understand or teach physics any longer in the way he wanted to understand and teach it. Albert Einstein, perhaps his closest friend, in his "In Memoriam"[13] explains with deep understanding that this was the quintessence of Ehrenfest's depression. Other factors contributed. The coming to power of Hitler and the surge of anti-Semitism in Germany must have touched him deeply. Göttingen, for instance, was to him a spiritual home and a place of pilgrimage. To see that all that was going to be destroyed —and I do believe that he foresaw as clearly as Einstein and more clearly than many others how things would develop—must have robbed him of one of the few remaining certainties. Einstein also hints at a partial estrangement from his wife Tatiana. Half a century is a long time, and I believe that by now I can tell the little I know about the other woman in his life.

She was a nice woman. I met her in the late thirties; she was an art critic and was organizing a series of lectures on modern art and asked me to discuss the relations between art and science. She wanted to speak to me about Ehrenfest. So I went to see her one evening and she explained that there were things she wished to get off her mind and some mementoes she wished to get rid of. She told me how she had met Ehrenfest, that he had been interested in her work, that he had become her lover and that he had always been kind, considerate, and grateful. But also, that she had more and more been burdened by his despair. He had considered getting a divorce and starting a new life with her, but he had been unable to make up his mind. He was deeply attached to Tatiana; that she could understand, but what had really irritated her was that he kept worrying over minor practical and financial details. Finally, she could bear it no longer and she had said, "Do whatever you like, leave me or marry me, but stop dawdling." He had been thunderstruck, had stood there for a while gazing perplexedly at her; then, without a word, he went away, forever. "It was Tatiana who won," she added, though without bitterness.

She then gave me a bunch of papers. She had encouraged Ehrenfest to write down what he remembered about his early youth in the hope that this might help him overcome his troubles. It is these papers on which Klein based his chapter on Ehrenfest's childhood in Vienna. They give a vivid description of the environment in which he grew up and portray an intelligent and very sensitive child whose feelings were often hurt by playmates and by grown-ups. Yet the tale of his youth is by no means a tale of misery; Klein even speaks about an idyllic tone.

Yes, she was a nice woman and no one should blame her for having been unequal to the task of reconciling Ehrenfest with his own limitations.

My wife and I read the news of Ehrenfest's death in a newspaper on the train that brought us home from Copenhagen. I then understood why Ehrenfest had insisted so much on my coming back to Leiden and why he wrote the letter I mentioned. Whether I was able to live up to his expectations I do not know, but I did try. I gave lectures, of a kind, and I kept the colloquium going, in a way. But it is one thing to do reasonably good theoretical work under the guidance of a man like Pauli, and another thing to have to run a department, however small, on your own. When, in the autumn of 1934, H.A. Kramers, who had been appointed as Ehrenfest's successor, arrived in Leiden, I felt really relieved.

Hendrik Anthony Kramers: the name invites comparison with H.A. Lorentz. Even more curious, during the spring of 1916 Kramers taught for a while at a secondary school at Arnhem. Is it a meaningful coincidence in the sense of Jung? In any case, it is not a meaningless comparison, for Kramers was a man of considerable stature. Like Lorentz he was well known and highly esteemed in the international world of physics, like Lorentz he was for many years the obvious leading figure among the physicists in his own country. Undoubtedly, Lorentz was the greater physicist and the influence of his work has been more profound and enduring than that of Kramers, but Kramers's interests covered a wider field. Lorentz was universally gifted and he would have succeeded in many fields of study, but he confined his activities consciously and conscientiously to the field of physics. "The physicist, and this holds for all of us, must restrict himself to reading in his way in the book of the world," he says in one of his many popular addresses.[14] He had an excellent command of English, French, and German, which stood him in good stead in his publications and as a chairman of international meetings, but it is unthinkable that he would have put a copy of Baudelaire's *Les Fleurs du Mal* among his physics books with the motivation that its spirit was somehow akin to that of theoretical physics. (I must confess that Kramers never entirely convinced me that this was really true.) Neither does one imagine Lorentz spending time on writing a beautiful Dutch translation of Mallarmé's *Les Fenêtres*. Kramers did; it was one of his favorite poems and I still remember how, at a festive meeting during the war years—it may have been on the occasion of his silver doctor-ate—he referred to the confiscation of laboratory equipment by the German occupying forces as "Le Vomissement impur de la Bêtise" (Stupidity's impure vomiting). Kramers was also a serious student of Shakespeare. Sometimes I visited him when he had to keep to his room—his health was not very robust—and then he might offer a possibly somewhat paradoxical interpretation of the play he had just been rereading; he was not always in the mood for Shakespeare, though, and he also introduced me to *Tristram Shandy*.

His older brother, Jan, was teaching Persian, Turkish, and Arabic at Leiden University. He was a well-known expert on the history and culture of Islam, and Hans Kramers (as H.A. was usually called) was interested in such problems too. He also liked to discuss problems of Christian theology, usually in a mildly ironical vein, and he pre-tended to be really shocked when in youthful ignorance I committed the well-known blunder of confusing immaculate conception and virgin birth. I do not share Kramers's theological interests but a bit of knowledge can always come in handy. Many years later, during a "déjeuner" in Paris, the French mathematician Paul Montel told me he had often won a box of candy from his partner at a dinner by betting that she would be unable to give a correct definition of immaculate conception. He did not propose a bet to me, but was clearly surprised when I rattled off such a definition. Kramers was certainly not a dogmatic atheist like, for instance, Dirac in his younger years, whose attitude was summed up by Pauli in one fa-mous sentence: "Our friend Dirac has a religion; and the main tenet of that religion is: 'There is no God, and Dirac is his prophet.' "[15]

Music was important in Kramers's life; his main instrument was the cello and he liked to play string quartets, but he played the piano too and was a competent accompanist. In an earlier chapter I told how a popular song became related to my first days in Denmark. Kramers told me about a similar experience but at a considerably higher level. In the early twenties a film that tried to explain Ein-stein's theory of relativity to the general public was shown in a Danish cinema. That was long before the arrival of sound films, and Kramers had been hired to provide the spoken word. As a musical introduction—in those days the larger cinemas had full-fledged or-chestras—he had chosen Beethoven's *Egmont* Overture. And so this overture and relativity remained linked together, and whenever and wherever he heard it performed, at the final chords he would feel that he had to get up and step forward—in full evening dress—and begin his allocution: "Mine Damer og Herrer, Relativitetsteorien ..." (The money he earned he invested in a sewing machine for his young wife.) The difference between our respective musical levels was also apparent in our disagreement about barrel organs. Barrel organs—not the small portable type, but big installations on a horse-drawn wagon—used to be a prominent feature of Dutch street life. (To a certain extent they still are, especially in Amsterdam, but they are by now fully motorized.) Once or twice a week a barrel organ would play right under the windows of the Institute for Theoretical Physics. Maybe their program had to be approved by the police: the police station was right across the street. I was always amused by the blatant vulgarity of this kind of music and I even suggested that it might be good fun to arrange classical music for barrel organs, but Kramers shuddered at this blasphemous thought and he really suffered until the instrument of torture had gone on its way. I mentioned this trait in an after-dinner speech on the occasion of his receiving the Lorentz Medal in 1947: somehow his abhorrence of barrel organs was also manifest in his way of doing physics. He would despise cheap success obtained by glib application of known principles.

By the way, I often use the barrel organ as an example to explain the difference between hardware—the machinery itself—and software. Preparing the perforated plates that are used for letting the organ play a given piece of music is equivalent to writing a program (in machine code) for a computer. To my knowledge, the equivalent of a compiler, which would automatically convert a standardized score into the instructions for the organ, has never been developed.

The painter Toon Kelder,* who was one of Kramers's lifelong friends, once explained to me why he painted Don Quixote: in his opinion he was the noblest of all heros because there was never the danger that the pure nobility of his attempt would be spoiled by the degrading influence of success. It seems incongruous to apply this metaphor to the work of a physicist: I do not want to imply that Kramers was fighting windmills and it would be foolish to deny the

*T. Kelder, 1894–1973, moved in the course of his life from rather sensual impressionism to severe black-and-white abstractions and to metal sculptures. His work is uneven in quality but the best of it—including some portraits—is outstanding.

importance of success. All the same: the work of Kramers has a special nobility. As I wrote in the *DSB*:

He tackled problems because he found them challenging, not primarily because they offered chances of easy success. As a consequence his work is somewhat lacking in spectacular results that can easily be explained to a layman; but among fellow theoreticians he was universally recognized as one of the great masters.

Let me come back to the comparison with Lorentz. It is a striking fact that the mathematics used by Lorentz is always fairly elementary. In his work you will find few intricate applications of special functions or of general theorems in the theory of analytical functions, no general theory of groups, and so on. The tools he needs he always has at his command, but somehow he succeeds in analyzing and formulating his problems in such a way that fairly elementary calculus is sufficient.* In the papers of Kramers there is more mathematical erudition, elegance, and ingenuity. "You and I have a certain mathematical vein," he once said to me. "I don't want to say it's running very rich, but it is there. I do not find it in Lorentz."

Let me now supplement these somewhat haphazard recollections and impressions by a short biography. H.A. Kramers was born on 17 February 1894 in Rotterdam, where his father was a well-known physician. He began his studies at Leiden in 1912; in 1916 he passed the "doctoraal" (See Chapter 3 for a description of the Dutch degree system).

In 1916, in the middle of the First World War, he set out on a rather adventurous trip to Denmark, where he arrived unannounced and, of course, completely unknown. However, it soon became evident that he was a great asset.† Especially his mathematical knowledge and virtuosity were an invaluable aid to Bohr.

Kramers's first major work consisted of a systematic application of Bohr's correspondence principle to the calculation of the intensi-

---

*I only know of one case where he points out a rather profound mathematical problem (Wolfskehlvortrag, *Collected Papers* VII, p. 236), namely the problem of showing that the asymptotic expression for the number of modes of vibration of a cavity per energy interval is independent of the shape of that cavity. In a doctoral thesis one of his students had confirmed this for a few cases in which elementary solutions could be written down, but he himself did not go beyond that. It was Hermann Weyl who took up the challenge.

†Niels Bohr has beautifully described this phase of Kramers's life in a lecture at a memorial meeting held at Leiden in May 1952. It is published in *Ned. T. Natuurk.* 18 (1952):161, and is followed by an article in which I describe the later work.

ties of spectral lines. It was published by the Royal Danish Scientific Society (Viderskabernes Selskab) and served as the thesis with which he obtained his doctor's degree at Leiden in 1919. Martin Klein writes (in connection with Ehrenfest's thesis), "Few doctoral dissertations are major contributions to science." Kramers's thesis definitely was, and it established him at once as one of the foremost experts on the older quantum theory.

He continued to work in Copenhagen, first as assistant, later as lecturer at Bohr's Institute, and between 1920 and 1925 there appeared a number of important publications covering a wide field of subjects. Heisenberg, when he first came to Copenhagen, was deeply impressed not only by Kramers's competence in physics but also by his *savoir faire*, by his fluency in Danish and in other languages, by the ease with which he moved in various circles, by his abilities as a musician (it was his musical interests that brought him to meet Anna Petersen, a promising young singer, whom he married in 1920). Heisenberg felt he would have to work hard to equal him. He did. Pauli in later life told a different story. "Kramers was good; I was better," he said with characteristic modesty.

In 1924 Kramers published his papers on dispersion theory, that is on the theory of the scattering of light by atoms. This theory was further worked out in a joint paper with Heisenberg that appeared in 1925. That paper became the starting point of Heisenberg's approach to quantum mechanics that later developed into matrix mechanics. In 1926 Kramers moved to Utrecht, where he had been appointed to the chair of theoretical physics. Perhaps this move partly explains that his further contributions to the creation of the new quantum mechanics were, at first, not as striking as might be expected. All the same, the papers he published in the late twenties and early thirties show a masterly command of the new theory. His textbook, of which the first part appeared in 1933 and the second in 1938, contains both a careful analysis of principles and a wealth of elegant mathematical detail.

His later papers can conveniently be classified under four headings: formalism of quantum mechanics, paramagnetism and magneto-optical effects, statistical and kinetic theory, and quantum electrodynamics.[16]

Kramers came through the occupation years unscathed. Of course, apart from a few visits by German colleagues—about which more later in this book—there were no contacts with scientists abroad, but, until the last winter of 1944–1945, scientific life in the Netherlands went on. Leiden University was closed, but work at the Kamerlingh Onnes laboratory continued and Kramers enjoyed his contacts with the experimental physicists there. An important paper with Kistemaker is one of the results of those years.

After the war he played a prominent role in organizational matters, both at home and internationally. In 1946 he was chairman of the Scientific and Technological Committee of the United Nations Atomic Energy Commission, and he presented a unanimous report on the technological feasibility of control of atomic energy. From 1946 to 1950 he was president of the International Union of Pure and Applied Physics. He was also a member of the steering committee of the Solvay Conferences. He died of lung cancer in 1952.

Kramers and I had many discussions together; he taught me many a useful mathematical trick, and I am proud that he counted me among his friends. Yet I must admit that his influence on my own work has been less than might be expected, and we never did a solid piece of research together. Somehow his approach to problems was different from mine: to me it seemed often somewhat indirect and complicated. Perhaps I was just too stupid. However, when Schüler, remembering our meeting in Berlin, informed me about some new results and again asked me whether I could explain them quantitatively—he himself had found a qualitative interpretation—Kramers encouraged me to go ahead at full speed, and this led first to a short publication and then to a prize essay.

Teyler's Foundation is a venerable institution in Haarlem, the Netherlands. It was created by Pieter Teyler van der Hulst (1702–1778), a wealthy merchant who was interested in theology and natural philosophy and who was a collector of art and of various curiosa. He left his collections and most of his fortune to this foundation. The buildings of the foundation house a museum of art, and also a collection of old physical instruments, foremost among them a really large electrostatic machine designed by Van Marum*; a laboratory, now

---

*Although Martinus van Marum (1750–1837) did not make very fundamental contributions to science he was an interesting polymath, and he was instrumental in introducing many new ideas, like those of Lavoisier, in the Netherlands. The study of his life and work in five volumes (published on behalf of the Hollandsche Maatschappy der Wetenschappen by H.D. Tjeenk Willink & Zoon, Haarlem) gives an interesting picture of scientific life in the Netherlands and elsewhere during his lifetime.

no longer active; a lecture theater; a library; and (a post–World War II addition) two organs (one with pipes, the other electronic) tuned according to the thirty-one tone system advocated by Christiaan Huygens and revived by A.D. Fokker (1887–1972). Here H.A. Lorentz was conservator from 1911 to his death, here he even had the possibility of doing some experiments, a possibility he had asked for but that had been denied him at Leiden.

One of the activities of Teyler's Tweede Genootschap—a kind of subcommittee of the foundation—was to offer a prize for an essay on a specified topic. Soon after I had done the calculations on Schüler's results I mentioned, they chose as the theme for a prize essay "The Interaction between Electrons and Nuclei," and I started at once to extend my work. I wonder whether Kramers suggested the theme, knowing that I was working in that direction. If that was so, it would be one more reason for me to be grateful to him. Well, I did get that prize, which meant a fair-sized gold medal, a somewhat solemn award ceremony, which my father thoroughly enjoyed, and a certain amount of local prestige.

More important was the work itself, as I explain in Appendix B.

I was slightly sorry not to have been the first one to publish the formula for the influence of the magnetic moment of the nucleus on the energy levels of the electrons. The effects Schüler had found were due to the influence of an electric quadrupole moment of the nucleus. In more popular terms this means that the nucleus is not exactly spherical but can be either elongated or flattened. This time I was the first to derive quantitative formulae for this quadrupole interaction, and to calculate values of the quadrupole moment (which is a measure for the degree of flatness or elongation) of the nuclei of europium and cassiopeium (lutetium). The prize essay was originally written in Dutch—it had been written in longhand by my wife—but after the award ceremony I translated it into English; it was then published by Teyler's Tweede Genootschap and turned out to be rather useful to workers in this field, although it was not very accessible. In 1963 it was reprinted as a paperback by Freeman and Co. of San Francisco.

In the meantime I had become interested in low-temperature physics. I began to assist my wife, who was working in the cryogenic laboratory, and together with C.J. Gorter I worked on the thermodynamics of superconductors.

In 1936 W.J. de Haas offered me the position of senior assistant ("conservator") in his laboratory, and during the next six years I tried my hand at experimental physics. My next chapter will deal with that period.

6

Low Temperatures

Kamerlingh Onnes

On 12 January 1807 a barge carrying gunpowder from the factory in Amsterdam to the armory in Delft made a stop in Leiden and was moored at the Rapenburg in the middle of the city, not far from the university. Somehow the cargo caught fire and the explosion and conflagration that followed destroyed more than five hundred houses, a sizable section of the city. It would take until the end of the century before the last remnants of the ruins were cleared away and replaced by a pleasant park, but, in the meantime, this drastic clearing project had provided room for an expansion of the university. Laboratory buildings were built and gradually extended. Anatomy, physiology, and physics were housed there. Later the medical activities moved to another site—the other side of the railway tracks—and today physics buildings occupy a whole block. Even that is now too small: part of the physics department is to be found in a new project outside the city, a common and undesirable feature of many university towns.

The creation of the Leiden Laboratories with their once unique and still honorable standing in low-temperature physics was to a large extent the work of one man, Heike Kamerlingh Onnes.

From 1885 onwards his scientific papers and those of his collaborators, usually first published in the *Proceedings of the Royal Academy* in Amsterdam, were reprinted and circulated as the "Communications from the Cryogenic Laboratory at Leiden" and these "Leiden Communications" contain a major part of the early history of low-temperature physics. There also exist two commemorative books. The first one, with an introduction by J. Bosscha, was published in 1904, on the occasion of Kamerlingh Onnes's silver doctorate; the second one, with an introduction by H.A. Lorentz, in 1922, on the fortieth anniversary of his inaugural lecture at Leiden. These two volumes contain excellent surveys of the development of the Leiden Laboratory. In Lorentz's collected papers one finds the text of his address at Kamerlingh Onnes's funeral and of a lecture on his doctoral thesis. On the basis of such sources there arises the picture of a man with great vision and remarkable singleness of purpose, rewarded by outstanding success; of a great manager, exacting but benevolent; of a man who realized his social responsibilities and who took both the interest of Dutch industries and that of his technicians and apprentices to heart. This is also confirmed by the article in the *DSB*.

All the same, I feel that Kamerlingh Onnes should not be idealized. But here I must add a word of warning. I believe I am more or less competent to judge the importance of Kamerlingh Onnes's scientific work, but what I know about the man himself is from hearsay, is based on stories told by former students and collaborators, on anecdotes that were still circulating among elder technicians in the cryogenic laboratory. From such admittedly incomplete, superficial, and uncheckable information there results a more human but far less perfect image, and part of that image I do not entirely like.

I dislike Kamerlingh Onnes's social prejudices as well as his avowed approach to physics. Since this book is not an attempt to write an objective history of science but an account of personal experience and opinions, I shall write the way I feel. The reader should be aware of the fact that this may not do justice to the real Kamerlingh Onnes.

Heike Kamerlingh Onnes came from a prominent family of merchants and manufacturers in which there also ran an artistic vein. A younger brother became a well-known painter (whose son became, in my opinion, an even better painter) and Kamerlingh Onnes himself was, in his youth, interested in poetry. Did he willfully suppress his poetic tendencies? His inaugural lecture at Leiden (see below) almost suggests he did. Yet his narrative describing the first liquefaction of helium is for all its factual precision a beautiful piece of prose. He was born on 21 September 1853 at Groningen, went to school there, began his studies at Groningen University in 1870, and ob-

tained his doctor's degree in 1879, having in the meantime spent
three semesters at Heidelberg, where he studied with Kirchoff and
Bunsen. In his thesis, *Nieuwe Bewijzen voor de Aswenteling der Aard*
(New Proofs for the Axial Rotation of the Earth) he studied the
influence of the rotation of the earth on the motion of a short pendu-
lum and arrived, not surprisingly, at agreement between theory and
observation. This thesis shows considerable mathematical ability—he
systematically applies the methods of Hamilton and Jacobi—and also
considerable skill in designing and handling apparatus, but the
choice of subject always struck me as rather pedestrian, not to say
dull. One could expect what one would find and the work did not
even lead to a striking demonstration experiment. After getting his
doctor's degree, Kamerlingh Onnes was a teacher of physics in Delft
for some years and in 1882 he was appointed to the chair of physics
at Leiden. He held that chair for forty-two years.

His inaugural address at Leiden, given on 11 November 1882, is
entitled "The Significance of Quantitative Research in Physics," and
contained a clearly formulated credo to which he would remain true
all his life. It is also remarkable because it shows an awareness of what
is today so much discussed as "social relevance." Let me try to translate the rather solemn opening phrases:

> Physics owes its fecundity for creating means for material well-being and
> its preponderant influence on our metaphysical views to the pure spirit of
> experimental philosophy. It can maintain its important share in the thinking
> and working of our present-day society only if by observation and ex-
> perimentation it continues to wrest from the unknown ever new territory.
>
> However, the number and the means of those institutions that offer
> physics an opportunity for doing this are lagging far behind the great impor-
> tance of the societal interests to the furtherance of which they are dedicated.
> Therefore, someone who accepts the momentous task of forming future
> practitioners of physics and of administering such institutions has to be
> doubly serious in rendering account of his ideas concerning the require-
> ments of experimental investigations in our time.
>
> It may be that he knows no other motivation for his working and striving
> than a poetical thirst for truth; fathoming the nature of things may be his
> main goal in life, yet the courage to accept a position that gives him the
> opportunity to do this can stem only from the conviction that he can be
> useful by adhering to certain definite principles.
>
> According to my views, aiming at quantitative investigations, that is at

establishing relations between measurements of phenomena, should take first place in the experimental practice of physics.

By measurement to knowledge [*door meten tot weten*] I should like to write as a motto above the entrance to every physics laboratory.

"Door meten tot weten." Here one can start philosophizing to what extent the structure of a language has an influence on the way of thinking of native speakers. The vicissitudes of the Dutch language —the Low-Dutch language, as Maxwell called it—have made "meten" (to measure) and "weten" (to know) rhyme. (The corresponding words in High German—"messen" and "wissen"—do not.)

Hence the slogan impressed itself on the mind; it has been the bane of much good physics. For although it is certainly true that quantitative measurements are of great importance, it is a grave error to suppose that the whole of experimental physics can be brought under this heading. We can start measuring only when we know what to measure: qualitative observation has to precede quantitative measurement, and by making experimental arrangements for quantitative measurements we may even eliminate the possibility of new phenomena appearing. (Lenard's set-up was better for certain quantitative studies than Röntgen's, so he did not discover X-rays.) It is also the task of experimental physics to create new circumstances, so that new phenomena may arise: low temperatures, high pressures, high magnetic fields. Fortunately, Kamerlingh Onnes was too good a physicist to adhere rigorously to his own narrow-minded precepts. He opened up a whole new field of research, and with his assistants, found at least one qualitatively new and most surprising phenomenon: superconductivity. The best of his work belies his dogma. (At least, so it seems to me, but he himself might have maintained that after all it was the desire to perform measurements that pointed the way.)

Systematically and thoroughly Kamerlingh Onnes built up his equipment. Originally some other groups in the laboratory were working on magnetism and on magneto-optics, and it was there that in 1898 Zeeman discovered the Zeeman effect. After Zeeman had left for Amsterdam and some of the other members of that group had also moved to other universities or Delft, such work either came to a halt or was incorporated into the main program. At first Kamerlingh Onnes was not pioneering. Air had been liquefied for the first time in 1877. It was 1892 before Kamerlingh Onnes's installation went into operation, but then it produced regularly several liters of liquid air (or liquid oxygen) an hour. In 1898 Dewar at the Royal Institute succeeded in liquefying hydrogen by the so-called Linde-method (Kamerlingh Onnes had already predicted in 1896, on the basis of the law of corresponding states, that this should be possible); it was 1906 before the hydrogen liquefier at Leiden worked; then it produced four liters an hour.

Two years later on the tenth of July, helium was liquefied, and this time Kamerlingh Onnes was clearly ahead of all his competitors, Dewar first among them.

From a purely scientific point of view the liquefaction of helium was perhaps not such a surprising thing: after the liquefaction of hydrogen few physicists would doubt that the extrapolation to helium was sound. But given the general state of technology and of the usual equipment of research laboratories in those days, it was a major technical feat. Certainly, some pumps and compressors were commercially available, but their quality was often hardly sufficient. In the early stages of the work the laboratory even had to produce its own electric energy, and most of the equipment had to be built in the laboratory workshops. And here Kamerlingh Onnes had a real stroke of genius: he founded a school for instrument-makers and glassblowers. The pupils followed theoretical evening courses—partly taught by some of his scientific assistants—and they did practical work in the laboratory workshops. This also gave Kamerlingh Onnes the possibility of paying his chief technicians more than he could have done otherwise.

He got a first-class glassblower from Germany, Herr Glasbläsermeister Kesselring, who started a tradition of glassblowing in the Netherlands. He also found his famous technical factotum, Mr. Flim.

Flim was a remarkable man indeed. He came to the laboratory in 1901 but he was still going strong when I worked at Leiden, so I knew him fairly well. He had little theoretical knowledge—he had been trained as a mechanic—but he was not only extremely skillful with his hands, he was also an excellent designer of apparatus and a good organizer. He usually spoke in short, clipped sentences. If you had some new plan and asked for his opinion, he might answer, "I'd say, madness"—and then you had better not try, at least not at Leiden. He had been largely responsible for the design and the building of the hydrogen and helium liquefiers, and Kamerlingh Onnes, in his account of the first liquefaction of helium, gives him fair credit. It was sometimes said that Kamerlingh Onnes would never have liquefied helium without Flim, but that is only partly true. Being the man he was, Kamerlingh Onnes would almost certainly have found somebody else to assist him; it would never have occurred to Flim to liquefy helium without Kamerlingh Onnes.

There can be no doubt that by creating this school Kamerlingh Onnes killed several birds with one stone. He provided the laboratory with excellent technicians, with well-equipped workshops, and with a number of apprentices that came in handy for many jobs—"de blauwe jongens" (the blue boys) we used to call them later, from the color of the overalls they wore. He also created satisfactory positions for these technicians and helped many gifted boys to find satisfactory positions in industry or in other laboratories later on. And he helped to raise the general level of instrument-making and glassblowing throughout the country.

He was a paternalistic dictator and, as far as his technicians were concerned, he got away with it. One of them, L. Ouwerkerk, who came to the laboratory in 1915, once explained to me: "You were called to his room around New Year, expecting a raise and feeling you were entitled to it; he sent you away without a penny and still you felt happy about it. I still don't quite understand how he did it." When free Saturday afternoons were introduced in most factories and offices and his technicians requested him to follow suit, he bluntly refused. He did not approve of that kind of revolutionary tendency. The laboratory workshops—and the laboratory—continued working Saturday afternoons until the early thirties.

He was very class-conscious and believed in a stratified society. I have some evidence that he even treated the great Lorentz somewhat condescendingly: he did not come from the same social circle. When Flim wanted to send his boy to the HBS (a school preparing for academic studies) he advised strongly against it. He should give the boy a sound technical training but not send him to a school that prepared for the university and not try to let him enter circles where he did not belong. According to the story, for once Flim rebelled and said, speaking in his usual way, "I'd say: my boy, my money, my decision." The son later studied medicine and became a well-esteemed general practitioner in Leiden. The story that Mrs. Kamerlingh Onnes fainted when she heard he was going to marry the daughter of a professor is apocryphal but instructive. But young Flim came to a tragic end: he was shot by the Germans as a hostage during the occupation.

Still, within the limits of his social prejudices, Kamerlingh Onnes might be said to have been generous to his technicians.

What about his relations with his students and assistants? I gathered that he was not inaccessible and that he liked to establish some social contacts. He was also willing to make allowances for youthful pranks, as is shown by the following story. (I cannot vouch for its authenticity, but the man who told it to me knew Kamerlingh Onnes well.) Kamerlingh Onnes's garden bordered on the Galgenwater, a canalized branch of the Rhine, across from where the students' rowing club had its shed. One year, when the crews had started their spring training, they established the custom of having a short swim afterwards, and they did not bother to put on bathing trunks. Mrs. Kamerlingh Onnes thought this highly improper and told her husband to do something about it. So he wrote a letter, in which he expressed his interest in their training program and approved of their swimming, but he asked them to have, all the same, some pity. He had a weak stomach and two young kitchen maids, and since the students had begun their training, food was more often burnt than not.

However, he did not treat his students and assistants as younger colleagues and as scientists in their own right. As a result, he was respected and feared rather than admired and loved. The order of things he established was authoritarian and would not be acceptable today, but within that order he acted correctly and gave recognition where it was due. He considered it evident that he determined the program of the laboratory and that all the results should be published under his name; only rather senior physicists could be co-authors. It is instructive to look, in this connection, at the discovery of superconductivity. After the liquefaction of helium Kamerlingh Onnes began systematically to investigate the properties of matter at the low temperatures that had become available. Measuring electric resistance as a function of temperature was part of the program, and Kamerlingh Onnes was well aware of the theoretical importance of such measurements. When experiments showed that—contrary to then-current expectations—the electrical resistance of gold and platinum was much lower at liquid-helium temperature than at higher temperatures, he even published some theoretical considerations that might explain this phenomenon (and that contain some elements of today's theory). The two physicsts designated to carry out the experiments were G. Holst, who had obtained a degree in physics and mathematics at Zürich and had come to the laboratory as an assistant and to prepare a doctoral thesis, and Dr. C. Dorsman. At the end of the first publication on electric resistance at helium temperatures, which does not yet contain the discovery of superconductivity, Kamerlingh Onnes writes: "I gratefully record my indebtedness to Dr. C. Dorsman for his intelligent assistance during the whole of the investigation and to Mr. G. Holst, who conducted the measurements with the Wheatstonebridge with great care." That acknowledgment is repeated in the papers announcing the discovery of superconductivity, and two years later, Kamerlingh Onnes ends a paper as follows:

Having completed the series H of my experiments with liquid helium I wish to express my thanks to Mr. G. Holst, assistant at the physics laboratory, for the devotion with which he has helped me, and to Mr. G.J. Flim, chief of the technical department of the cryogenic laboratory, and Mr. O. Kesselring, glassblower to the laboratory, for their important help in the arrangement of the experiments and in manufacturing the apparatus.

In 1914 when Holst had already left Leiden there appear some papers in which Holst is co-author, and finally, in a confidential document in which Kamerlingh Onnes nominates Holst for membership in the Royal Netherlands Academy—Holst was elected in 1926—he emphasizes Holst's role in the discovery of superconductivity.

Had anyone suggested to Kamerlingh Onnes that Holst should have been co-author from the beginning and should be regarded at least as co-discoverer, he would probably have been highly surprised. After all, he himself had created the possibility of carrying out measurements at helium temperatures—to that, Holst had not contributed. He himself had given instructions to measure the electrical resistance of mercury, partly because on the basis of a theoretical speculation, he counted on the possibility that it might decrease rapidly at very low temperatures. Under these circumstances, no competent experimental physicist given that task could possibly have failed to discover superconductivity. That Holst was very competent he did not deny; for that he had given him ample and well-deserved credit.

The argument sounds rather convincing, yet by today's standards Holst ought to have been co-author. How did Holst himself react to this situation? We shall have to come back to that point later on.

Mrs. de Haas in her book on Lorentz tells an unfortunate story about two rooms, which occurred around 1906. I quote:

A new addition to the laboratory was constructed which contained a lecture-room, a room for Lorentz, one for his assistant and two small laboratories for experimentation, which were intended for Lorentz's personal use. This accommodation, which he had greatly lacked and had asked for, was now given him. He liked to do experiments himself, just for the sake of pleasure. Probably through a mistake those two rooms were "temporarily" added to the large laboratory. I remember very well how disappointed my father was at the result of this "administrative" measure. However, the matter was not discussed. My father preferred, rightly or wrongly, to keep his peace of mind rather than to create a disturbance unless it were strictly necessary.

I am inclined to believe that Kamerlingh Onnes just pinched those two little rooms; in any case it would have been easy for him to redress the matter.

As I wrote before, I am prejudiced. So let us not forget the enormous merits of Kamerlingh Onnes. In an age when in many laboratories technology did not go much beyond string and sealing wax and home-blown glassware he created an almost industrial organization. It was "big science" *avant la lettre*. To achieve what he achieved required not only scientific vision and technical competence, energy, perseverance, and diplomatic skill, but even a measure of ruthlessness.

Until 1923 Leiden was the only place in the world where liquid helium was available. Then McLennan at Toronto followed Kamerlingh Onnes's example. Liquefiers were also built at the Physikalisch Technische Reichsanstalt in Berlin and at the Bureau of Standards in Washington. In the thirties several new and very active low-temperature centers were started. Kapitza built a novel type of liquefier at Cambridge and started the Mond Laboratory. Then he moved to

Russia and started an important group there.*

Still another type of liquefier was used at Oxford at the Clarendon Laboratory by Mendelssohn and by Simon and Kurti, who had had to leave Germany because of Hitler's anti-Jewish measures. F.A. Lindemann, later Lord Cherwell, who had obtained his PhD in Germany with Nernst in 1904, was then professor of physics at Oxford. In his role of wartime advisor to Churchill he has often been criticized,[1] but there can be no doubt that he did a magnificent job in bringing the moribund Clarendon Laboratory to life again. Finding positions at Oxford for refugees from Germany was an act of human kindness that also contributed in no small measure to Britain's scientific strength. There was one more low-temperature laboratory: that of Giauque at Berkeley, California. Liquid helium was no longer a Dutch monopoly but it remained an oligopoly until after the Second World War. Today it is almost a household commodity.

Kamerlingh Onnes died in Leiden on 21 February 1926. He was buried in the churchyard of the nearby village of Voorschoten. His technicians were supposed to follow the cortege on foot, of course in black coats and tophats. Once outside the city limits the horse-drawn hearse went at a rather lively pace, and the poor technicians arrived sweating and panting at the graveyard. Then one of them, wiping his brow, smiled and said, "Just like the old man; even when he is dead he keeps you running."

This is true also in a metaphorical sense. Many generations of physicists have been working and are still working in the field he opened up.

De Haas and Keesom

After Kamerlingh Onnes's retirement the cryogenic laboratory was directed by two successors, De Haas and Keesom. They were very different characters and they did not get on too well with each other.

Wilhelmus Hendrikus Keesom, born 21 June 1876, son of a sheepfarmer on the island of Texel, must have been a paragon of a clever and diligent schoolboy and student. He studied at Amsterdam, where he followed the lectures of Van der Waals, and he passed his

---

*I am not sufficiently familiar with details to tell the story of the circumstances that induced—or forced—Kapitza to stay in Russia in 1934 when he was there on a visit.

examinations with clockwork regularity and always *summa cum laude.* In 1900 he became assistant to Kamerlingh Onnes at Leiden, and although he got his doctor's degree at Amsterdam—in 1904—the work for his thesis was done at Leiden. He might be said to be—at least in the earlier stages of his career—a typical product of the "door meten tot weten" doctrine. However, he was not only an experimental physicist. His knowledge of kinetic theory and of thermodynamics was considerable, and although his theoretical work was not highly imaginative it was extremely solid. Kamerlingh Onnes relied on that when he designed his helium liquefier and in many other cases; he mentions this specifically in his Nobel lecture. Keesom stayed at Leiden until 1917, when he became teacher of physics and physical chemistry at the veterinary school at Utrecht. A year later that institution acquired university status and Keesom almost automatically became full professor. It is told that his father was rather surprised when his son was appointed at Utrecht. "Willem at the veterinary school?" he is rumored to have said. "He hardly knows the difference between a sheep and a pig." At Utrecht, Keesom did valuable work on X-ray diffraction by liquids and gases—probably not his father's idea of veterinary work. In 1923 he returned to Leiden and resumed work in low-temperature physics. Keesom was not a brilliant orator nor a polished society man. His heavy bulk—in later years he must have weighed over 125 kilos—interfered somewhat with his respiration, and his talks were interspersed with little puffs.

In his younger days he liked to go skating, using the typical Dutch skates: steel runners mounted in wooden blocks that were fastened under the shoes with straps. But he had to give it up. "Did you go skating this weekend, professor?" one of his students once asked him. "No, I did go to the skating-rink *pff* but there was no one to tie on my skates *pff* so I had to go home again." His sense of humor was slightly primitive but it was certainly there. It is told that when at a meeting a colleague, who has getting rather hard of hearing, complained that it seemed to him people spoke less distinctly every year, Keesom replied, "Yes, to me it seems the chairs are getting narrower every year." He was a pleasant enough companion to drink a glass of beer or wine with. During my stay at Zürich there was an international meeting at which Keesom was present. At a reception at Pauli's home I had to serve the drinks. When somewhat later, at some gathering in Leiden, he saw my future wife, he went to her and said,

"Your betrothed *pff* handles the bottle *pff* quite expertly." I wonder whether it was partly due to this preliminary lubrication that later —unlike my predecessor Wiersma—I was never aware of any friction between Keesom and myself.

Wander Johannes de Haas, born 2 March 1878, was the son of a schoolmaster. After leaving the HBS in 1895 he first started to read for the examinations for notary public, but after having passed two of the required three and after having worked for a short period in a notary's office, he decided that this career was not to his taste and he went to Leiden to study physics. He was assistant to Kamerlingh Onnes from 1905 to 1911, and in 1912 he obtained his doctor's degree with the thesis: "Measurements Concerning the Compressibility of Hydrogen, in Particular of Hydrogen Vapor at and below the Boiling Point" (again a "door meten tot weten" thesis). In the meantime he had married Lorentz's eldest daughter, Geertruida Luberta, herself a theoretician of merit, and in 1911 the young couple had moved to Berlin. There he worked first with Professor H. du Bois, a specialist on magnetism, a subject that was more to his taste than the accurate measurement of isotherms and vapor pressures. From 1913 to 1915 he was "Wissenschaftlicher Mitarbeiter" at the Physikalisch Technische Reichsanstalt. It was there that he, together with Einstein, did the work on the phenomenon that is still known as the Einstein–De Haas effect, and although it is arguable whether he really found the effect he set out to discover, there can be no doubt that his experiments stimulated and showed the way to later researchers in this field.

This may seem a nasty remark. The point is that he announced that he had found an effect roughly twice as large as that found by later and far more accurate measurements. That shows that he had not completely eliminated all spurious effects, although he was well aware of them (he discussed them at the third Solvay Conference in 1921). But if a spurious effect is as large as the real effect, can one then claim to have proven the existence of the real effect?

In 1917 he became professor at Delft, in 1922 at Groningen; in 1924 he returned to Leiden.

The division of activities between De Haas and Keesom was roughly as follows. Keesom was responsible for the cryogenic plant and would carry out research on the properties of helium and other gases and also on the thermal properties of solids. De Haas was to study electrical, magnetic, and optical properties of matter at low temperatures. On the whole that was a very sensible agreement and the arrangement worked reasonably well. It would have worked even better had there been a close and friendly collaboration between De Haas and Keesom, but that was not the case. I shall not try to trace the nature of their disagreements. During the period I worked at the Kamerlingh Onnes Laboratory, I always tried to behave as if they did not exist. Today I admit that they did exist but I shall not go beyond that.

Let me tell a bit more about De Haas and Keesom as physicists. Keesom was solid and reliable; De Haas was imaginative and whimsical. His flights of fancy were sometimes disconcerting, often constructive, and always amusing. "Do you believe in the reality of lines of force?" he might suddenly ask. "We need a complementary mathematics," he said another time. I did not at once get the point so he continued, "—a kind of mathematics where everything that is now complicated and difficult is simple and easy; but of course simple arithmetic will get very complicated. That is what we need." Or he might be gazing at drops of water running down a misted window and say pensively, "When one sees all those drops running down one just wonders what it is all about."

De Haas's knowledge of theoretical physics, or rather his grasp of the mathematical formulation of physics, was slight, far inferior to that of Keesom. I have never seen him handle any but the most elementary formulae. He made up for that by a remarkable intuition that often enabled him to spot the weak point in a train of thought even though he did not follow the mathematics. A similar intuition guided him in his critical evaluation of experimental results.

The first experiment I tried to do when I came to the laboratory was to see whether the stopping power of tin for alpha particles changed when the tin cooled down and especially when it became superconductive.* I thought at first that I had found an effect. I showed the curves I had obtained to De Haas, who eyed them doubtfully. "I am convinced these results are right." I said, to which he replied, "You have convictions? Dangerous thing for an experi-

---

*The idea that the stopping power should increase with increasing conductivity was suggested by a theoretical paper of C.F. von Weiszäcker, *Ann. Physik* 17 (1933): 869. A more refined theory, which shows that no such effect is to be expected, is due to H.A. Kramers, *Physica* 13 (1947):401–412.

menter to have. Better get rid of them." Of course, he was right: the effect turned out to be spurious.

Keesom believed in laying out extensive programs for measurements; De Haas delighted in little experimental tricks. His health was frail and he was often unable to give regular lectures. That was a pity, for if he was in the right mood he could design most ingenious lecture demonstrations and present them with an inimitable kind of dry humor. He could also tell very instructive stories. A typical De Haas yarn is a story about precision measurements at Groningen. Haga,* his predecessor there, carried out accurate measurements of currents with a tangent galvanometer. About this, Curtis of the Bureau of Standards writes in 1937:

Haga and Boerema claimed an accuracy of 100 parts in a million, which is probably the limit of accuracy to be obtained by this method. The method is, however, of great historical importance, since for nearly half a century following the first description in 1837, it was in regular use in all electrical laboratories. It was discarded when laboratories found that the magnetic field from commercial circuits affected its reading.[2]

But Curtis does not mention the particular difficulty encountered by Haga. He had insisted on a completely iron-free building, and as a further precaution it was surrounded by a little park. One summer it was found that an hour or so after sunset there occurred irregular perturbations in the magnetic field, especially when the weather was fine. At first it was not clear where the precautions had failed—or was it a new meteorological effect? Finally the cause was traced: a regiment of horse was stationed in Groningen; on fine evenings the men would take their girls into the park after dark—and they wore spurs.

De Haas could be a shrewd diplomat—if he wanted to be—but sometimes his methods were bizarre. C.J. Gorter told me that when he was writing his thesis and wanted to discuss what he had written with De Haas, he was always friendlily received, but instead of talking about the thesis De Haas would begin to enlarge upon quite different themes—for instance, on the relative advantages of chicken farming as compared to a career in physics, and he asked Gorter for his opinion. Next time he would then ask whether he had thought things over, but when Gorter got going with his ideas and results on

*H. Haga (1852–1936) had been a classmate of Lorentz in Arnhem. He was professor of physics at Groningen from 1886 to 1922.

magnetism, De Haas cut him short. It was not magnetism he was thinking about, but chicken farming. "It might be risky after all," he said. "I've been told those birds may catch the pip, and then where are you?" The reason for these delaying tactics was that De Haas felt that Wiersma, who was several years older than Gorter and who had taken much organizational work out of his hands, should get his degree first. But Wiersma, besides being rather occupied by such tasks, also had, as I explained, a tendency to be held up by trifles. And so Gorter had to be slowed down until Wiersma was ready. The argument was understandable—and Gorter probably wrote a better thesis because he had to spend more time on it—but I have still to find a manual on research management that recommends a discussion of disease of poultry as a suitable procedure for dealing with overactive physicists.

Once I found De Haas in his office diligently pouring water from a variety of teapots. This turned out to be a very rational occupation, however: the Dutch patent office had called him in as an expert to decide whether a patent application for a non-dripping spout was anticipated by an earlier invention. Somehow he was also consulted on the quality of tiles for a bathroom at the palace of Princess Juliana and her consort. He had them subjected to a large number of rather extreme temperature cycles to make sure the glaze would not chip. "His Royal Highness should not get any chips in his bottom," he said rather disrespectfully.

His own inventions did not always work. He had an idea for drying hay by putting it with preheated iron balls into a kind of concrete mixer. Then, when the hay was dry, the balls had to be got out by means of a magnetic field. It seems the drying went more or less all right, but in a magnetic field the hay would be caught between the balls and it was very hard to pull it out. (At least, that was my impression, but De Haas was secretive about his invention.)

In 1942 Keesom published *Helium,* an impressive book in which everything then known about helium is reported and discussed. It shows Keesom's talents at their very best: his diligence—an enormous amount of work must have gone into the preparation of the manuscript—his accuracy, his theoretical understanding. De Haas would never have had the patience to write such a book. He once signed a contract with the firm of Julius Springer to write a book on superconductivity, with me as co-author. We worked a few evenings on it but it was a hopeless undertaking. He was completely averse to drawing up even the vaguest kind of outline and did not consider it necessary to collect material. He just began to dictate what he remembered about one particular—and not very relevant—experiment. After that he lost interest. On the other hand, unlike Keesom, he liked to pursue some personal experiments. They were usually experiments relating to very fundamental questions. Let me give an example: A question that had been on his mind for a long time was whether the negative charge of an electron is exactly compensated by the positive charge of a proton, or in other words, "Is a neutral hydrogen atom really exactly neutral?" He had raised this question already at the third Solvay Conference in 1921.[3] Later he set up an experiment to determine if there was a "charge deficit." It speaks highly for his experimental skill and objectivity that he found no indication of such a deficit whatever. (Later such experiments have been repeated by other experimenters with more elaborate techniques and far greater precision but with the same negative result.) In a similar vein he once started experiments to ascertain that a current of electrons and a current of ions transporting the same charge did give rise to exactly the same magnetic field.

When in the late thirties De Haas heard about uranium fission and about the first—unsuccessful—attempts of Joliot to get a chain reaction going, he at once realized the immense importance this might have and he induced the Dutch government to buy a sizable quantity of uranium oxide. One day a consignment arrived at the Leiden laboratory, and De Haas himself supervised the unloading. The material came in small barrels, which were distributed over the basement. "This one here," said De Haas, "and that one there, and that one behind that column, and don't put them on top of one another, for then they pop." Of course, making an atomic bomb takes a bit more than piling up impure uranium oxide, and "they pop" is hardly an adequate description of a nuclear explosion, but this weird kind of humor was characteristic of De Haas. You were never quite certain whether he was not partly in earnest.

This uranium oxide—I believe that all of it was later transferred to Delft—was kept hidden through the occupation years. It later served as a starting point for the Dutch-Norwegian collaboration on atomic energy, in the organization of which Kramers was going to play an important role. The Netherlands could supply uranium, the Norwegians had heavy water.

There was one more major difference between De Haas and

Keesom. Keesom was a devout Roman Catholic. I remember the day when P.H. van Laar got his doctor's degree. Van Laar was a Roman Catholic priest,* a man of considerable erudition and a most loyal colleague; he was also a capable experimental physicist and his thesis was an excellent piece of work. Keesom in his "judicium" praised it highly. He ended with a humble request: would Van Laar, when he celebrated mass on the anniversary of that day, remember him in his prayers? It was unusual, but moving, to hear a thesis supervisor address a young doctor in that way.

With De Haas you never knew what he really thought about such matters. A few years before his death—he must have been about eighty by then—I paid him a visit. "I have always been a heathen," he said, "but now I don't have much longer to live and I have in-structed Woltjer to convert me.† But he is not making any progress. He reads from the Bible beautifully but he is not making any prog-ress. [De Haas usually repeated his phrases several times.] He should hurry up."

Keesom and De Haas were very different persons. All the same, comparing the work done under their respective supervision and especially the theses of their graduate students, we find somewhat to our surprise that there is not really any great difference. Keesom may in principle have been more interested in systematic series of mea-surements than De Haas, but the curious properties of liquid helium and of superconductors, as well as the new insight into the behavior of electrons in solids provided by new quantum methanics, drew him automatically into more adventurous work. De Haas was interested primarily in strikingly new phenomena, but the facilities and tradi-tions of the Leiden laboratory and the need to find suitable themes for his students encouraged a type of work that was not his own first choice. "I've already carried out so many measurements," he used to sigh, "or rather, I've made others carry out so many measurements!" But, willy-nilly, he continued to let his students carry out measure-ments.

If we look in retrospect at the contribution from the Kamerlingh Onnes Laboratory in the thirties, they present a unified picture in

*He later held a chair of Thomistic philosophy at Leiden.

†H.R. Woltjer had been "conservator" in the Kamerlingh Onnes Laboratory. In the early thirties he went to Bandung in Indonesia as a professor; he survived the Japanese occupation and returned to the Netherlands. He was a practicing Calvinist.

which the contributions by Keesom and his school and by De Haas and his school complement each other in a harmonious way—so let us forget their passing discords.

Gorter

Keesom and De Haas were succeeded by C.J. Gorter, whose name has already occurred several times in my narrative. He was my senior by two years and had a decisive influence on some of my work. I regard him as the most outstanding Dutch experimental physicist of his generation. His death occurred while I was writing this part of my book, but his scientific activity had come to an end several years before: he died a patient in a mental home. There is a gross contrast between the tragedy of his final years and the image of the lively, brilliant, and energetic man that was such a well-known figure in the international world of physics.

Cornelis Jacobus Gorter, son of a civil servant, was born in Utrecht on 14 August 1907; he went to school in The Hague—at the Nederlands Lyceum, my father's school—came to Leiden in 1924 to study mathematics and physics, and obtained his doctor's degree on 1 March 1932 with a thesis (in German) *Paramagnetische Eigenschaften von Salzen* (Paramagnetic Properties of Salts). It was a competent synopsis of experimental results and their theoretical interpretation based only partly on measurements of his own. It should be remembered that at that time Van Vleck's famous book had not yet appeared.

Between 1931 and 1936 he held a position at Teyler's Foundation in Haarlem; from 1936 to 1940 he was lecturer at Groningen University; in 1940 he was appointed as successor to Zeeman at Amsterdam. He returned to Leiden in 1946 as successor to Keesom and after 1948 also to De Haas, so that he became the sole head of the Kamerlingh Onnes Laboratory.

For Gorter's own work, the years from 1932 to 1946 were undoubtedly the most fruitful ones. One is sometimes inclined to regard him as the man who *almost* discovered nuclear spin resonance, who *almost* was the first to orient nuclear spins, and some of his utterances in later years may have given support to that view. Certainly, he has been close to results that would probably have earned him a Nobel Prize, but by emphasizing that fact one does scant justice to what he *did* achieve. He was the first to carry out systematic work on paramagnetic relaxation and in that field he obtained many striking results. In the preface to a monograph on this subject published shortly after the Second World War,[4] he writes, "Paramagnetic relaxation is not an important chapter in modern physics." I do not agree; electron-spin resonance and nuclear-spin resonance have become very important chapters of modern physics and one cannot study resonances without paying attention to relaxation. Gorter was well aware of that, and he realized from the very beginning the possibility and importance of research on resonance. Unfortunately the technical means at his disposal were inadequate and his strength did not lie in creating an entirely new experimental technique.

Gorter was certainly one of the pioneers of the investigation of matter by radio waves. It is in no small measure thanks to him that also in later years Dutch physicists have maintained an honorable position in this field. As to his influence outside the Netherlands, let me only mention that the application of high-frequency fields to atomic and molecular beams by Rabi and co-workers was undoubtedly at least in part inspired by Gorter.

In the thirties we did some work together on superconductivity. After Meissner and Ochsenfeld's discovery, we formulated the thermodynamics of superconductivity, and somewhat later we showed that many of the properties of superconductors could be interpreted in terms of a simple two-fluid model. In both cases the initiative came from Gorter, who also applied the two-fluid idea to liquid helium II.

When Gorter returned to Leiden he had already laid the foundation for much of the work that would take place under his direction and would result in numerous publications, among them more than sixty doctoral dissertations. He stayed at Leiden until his retirement in October 1973. That does not mean that he was always in Leiden. Already in his student days he was a great traveler—a bicycle trip to the north of Norway, undertaken together with the astronomers Bart J. Bok and Gerard P. Kuiper to observe a solar eclipse, became famous among his fellow students; unfortunately it rained at the critical moment—and the love of traveling stayed with him throughout his active life. Conferences, visiting chairs, visits to other laboratories, and committee meetings took him all over the world, and everywhere he was a most welcome guest. In this way he contributed much to international scientific cooperation. A journey to the eastern states of the Soviet Union offered to him by his Russian colleagues became a crowning adventure of his later years.

I shall not enumerate the many distinctions he received, nor the many committees and organizations on which he served, often as chairman. He seemed indefatigable (as a matter of fact he occasionally took me to task for my laziness) until during the last years before his retirement the first symptoms of his decline became apparent.

Gorter was an outstanding physicist. I have also known him as a good friend. Ambitious, certainly, but also cordial and trustworthy and endowed with a disarming sense of humor. He was a worthy successor to a great tradition that began with Kamerlingh Onnes and that had been continued by Keesom and De Haas.

Low-Temperature Physics, Personal Aspects*

It was the special structure of the Kamerlingh Onnes Laboratory that saved me from becoming a total flop as an experimental physicist. Operating the liquefiers, and distributing and handling the liquid gases, was entirely in the hands of the technicians under the able direction of Flim, in whom Keesom had full—and fully justified—confidence. That obviated the need for any technical skill in that direction. Once you had liquid helium, there were many experiments that could be done with rather simple means, and if I needed more complicated apparatus there were well-equipped workshops and excellent craftsmen.

Of course, the arrangement had its drawbacks. Technicians without much scientific background, even technicians as competent as Flim, have a tendency to become conservative. Kamerlingh Onnes they had respected and rightly so. Now they felt there was no reason to make changes in the cryogenic practices. Helium was liquefied once a week—you could count on that—but any proposals to make liquid helium available more frequently were not well received.

Helium days were therefore days of importance. Each time only a limited number of cryostats could be filled, and one might have to wait several weeks before one's turn came. This ensured that experiments were carefully prepared: if something went wrong you might have to wait again for several weeks until you got another

*A short survey of the scientific aspects of low-temperature physics is given in Appendix C.

chance. Such careful preparation was not a bad thing, but a less desirable feature was that the experiments were not always sufficiently often repeated and checked. Also, one did not dare to sacrifice a helium run for some preliminary qualitative observations. To put it in a different way: the system encouraged measurements but discouraged experiments.

All the same, on the whole I remember with pleasure the ritual associated with liquid helium. When you came in, in the morning, the technicians had already started. You inquired tactfully how things were going, careful not to be in their way. Then you retired to your room and waited, and at a certain moment the Dewar with the precious liquid was solemnly carried in and your work could start. If you were lucky you got a refill in the afternoon. On such days you would be busy for eight or ten hours, taking hundreds of galvanometer readings and bridge settings and so on. Temperatures were varied by changing the vapor pressure, that is by pumping away the evaporating helium more or less rapidly. They were measured by determining the height of a mercury column with a cathetometer, an instrument almost unknown to the present generation of young physicists. Usually you had organized a little team to help you with these and other auxiliary measurements or possibly with the task of keeping currents constant by means of rheostats. The art of automatic registration and control was not highly developed in those days and certainly not at Leiden. De Haas did not hold with electron tubes at all and was not convinced that measurements involving electronic amplifiers could ever be reliable. The high-gain low-frequency amplifier built for my own experiments on magnetism made experts in the field roar with laughter. It was a simple, straightforward three-tube resistance-coupled amplifier, but I had heard stories about "motorboating," spontaneous low-frequency oscillations caused by various kinds of spurious feedback. So I simply put each tube, with its own dry battery, into a separate metal compartment of generous dimensions. Ridiculous or not, it worked well.

As I wrote before, I had already helped my wife, doing simple auxiliary measurements and providing food and coffee. (For many years free tea had been served on helium days, and a modest sum taking care of that appeared on the laboratory budget, until the government audit department—perfectly in line with the principles later so eloquently described by Parkinson—put an end to this extravagant luxury.) By the time I had entered the laboratory full time she had already given up her work, and now she would drop in to feed me. However, she was not always able to do this. Our first two children were born during the night or morning preceding a helium day—and at our home, as was then still the prevailing practice in Holland. So after I had seen that everything was all right I hurried on my bicycle to the laboratory. I do not think I made any important discoveries on those days.

These years in the Kamerlingh Onnes Laboratory were, on the whole, very happy ones. In a former chapter I related the difficulties I had had with the "practicum," the prescribed classroom experiments. Now I found, almost to my surprise, that I was not so bad, as soon as I could set up my own experiments. I found it amusing to design things to be made in the workshop and I even made some myself. I also acted as a general theoretical counselor, and although I did not do very profound theoretical work I developed some useful ideas. The most important ones I mention in Appendix C.

I should add that I was a poor executive. A conservator was also supposed to look after instruments and see to necessary maintenance and repairs to the building. That part of my duties I fulfilled badly, although my negligence did not lead to major catastrophes. I do hope that I have to a certain extent made up for my shortcomings, on the one hand by my own scientific work and my advice to others, on the other hand by avoiding, or appeasing, conflicts between various groups and persons in the laboratory.

In my personal affairs I was throughout my life an equally bad executive. Fortunately, my wife is far more competent than I in financial and other practical matters, so at home I left such things to her. This was well known in the laboratory, as is shown by the following episode. For small repairs and purchases I could fill in and sign a certain kind of form that enabled a technician to have the work done by an outside firm or to buy the thing in question. Once a technician had asked for such a form and was waiting while I was writing it out. "What is the date?" I asked and he said, "Twenty-second"—and added, to tease me with my not knowing the date, "February, 1938."* By that time I had arrived at the signature, and reacting to his teasing I asked, "And what is my name, please?" The

*The actual date may have been quite different: I remember the story, but not the date. It is irrelevant.

answer came promptly, "Mr. Jonker"—my wife's maiden name.*

A pleasant feature of those years was the friendly contacts with other low-temperature laboratories and especially with the Mond Laboratory at Cambridge. The Mond Laboratory was created by P.L. Kapitza under the auspices of the Royal Society and the general supervision of Rutherford. (From a formal point of view this may not be a correct formulation but I believe it corresponds to reality.) It was an elegant white-brick building erected in the inner court surrounded by the Cavendish and other laboratories. In the entrance hall a stone carving by Eric Gill represented Lord Rutherford. It had been highly controversial and many people objected to it.[5] It had finally been tolerated, not least because Niels Bohr, whom Kapitza had called in as an expert—on Rutherford rather than on modern art —had said he liked it. Eric Gill had also carved a crocodile in the brick façade; that was a joke of Kapitza's, who liked to refer to Rutherford as the crocodile (according to Gamow an apt allusion).

When Kapitza did not return from the USSR after 1934, Cockcroft was put in charge. He had collaborated with Kapitza on the production of strong magnetic fields and also on the construction of liquefiers, and although his main interests were no longer in low-temperature physics, being the man he was, he of course kept things going. Moreover, the younger people at the Mond Laboratory, Allen and Misener—who had come from Toronto where they had worked with McLennan—and Shoenberg, were perfectly able to look after themselves. Yet Lord Rutherford felt that closer contacts with the larger center at Leiden were desirable. He first invited De Haas to give a number of lectures, and I accompanied him on that occasion. One of his lectures dealt with adiabatic demagnetization and Wiersma had provided him with a large number of slides, some of which were rather irrelevant; I was worried by this but De Haas never turned a hair and did not even take the trouble to remove the useless slides before the lecture. He would ask for the next slide and

*In the Netherlands a woman keeps her own name when she marries. That is to say, her passport is still in her maiden name with the additional "married to [say] Jansen," and she has to sign very official documents that way. Of course, she will be known and addressed as Mrs. Jansen, but in writing one often uses a double, hyphenated name: her husband's name followed by her maiden name. I have been told that people at the Clarendon (Oxford) were once thoroughly puzzled by a series of papers by W.J. de Haas, J. Voogd, and J.M. Jonker, by W.J. de Haas and J.M. Casimir-Jonker, and by H.B.G. Casimir respectively until some bright boy arrived at the correct solution.

if he found nothing to say about it he would simply say, "And that is also a slide; next one please." Not a perfect way of lecturing perhaps, yet such crazy touches make a lecturer memorable; a perfectly smooth lecture is often soporific and easily forgotten.

It was agreed between De Haas and Rutherford that I would be invited to come to Cambridge from time to time, and on the basis of that arrangement I spent a month at Cambridge in the autumn of 1937 and again in the spring of 1938, when I gave a number of lectures on magnetism and very low temperatures. They were published in somewhat extended form as a Cambridge tract in 1940 and later reprinted as a Dover paperback.

I loved Cambridge and the charming countryside outside the city. Rupert Brooke's "Grantchester"—written in a nostalgic mood, while "sweating, sick and hot" in Berlin—may not be great poetry, but lines such as

> To smell the thrilling-sweet and rotten
> Unforgettable, unforgotten
> River-smell, and hear the breeze
> Sobbing in the little trees. . . .
>
> Say, is there Beauty yet to find?
> And Certainty? and Quiet kind?
> Deep meadows yet, for to forget
> The lies, and truths, and pain?

convey exactly my own nostalgia. There were the beautiful colleges and the informal formality of a dinner at high table and a glass of port in the senior combination room afterwards. And I discovered the art of leisurely working very hard.

Whenever I come to Cambridge these days I feel again that same charm, but would I have liked to live there permanently? Would I not have felt stifled by college politics of the kind so ably described by C.P. Snow in books like *The Masters* or *The Affair,* or would I have been able to avoid them? And how do ancient university traditions integrate with modern contestation by mass-produced individualists? I wonder.

Perhaps Cambridge does manage to achieve a synthesis between the old and the new. On 12 June 1967 my wife and I, invited by some friends, attended the May Ball at Trinity College. I do not think that the discrepancy between the name of this festivity and the date implies that the organizers did not yet recognize the Gregorian Calender, recently—i.e. in 1752—introduced in England, but of course in Cambridge you never know.

A feature that at once caught the eye was the enormous diversity in attire. Senior statesmen like myself appeared in the garment known in England as white tie, their spouses in formal long evening gowns. Most men of younger generations came in dinner jackets, adding color by a great variety of frilled, fluted, or pleated shirts, often of brilliant hue. The girls' dresses displayed even more imagination, ranging from traditional long gowns to pajama trousers combined with a bikini bra, but at least in one case a length of black string, worn as if it were a scarf, was added to make the ensemble more dressy.

This crowd milled through the corridors and the grounds of the venerable college, which had been adapted to the occasion. Bars served drinks in several places, tents in the grounds gave shelter to a choice of bands, and in the main court a scaffolding had been erected on which no one less than Françoise Hardy was to give a performance. A minor accident occurred when one young man, trying to climb to a vantage point, threw down a loudspeaker, which damaged and slightly fractured the nose of one of the ladies in our party; however I am happy to say that she suffered no permanent disfigurement.

The diversity of dancing styles fully matched that of dress. More or less classical ballroom dancing was by no means absent, but a new style where the partners only from time to time reach hands, or embrace or what have you, but mainly carry out complicated gymnastics separately—in a way reminiscent of courting sticklebacks—prevailed, and there were many variations. I was particularly impressed by a young couple that behaved as follows. When the music started, the girl clasped her arms around the boy's neck and lifted her feet. Throughout the dance he never moved his feet at all: he just stood there, swaying slightly with the rhythm. Finally the music stopped, she put her feet on the ground again, and that was that. As a choreographic performance it struck me as somewhat unimaginative but it must have provided marvelous training for the muscles of neck and shoulders.

The weather was chilly and in the small hours of the morning a cold gray fog began to rise from the river. Bonfires were lit and groups gathered around them, chatting and warming their feet.

Close to the largest fire sat a girl in a long, wet, white dress. The fire was quite hot, but she was shivering although her boyfriend had put his dinner jacket around her shoulders. He had taken off his shoes and had put them where they almost sizzled; he had taken off his trousers and tried to dry them over the fire. The couple had been trying their luck on the river, but their punt had overturned, hence their plight. The boy looked sad and the girl kept muttering: "I won't drink champagne again, I hate the stuff; never again, no champagne again." It seemed a dubious overture to future marital bliss.

I am no dancer, although with a sufficiently lightweight girl I might have been able to imitate the performance I described above. But I was glad to see that old Trinity could unshakenly accommodate this curious medley. And if some ancient dons turned in their graves, or even if their revolted ghosts walked the grounds, their protests went unnoticed amid the general din and turmoil.

John D. Cockcroft, Ernest Rutherford, and J.J. Thomson

John D. Cockcroft (27 May 1897 to 18 September 1967) was an admirable man. Before coming to Cambridge he had been trained as an electrical engineer at the Technological College of Manchester University and in the Research Department of Metropolitan Vickers, and his engineering background served him well in his pioneer work as a physicist. Together with E.T.S. Walton he built a high-voltage generator of the cascade type; the circuit they used had already been described by Greinacher, but this was the first time it was used in a spectacular way, and the name "Cockcroft-Walton generator" is well deserved for this type of apparatus. In 1932 they showed that the lithium nucleus is split when bombarded by fast protons. This was the first time a nuclear transformation was brought about by artificial means, and it was the beginning of a new era of nuclear physics in which "string and sealing wax" were rapidly being replaced by powerful and technologically refined apparatus. Soon a number of other nuclear reactions were found. In 1951 Cockcroft and Walton received the Nobel Prize for their work.

When I came to Cambridge, work on nuclear reactions was being continued mainly by others. Cockcroft himself was working on expanding and modernizing the Cavendish Laboratory. Building a cyclotron was one of his main concerns, but he also kept an eye on the installation of a new cascade generator by Philips of Eindhoven. I had, at the time, no special relations with the Philips company, but seeing that generator I felt some national pride. It looked both powerful and elegant compared with the old Cockcroft-Walton generator, that looked rather "as if someone had made it himself." More important: it worked well. It rendered valuable service throughout the Second World War. Later it was donated to the University of the Witwatersrand at Johannesburg, South Africa, where it was in operation for many years until it was finally dismantled.

Robert Spence writes in his article on Cockcroft in *DSB:*

Although habitually economical in the use of words, he was always genial and approachable yet made firm, impartial and prompt decisions that were accepted almost without question.

I could not agree more and I can only add that this also held for his letters: you always got a prompt answer, perhaps only one short sentence, but it told you exactly what you wanted to know. For instance: in the preface of my little book, *Magnetism and Very Low Temperatures,* I wanted to state that it was based on lectures given on the kind invitation of Lord Rutherford, but I had to make sure that this was correct and that there was no other person or organization that should be mentioned. So I wrote to Cockcroft; his answer ran to the best of my recollections as follows:

> Dear Casimir,
> The invitation was certainly Rutherford's.
> Regards,
> Cockcroft.

Once I had to wait a few weeks. I had written a popular article—in Dutch—and had asked for a picture of the Cambridge cyclotron. This time the letter was twice as long. It said:

Excuse the delay. We are rather busy preparing for Mr. Hitler.

Cockcroft was a fellow of St. John's, and one of his duties there was to look after repairs of the old buildings. He was quite an expert on bricks, and during the weekends he often drove around the neighboring countryside looking for ruins or dilapidated barns with bricks from the right period that he could buy at small cost.

During the war years he rendered valuable service to his country in radar—his main assignment—and in nuclear weapons. After the war he first worked in Canada—the first Canadian research reactor at Chalk River was built under his direction—and then returned to England to become director of the Atomic Energy Research Establishment at Harwell. In 1959 he returned to Cambridge as master of Churchill College.

Lord Rutherford I hardly met. He had made the arrangements for my visits but told me to discuss all details with Cockcroft. He died during my first visit, on 19 October 1937, after an operation for a strangulated umbilical hernia. I assisted at the solemn funeral service at Westminster Abbey on the twenty-fifth of October, where Niels Bohr was one of the mourners. This Anglican funeral service was entirely impersonal, the name of the deceased was mentioned only once: ". . . our brother Ernest who has passed away." I felt then, and still feel, that there can be great consolation in ritual. It expresses better than words that all of us partake of the same unavoidable human destiny, and that we have to accept the death of a great man just as innumerable people before us had to accept, and innumerable people in the future will have to accept, the death of someone dear to them. At the same time, celebrating the funeral service in Westminster Abbey and placing the remains of Rutherford close to those of Newton in that venerable building expressed more eloquently than any wordy glorification the grateful admiration of an empire for one of its greatest sons.

There can be no doubt about it. Ernest Rutherford was the leading figure in nuclear physics from its early beginnings, after the pioneer work of Becquerel and the Curies, right up until his death. He died only a few years before a final breakthrough: the discovery of nuclear fission. He himself had never believed that it would be possible to extract energy from the nucleus, and now, all of a sudden, a result in the field of pure research, to which he had devoted a lifetime of almost unparalleled scientific effort, became a major factor in the strife between nations and will quite possibly become the main agent that will destroy civilization as we know it. How would he have felt about this new development? His death might have been avoided; he was still vigorous and in good health shortly before, but he had been suffering from a rupture for quite some time and instead of submitting to a rather harmless operation he had preferred to live on with it. One day the unavoidable happened, the rupture became strangulated and a surgical operation, probably carried out after unnecessary delay, failed to restore the intestinal function.

In those days I got thorough lectures on the surgery of ruptures during lunch or dinner. I was staying as a paying guest in the home of a retired surgeon, Mr. Griffith. I later learned from one of his cronies that he had been a very well-known surgeon and had earned quite a bit of money but had lost most of it in farming adventures—he certainly had some rather special ideas about pig farming. He was a great believer in the medical virtues of suet, and suet pudding figured prominently on the bill of fare in his home; I have to admit it was nourishing. He explained to me that neither the Dutch nor the Danish had the slightest idea how to raise pigs with plenty of suet; he claimed to know the secret. As a consequence, his wife—an admirable woman—had to take in guests and to go shopping on her bicycle, so that they could stay in their big, old-fashioned, and pleasant house.

His usual table talk was about his operations, and his stories invariably ended with "heals beautifully." One of his major feats seems to have been the excision of the tongue of a young woman: I heard the story at least twice. First he had nicely split the jaw and bent the two halves outwards so that he could get at the tumor. "But then I had to fit the two halves together again. How do you think I did it?" "Well, Dr. Griffith, I really wouldn't know. I don't suppose that plaster of Paris would be much use." "You are right, my boy, plaster of Paris would be no use at all. No, together with a dentist I had caps made to fit over the teeth, I put tie-rods between them, and then I could sew the two halves together again [all this with gestures and indications on his own jaw or mine]; healed beautifully." On another occasion he looked critically for quite a while at my hands. I wondered what was wrong with them; they looked reasonably clean to me. Then he said, "You would be no good as a professional boxer." "Well, Dr. Griffith, I suppose you are right. As a matter of fact I never considered that career. But why do you say so?" "Your hands are far too long. They would not stand a really heavy punch; what you need is a very short and broad metacarpus."

He was somewhat skeptical of newfangled ideas and he was not really convinced of the role of bacteria as a cause of disease. In surgery, doing a clean job was far more important than asepsis. Here his wife interrupted: "But darling, you did always boil your instruments, didn't you?" He looked at her indignantly: "Of course I did, of course. Nevertheless, a fast clean job is even more important." He also had a philosophical argument against the notion that lack of vitamins could be the cause of disease. "An absence can never be a cause," he said. He was a great admirer of the work of Theodor Schwann (1810–1882) and felt that more attention should have been paid to the centenary of his discovery of the cell as building block of animals. I am afraid he was not quite accurate in his interpretation of history. Schwann's work on yeast cells and fermentation appeared in 1837, some preliminary notes on the cell-theory of animals were published in 1838, but his most important publication, the *Mikroskopische Untersuchungen über die Uebereinstimmung in der Struktur und dem Wachstum der Thiere und Pflanzen* appeared in 1839. So it is not obvious which year one should choose, and anyway, to regard Schwann as the sole discoverer of the cellular structure of living beings seems to me an oversimplification of the history of cytology. But Dr. Griffith in his later years was a man of outspoken, clearcut opinions, who did not bother about minor factual details.

I think it was Mrs. Griffith who introduced me to another elderly gentleman of pronounced opinions—and of much greater stature: the almost legendary Sir J.J. Thomson, who had been Rutherford's predecessor as Cavendish Professor and had been master of Trinity College since 1919. In any case, my wife—who had joined me during the last week of October—and I once had tea at the Master's Lodge. Joseph John Thomson was born in 1856, came to Cambridge with a scholarship in 1876, and remained there until his death in 1940. He played a decisive role in the development of the atomic theory of matter, and under his leadership splendid work was done in the Cavendish Laboratory by many other physicists. Anecdotes concerning him related mainly to his absentmindedness—especially where his external appearance was concerned—and to his inability to handle delicate apparatus himself, although he was a master at designing it (a feature he shared with many prominent experimental physicists). I remember that at the Cavendish I came upon some old, rather dismal-looking glassware, covered with a thick layer of dust and provided with a label: THIS APPARATUS NOT TO BE TOUCHED, J.J.TH. Had he intended to handle it himself? (In the old days many famous laboratories had in common with the desks of theoreticians that they were extremely messy. Perhaps this is no longer true, and in any case messiness is neither a necessary nor a sufficient condition for good work.)

During our visit the conversation turned, after a while, to modern art, much of which was not at all to Sir Joseph's liking. When I said something in defense he answered, "Well, if you really like it, that is all right with me, but don't say you like it because that is fashionable these days. Always rely on your own judgment, in art as well as in physics. That's what I always have done." Remarkable advice from a man over eighty to one under thirty. It did make an impression: otherwise I would not remember, after all these years.

During the academic year 1938–1939 I gave lectures at Utrecht: I had to replace G.E. Uhlenbeck, who was spending a sabbatical year at Ann Arbor. He came back to Utrecht, but a year later left again for Ann Arbor, so he escaped passing the war years in the Netherlands. There may have been a connection between this decision of Uhlenbeck's and Goudsmit's decision not to accept an appointment to the chair of Zeeman at Amsterdam. We should be glad things happened the way they did: Uhlenbeck would probably have survived the occupation: whether Goudsmit would have been able to escape the Holocaust is doubtful. Now Gorter took Zeeman's place.

In the spring of 1939 I gave, moreover, a series of lectures at the Philips Laboratories in Eindhoven; this established a relation that would have a decisive influence on the course of my life a few years later. Also in 1939 I became a professor. The chair, an "extraordinary" one, had been vacated by G. Holst, the director of the Philips Laboratories. The title "extraordinary professor" caused considerable amusement among my English friends. Years later, after the war, I visited Cambridge again and had dinner at the home of Dave Shoenberg. He had told his children that they would have an extraordinary professor for dinner. I noticed that his boy was observing me rather carefully. Finally I heard him ask, in a clearly audible whisper: "Daddy, what is extraordinary about him?"

Laughable or not, these "extraordinary" chairs played—and continue to play—an important role in Dutch universities. Their main function is that they make it possible to appoint, usually on a part-time basis, persons whose special competence, or particular point of view, is not represented among the permanent academic staff. Unlike associate professors and assistant professors in American universities, these extraordinary professors were not—and are not—predominantly junior scientists. They might be directors of government laboratories or of museums, or a famous architect might be willing to teach one day a week. I think it is an excellent arrangement provided the teaching of the "extraordinaries" remains an extra: the main curriculum should be determined by and taught by the permanent staff. Universities should retain their independence, but if that condition is observed, having scientists from industry teach at a university will be mutually beneficial.

Sometimes, however, extraordinary chairs were used to give encouragement—and some reason for not going away—to younger people. That was what happened in my case. Or did Holst, who must have had a hand in the choice of his successor, already have in the back of his mind the idea that I might eventually become his successor in Eindhoven? It did work out that way, but one does not like to think that decisions one believed one took of one's own free will were really planned by others. Anyway, it is doubtful whether I would have come to Philips but for the special circumstances of the war years, and I do not think those could have been foreseen by Holst.

De Haas—who, I should emphasize, always treated me kindly and generously—reacted in his usual quizzical way. "Congratulations, but don't get ideas into your head: I have seen many professors in my time and some of them were complete asses. But you may find the title comes in handy when you are dealing with your landlord." The stupidity of professors was a favorite theme with De Haas. Once, when one of his sons had failed in an examination at Delft, he told me that he had addressed the professor concerned in the following way: "You might have disregarded a few errors; after all he is Lorentz's grandson, and Lorentz had to turn a blind eye to rather more when he recommended a blockhead like you as a professor." That such a conversation really took place I doubt. I think it more likely that this kind of railing was purely imaginary and served to cool his temper, and that he approached the professor in question—if he approached him at all—in a far more amiable and diplomatic way. It was a weak point in De Haas's diplomacy that, although he would omit scathing and sarcastic remarks during actual discussions, he had to get rid of such remarks somehow, so he made them to his students and co-workers. Sometimes they would leak out and reach the persons concerned, and then the effect could be rather harmful.

I gave my inaugural lecture on 28 April 1939, and after that I more or less regularly gave one special lecture a week. Otherwise the new appointment did not make much difference in my routine. Then came the war.

7

War Times

Prelude

In the evening of 18 September 1944 the British Second Army reached Eindhoven and passed through. Two days earlier American paratroops had begun to clear the way and now an endless stream of tanks and guns and lorries rolled by. But, of course, this was not yet the end of the war; that was forcibly brought home to us the next evening when the Germans tried to break through the lines and opened their attack with a heavy air raid. A week later the tragic issue of the battle of Arnhem was to rob us of all hope that the whole of our country would soon be liberated. That evening, however, we were exultant. It was an almost incredible relief to be all of a sudden out of the grip of an insane oppressor. But, with that feeling of relief, of deliverance, there came also a feeling of guilt and compunction. Here I had come through the occupation unharmed and having suffered no hardships worth mentioning. What had I done to help others? How many, more courageous than I, had been caught, had suffered, had died in a concentration camp or had been summarily executed! Jews had been rounded up and deported; we had little hope for their survival, even though the grim details of the extermination camps were not yet known. Could I not at least have saved one or two? I felt I had been a coward and an opportunist. True, I had not in any way, neither materially or spiritually, given support to the occupier. I do not think our wartime work in the Philips Laboratories had made any contribution to the German war effort, and I had partly worked, and had let other people work on things we were not supposed to work on. I had on occasion, for short periods, given shelter to people who had to hide and I had once or twice just escaped being arrested. It was not enough.

In my younger years I was not unduly afraid of physical danger, but I had been afraid of having to face human cruelty, of having to face the risks of being questioned and tortured. Moreover, I had a wife and three young children (one more daughter had been born in August 1943); it was my responsibility to get them through the war. It was not only my own safety I had been thinking of.

There were other factors: I had always tried, and I have tried also later in life, to avoid conflicts, to arrive somehow at mutual understanding and agreement. During the war that was a wrong attitude. One of my colleagues at Leiden, a young professor of theology, put it clearly: "With the Devil you cannot negotiate, with the Devil you cannot plead or argue; the Devil you can only fight." However, I was not cut out for the "illegal" underground work, I am bad at keeping secrets, I talk too much, I am a poor dissimulator. And finally, a good deal of German culture had gone into my idiosyncrasy: I had learned much from German books, German was the first foreign language I spoke fluently, my father had been strongly influenced by the German philosophers—he hated the Nazis perhaps most of all because they destroyed his image of Germany. That made it difficult for me to identify Germany—even Nazi Germany—with the Devil.

So I think my behavior can be explained and, perhaps, partly excused, but even today that does not entirely remove my feeling of guilt.

The foregoing lines will explain why I find it difficult to write about the war years and also why I do not feel entitled to condemn those Germans—a majority—who did not offer any active resistance to National Socialism, but just played it safe and muddled through. That is the reason I put this section at the beginning of this part of my narrative.

While I was in Zürich Hitler came to power. Under his reign of dictatorial terror Germany rapidly began to build up its military power and to prepare for its attempt to gain European leadership. Even more rapidly, whatever leadership Germany had in science and culture was lost. Alan D. Beyerchen has written a scholarly volume on *Scientists under Hitler*.[1] The gist of his story is that academic people in general and physicists in particular did, in no way, form a united front against the National Socialists. With a few exceptions, first among them Von Laue, they tried to go on with their work as usual and to stay out of harm's way. The laws barring Jews from academic positions did not give rise to any massive and collective protest. Also Walter Scheel, in the address previously mentioned, refers to the political immaturity and indiscriminating patriotism of most scientists.[2] Only Einstein saw clearly what National Socialism would lead to; resolutely and forever he severed all ties with Germany. On the other hand, according to Beyerchen's analysis the German physicists stood firm when it came to defending physics itself against assaults by Nazi party bigotry and by a rather cranky group around the once competent Lenard that advocated a special brand of "Aryan Physics."

I do not disagree with Beyerchen's conclusions, though I think he underestimates the risks that would have been involved in any form of political protest. Concentration camps were ready to receive one, and they were hell. Also, he may have been insufficiently aware of the despair and disarrangement prevailing in Germany in the early thirties. And he does not really take into account the force of German patriotism, a patriotism that, as Walter Scheel points out, was inclined to identify love for one's country and one's people with obedience to the state. But I am certainly not qualified to improve on his and others' studies. Here, as elsewhere, I can only relate some personal recollections and impressions.

Perhaps we should be grateful that the anti-Jewish laws relating to civil servants (and hence to academic staff) were instituted right at the beginning of the Nazi era and before the program of radical extermination was put into action. Thus the exodus[3] began in the spring of 1933, and all the Jewish physicists I know about were able to leave the country without too much difficulty.

In those days many, in Germany and elsewhere, still entertained a hope that the excesses of the National Socialists were a passing phase and would soon be followed by a return to reason. I myself in those early days even considered whether I should not try to swap positions with someone in Germany, like for instance Teller, who was about my age (he is only a year and a half older). Thank God such foolish plans came to nothing! But the exodus went on. Beyerchen enumerates no less than eleven Nobel Prize–winners in physics who had to leave Germany or German-controlled areas; Laura Fermi gives a more extensive, but by no means complete, list of emigrated physicists (but limits herself to those that went to the United States).[4]

The impact on the world of theoretical physics and mathematics was immediate and momentous.[5] Before the end of the year the University of Göttingen, cradle of matrix mechanics and of much modern mathematics, had been reduced to insignificance. (In experimental physics Pohl and co-workers continued to do solid, though not very spectacular, work.) Berlin, without Einstein and Schrödinger and without younger stars like Wigner, Von Neumann, Szilard, and Fritz London, was losing much of its already slightly tarnished glory. Sommerfeld continued at Munich as an outstanding teacher, but he did not find any more pupils of the caliber of a Pauli or a Bethe. Leipzig, under the leadership of Heisenberg, and with the active participation of the Dutch mathematician Van der Waerden, remained the most lively center for theoretical physics.

Soon German was no longer the main language of physics and mathematics. I do not recall many really important monographs or textbooks written in German after 1933. At Bohr's annual conferences there was a shift towards English. German emigrants courageously—though not without regret—abandoned their mother tongue, or rather, tried to abandon their mother tongue. Somehow it always remains embedded in the depths of our memory even if apparently forgotten. On one of my many flights to America after the war I was sitting next to a lawyer of German-Jewish origin. He had left Germany in the thirties, had done rather well for himself in the United States, and spoke English with complete fluency, although with an unmistakable accent. In those days one still made an intermediate stop at Gander on Newfoundland. On that particular flight there were some difficulties in coming in (I later heard from the pilot that the GCA of the airport had been out of order) and for what seemed a long time we kept circling around; sometimes we were in the clouds, sometimes we would get glimpses of endless snow-covered woods and lakes. The atmosphere got rather tense and suddenly my neighbor asked in a small, high-pitched voice, "Was macht denn der Flieger? Wird er bald landen?" (What is the pilot doing? Will he soon land?) I tried to comfort him, in German of course. Soon after, the pilot managed to pick up a radio beam, got within sight of the airfield, made an extremely sharp turn, and brought the plane safely down. The moment the wheels touched the ground my companion returned to English. I wonder whether he was aware of having spoken German in his moments of fear.

As for myself, after having more or less mastered the intricacies of German grammar I began a lifelong struggle with the irregularities of English pronunciation.

I did not visit Germany in those years: I met non-emigrated Germans only as visitors, to international conferences or otherwise. In that connection I remember the Van der Waals centenary conference that took place in Amsterdam on 25 and 26 November 1937. Among the participants were several emigrated Germans, for instance Fritz London and Franz Simon. There were also German physicists who had stayed in Germany. There were rather severe restrictions on taking money out of Germany in those days and, although we had some funds available for helping out, our German visitors had to live rather parsimoniously and chose simple hotels. Only one of them—not a well-known physicist—had a room booked at the Amstel Hotel, at that time certainly the most distinguished and perhaps the most expensive hotel. He was at once suspected of being the official "spy," for there was a probably not unfounded rumor that at every international conference there was at least one participant sent by the Nazis who had to report on the behavior of his countrymen. It was my impression that some of the Germans tried to avoid their former colleagues; others first looked carefully around before they engaged in conversation; but as soon as Sommerfeld spotted London he went straight up to him and greeted him most cordially. I had not met Sommerfeld before. He did not belong to Bohr's inner circle, and at Leiden he had the reputation of being a rather nationalistic Prussian. George Uhlenbeck told me recently that Ehrenfest had occasionally complained about Sommerfeld's "nostrification." This rather puzzles me: maybe Ehrenfest felt that Sommerfeld did not always sufficiently recognize the work of others and overemphasized the results of himself and his school, but the word nostrification was used in Austrian universities for "recognition of foreign diplomas." Perhaps Ehrenfest misunderstood the word. What about Sommerfeld's nationalism? The first draft of a letter to Einstein written in 1934—the letter actually sent was shorter—is revealing. Sommerfeld writes:

Moreover I can assure you that the misuse of the word "national" by our rulers has thoroughly broken me of the habit of national feeling that was so pronounced in my case. I would now be willing to see Germany disappear as a power and merge into a pacified Europe.[6]

And that brings me to my second meeting with Sommerfeld. He received the Lorentz Medal of the Dutch Academy in Amsterdam on 24 June 1939; in the evening my wife and I attended a small dinner party in the Hotel Oud Wassenaar, where he was staying.* It was then that Sommerfeld began to speak about the Nazi regime and I remember his words clearly:

When I learned about the measures against my Jewish colleagues my first reaction was "when this is possible in Germany, then I would rather die." But somehow life goes on.

He said it without any dramatic pathos: it was just an impressively sincere statement.

I do not think it had occurred to Sommerfeld to leave Germany. He must have felt that it was his duty to do what he could for physics and for his students, and he went on teaching even when he was past the age of retirement because no agreement could be reached on a successor.

The idea that one had to stay in Germany in order to defend some measure of scientific integrity and personal decency must have weighed heavily for many Germans who were definitely not Nazis. When Van der Waerden returned to Holland after the war he was therefore surprised to find that his behavior was criticized. He felt that it had been an honorable and even a courageous course of action to stay in Germany and to help people like Heisenberg and others to protect essential values of academic life. We could accept such an argument in the case of Germans, but we did not consider it the duty of a Dutchman to fight German madness on behalf of the Germans. A German remaining in Germany did not thereby condone the crimes of National Socialists; to a certain extent a Dutchman staying on in Germany did. Our criticism did not destroy old friendships, still less did it lead to real ostracism, and in 1948 Van der Waerden was appointed to a chair of mathematics at Amsterdam. Did he feel all

*The hotel was beautifully situated in a wooded area between Leiden and The Hague. It was a favorite haunt of Van Vleck's, but like many of our once-famous hotels it went out of business.

the same that he had had to suffer some injustice? In any case, when in 1951 he got an offer from Zürich he left the Netherlands, and he remained in Switzerland after his retirement.

Experimental physics was hurt to a lesser extent than theoretical physics and mathematics. Especially in the Kaiser Wilhelm Institutes, work went on somewhat as usual. Lise Meitner even preferred to continue her work at Berlin Dahlem as long as possible. As an Austrian she was not subject to the German laws, but after the Anschluss of Austria in March 1938 her position became precarious. She was whisked out of the country by the joint efforts of Otto Hahn and the Dutch physicist D. Coster. That was not long before the discovery of uranium fission. My countryman Peter Debye, who after 1934 directed the Institute for Physics—which was rechristened Max Planck Institute—even stayed on in his job until after the outbreak of war. Debye was not only a prominent physicist, he was also famous for his ability to talk authorities into accepting his conditions. It was sometimes said that a milli-Debye was a convenient unit of salary for an average physicist. And, as a rule, he did not worry about things he could not change, whether in physics or in daily life. His well-known remark about electrons in nuclei, a subject that at one time presented apparently insoluble problems, illustrates this point. He said: "It is like the new taxes—better not think about them." But in January 1940 the position became untenable even for Debye, and he left Germany for the United States.

For many of the older Jewish physicists, having to leave Germany was a painful experience. They were patriotic Germans, had served in the First World War, and I am convinced quite a few would have supported Hitler and his rearmament program had there been no anti-Semitism.

In the spring of 1939 an international meeting on magnetism was held in Strasbourg. It was the last international meeting before the war in which I took part. The conference was well attended, and most of the prominent figures in magnetism were there. Becker and Gerlach from Germany; Van Vleck from the United States; from France the venerable Henri Abraham, born in 1868 (who was to die in the hands of the Germans in 1943), and Louis Néel, then still a coming man; Gorter, Kronig, and Kramers from the Netherlands; and of course some German emigrants, among whom was Franz Simon (later Sir Francis) from Oxford. This is by no means an exhaustive list; they are just names I remember after more than forty years. The conference was well prepared: documents had been circulated well in advance and had been studied. Pierre Weiss (1865–1940) was a charming host and a tactful chairman. The weather was beautiful.

We all knew that war was threatening, and that threat was in weird contrast to the atmosphere of friendly cooperation and of gracious hospitality we were enjoying.

That feeling of contrast was enhanced by our surroundings, for whatever claims contending parties may have made, Strasbourg and the Alsace do present a synthesis of French and German culture. (A volume of poems in German published by the French Nobel Prize–winner Alfred Kastler under the title *Europe ma Patrie* is an eloquent manifestation of this characteristic.) Richard Becker told me he was deeply impressed by the apparently quite peaceful coexistence of the French and the German language. (It seems to me that the local patois was helpful in that respect: it helped people to take neither French nor German too seriously.) The Préfet du Département, the highest government official in the region, in a brilliant address, did not fail to point out that we were situated "entre la ligne Siegfried et la ligne Maginot" (between the Siegfried line and the Maginot line—Poor Maginot line, it was to offer but little resistance). It was during that conference I went for a walk with Simon. "It is strange," he said. "Last time I was here I helped to defend this country against the French. I was wounded twice, spent time in hospital. 'Das Vaterland wird euch immer dankbar sein,' hat man damals gesagt . . . Das hat man gesehen!" (The Fatherland will forever be grateful to you, one said then. That we have seen!) Simon became a most loyal British subject, he did much to build up physics at Oxford, but the feeling of having been betrayed by his native country rankled.

Perhaps this is the place to say something more about the enormous contribution emigrants from Germany and from Italy made to the development of science—and, often indirectly, to the development of technology—in England and even more in the United States. I have on occasion said ironically that the only really successful action of a developed country on behalf of an underdeveloped one was the action of Adolf Hitler on behalf of the United States of America, a remark that did not endear me to either the Germans or the Americans, but that does contain a good deal of truth. As a matter of fact, by excluding all Jews from academic positions Hitler provided the United States with a group of fundamental scientists that supplemented existing technical prowess and burgeoning theoretical capability in a most effective way. The subsequent sharpening of anti-Jewish measures and the horror of the concentration camps enhanced the loyalty of these immigrants to their new-found country. The nuclear bomb was the most obvious result, but the influence of the immigrants can be traced in many other developments. Destroying German science did not at once cripple German military strength: there remained enough knowledge to support existing technology and even to develop rockets and some well-engineered forms of navigational aids,[7] but in the long run the United States emerged as the leading power, scientifically, technically, and, partly because of that, politically. If it had been the intention of Hitler and his henchmen to bring about this situation they could hardly have chosen a better way to reach their goal. Will someone someday, in a madly distorted reinterpretation of history, pretend that Hitler consciously aimed at creating in this way a second line of defense against Russian aggression? Unlikely of course, but I have come across even more unlikely distortions of historical truth.

As I explained before, I had in my younger years the curious habit of destroying all letters and documents as soon as the matter they referred to had been dealt with. This is also true of the few letters I exchanged with German physicists between 1933 and 1940. Some of them might have been of historical interest, but the damage is beyond repair. I can only try to reconstruct as much as I can from memory.

In 1934 Planck published a paper on the so-called principle of Le Chatelier-Braun.[8] Now this matter had also been discussed by Ehrenfest[9] in 1911, and although Planck had quoted Ehrenfest, it seemed to me that he had not really noticed to what extent his own analysis had been anticipated by this earlier paper. So I wrote a letter to Planck which by any standards and certainly by German standards of those days must have been a very impertinent one, for after explaining my point of view I finished by writing: "Es scheint mir also Sie haben dem Ehrenfest Unrecht getan" (It seems to me you have done Ehrenfest an injustice). Planck, who must have understood that I had been profoundly shocked by Ehrenfest's death, sent me a most amiable answer. He began by saying that he fully understood that it was hard to accept criticism of the work of a man one had loved and admired. He then went on to explain that there were some essential elements lacking in Ehrenfest's analysis (*das sind doch wesentliche Sachen*). I felt ashamed. I still think that there was a bit more in Ehrenfest's paper than Planck had seen, but that does not really matter. His letter showed that he was a wonderfully generous man.

I exchanged some letters with Schüler on hyperfine structure. I had published a short preliminary notice on quadrupole interactions and in one of his letters he wrote me, "We are quite willing to believe your formula is right but please do publish a derivation." This may have been one more encouragement to write the prize essay mentioned earlier.

And then there was a letter from Heisenberg—that must have been around 1934. I had written him first a letter with critical questions on fundamental theory. Then, somewhat later, I had realized they were foolish questions and I wrote him apologizing for my first letter. His answer was surprising and went more or less as follows: "To show you that you have not been putting particularly stupid questions, I have a question myself." And then he went on to ask me whether I would be interested in a position in Germany. I must confess that I was tempted; I was living on a rather meager salary at the time and prospects in Holland did not seem too good. Also, I had a great and well-founded admiration for Heisenberg. I wrote to Bohr to ask his opinion, but it was Mrs. Bohr who answered. Niels had said that this was a decision I had to take myself, but that there were many things to be considered. Perhaps Bohr did not want explicitly to thwart Heisenberg's attempt to keep theoretical physics going. (There were not many young physicists of unquestionable "Aryan" descent who spoke German well and who had a reasonably good understanding of quantum mechanics.) But I knew how to interpret Bohr. So I declined to look into the matter any further. Again, I am glad things went that way. When somewhat later Bühl wrote me a similar letter the answer was easy. I just wrote him that I had received a similar letter from Heisenberg but that I had declined.

Now that Bühl letter is remarkable. As I related in Chapter 5, we had been acquainted at Zürich, and he must have known that I definitely belonged to the "relativity and quanta" crowd. Did he believe he could convert me? Or were his ideas then more moderate than they would become later on? I remember one sentence: "Ich würde diesen Brief nicht schreiben wenn ich nicht davon überzeugt wäre, dass Sie sich im heutigen Deutschland wohl fühlen würden" (I would not write this letter if I were not convinced that you would feel happy in the Germany of today). I took it almost as an insult.

It is curious how National Socialism in Germany turned shades of gray into contrasts of black and white. At Zürich it had seemed perfectly acceptable that an experimental physicist showed a certain abhorrence for formal mathematical theory. As a matter of fact a word of warning against too formalistic an approach to physics can be very useful. And if someone wants to rate Faraday higher than Maxwell he has my blessing . . . as long as he recognizes that Maxwell was a great man too. But, as I said, shades of opinion grew into sharp contrasts and playful debates became deadly struggles. We Dutchmen would soon witness a similar polarization in our own country.

Occupation

On 10 May 1940 the German armies invaded the Netherlands. Our armed forces were too weak and too poorly equipped to offer significant resistance, and on the fourteenth, after a vicious and militarily unnecessary air raid that destroyed Rotterdam, the Dutch army surrendered, Queen Wilhelmina and her Cabinet* having left the country the day before.

At Leiden these days were tense but uneventful. However, on the third day I was summoned by De Haas. He wanted to go to The Hague; would I please come along and drive his car? He himself did not feel too sure as a driver and the traffic might be difficult. So we got going. Between Leiden and The Hague, a distance of ten miles, we had to pass many military posts, but De Haas had the necessary permits; sometimes the soldiers amused themselves by letting us pronounce one or other shibboleth—usually a word involving plenty of Dutch *g*'s (akin to Scottish *ch* in *loch*)—but that did not give us any trouble. It turned out that De Haas wanted to see the minister of defense and the commander-in-chief. He felt he could give them some useful advice on arranging the defense of the country. What kind of pipe dream he was cherishing he did not explain to me; his plans may have been ingenious, but they must have been utterly

*In the British sense—i.e., the prime minster and the other ministers.

impracticable under the circumstances. Of course, he did not find the gentlemen concerned at either their homes or at their offices, so after a useless search and half an hour or so spent in an air-raid shelter —an alarm had been given, but it came to nothing—we returned to Leiden.

The first months of the occupation were less horrible than we had feared. Daily life soon returned to normal. The occupying troops were, on the whole, well behaved, there was little rape and little pilfering. There was also a slight feeling that England and France had let us down—many of us had hoped that the RAF would send powerful assistance as soon as we were attacked, little realizing how weak England really was. This did not render the majority of the people pro-German, but for the time being many were inclined to accept the unavoidable with equanimity.

Soon things began to change. A civilian government was installed, and gradually Nazi rule was established. At Leiden University this had serious repercussions. On the twenty-sixth of November a number of Jewish professors were dismissed, among them Eduard Maurits Meijers (1880–1954), who was then Holland's most prominent lawyer and legal scholar. He had been a professor since 1910 and was also universally respected and admired as a teacher. On that same day R.P. Cleveringa, the dean of the Faculty of Law, took over Meijers' lecture hour and in a justly famous speech voiced his protest against this infamous and illegal measure and paid tribute to the man they had so scandalously dismissed. Cleveringa was imprisoned—he had packed a suitcase before the lecture—but was treated with some respect, and was later released and survived the occupation. So did Meijers, and his family, although they were sent to Theresienstadt.* Cleveringa's speech was followed by a students' strike, and the university was thereupon closed and remained closed for the rest of the occupation, except for a short period during which examinations could be taken.

At the Kamerlingh Onnes Laboratory things went on much as usual. There was no German interference with the work. As to daily life, food was rationed, but rations were more or less adequate. If it had not been for the constant political threat and the tightening of anti-Jewish measures, life would have been supportable.

*Theresienstadt was partly a transit place for extermination camps, but there was a section for "privileged" Jews.

The eighth of February is the *dies natalis* of Leiden University. On that day there used to be a traditional dinner party for all the professors. I attended for the first time in 1940 and enjoyed the oratorical skill of the speeches after dinner. During dinner too, for in Holland it is customary at an official dinner to have speeches between courses. (Another difference from British dinners: there the toast to the Queen is proposed at the end of the meal: it means that then you are allowed to smoke. In Holland the toast to the Queen is at the very beginning of the meal: it means you are allowed to drink. During the occupation years, however, such a toast meant a lot more than that.)

Of course official *dies natalis* dinners were out of the question after Leiden University had been closed, but a small circle of professors to which I belonged did manage to celebrate all the same. In 1941 and 1942 we met at the home of the geologist B.G. Escher. Later, when some of the members had been exiled from Leiden or were in hiding, a place in the east of the country was selected. I remember specifically the dinner in 1942; it was a rather cold winter and the streets were covered with snow. One of the most prominent members of our group was the great historian J. Huizinga (1872–1945), who was then in his seventieth year. It was somewhat doubtful whether he would be able to walk all the way to Escher's house, which was in a village some distance from Leiden. Other transport was either unavailable or would have involved too great a risk of being found out. One of the younger professors—the same one who had warned me not to try to come to terms with the Devil—found an old-fashioned "push-sleigh," consisting essentially of one fairly comfortable seat semi-enclosed by a wooden body and mounted on runners. This type of sleigh, that can already be found in seventeenth-century paintings, is not built for speed, and it should be pushed or pulled by a man, not drawn by a horse. The contraption was in need of repair, but I could restore it sufficiently for our purpose. In the evening we picked up Huizinga at his home; his much younger wife—he had been widowed in 1914 and had remarried in 1937—wrapped him up in blankets and provided him with one or two hot-water bottles and off we went. It was quite a pleasant run and it earned us that evening the nicknames of Cleobis and Biton, as was to be expected in this learned company. Fortunately—for our ideas about happiness do not entirely coincide with those of the ancient Greeks—that night we did not share their fate.*

Such a feast provided a much-needed temporary relief from the grim reality. The Germans, or rather their even more despicable Dutch henchmen, wanted to turn Leiden, the oldest and most representative of Dutch universities, into a National Socialist institution; the future was very uncertain. Also, there were some difficulties in my relations with De Haas. He had been absent for a long time and I had been directing the research program. Now he was back, and he did not agree with the plans. Of course, he was entirely within his rights and he never interfered in any way with my own work, yet it was somewhat unpleasant for me to see my proposals countermanded without any relevant discussion. Also, in his dealings with the authorities De Haas pursued a line of his own, and that put me occasionally in an awkward position at faculty meetings, where De Haas rarely turned up and I had to defend his point of view. In Chapter 6 I mentioned that I had lectured at the Philips Laboratories. Moreover, I had two brothers-in-law working there: my sister's husband J.W.L. Köhler—who later developed the Stirling refrigerator—and E.J.W. Verwey, husband of my wife's sister and in later years my co-director at Philips. They suggested that I should in any case see Holst, the leader of the laboratories, and find out whether there would be a place for me, if I should have to leave Leiden. Holst received me enthusiastically and at once offered me a position. I decided to accept it, and in April 1942 I moved to Eindhoven, where I stayed with Philips for thirty years. I had arranged with Philips that I could keep my "extraordinary" chair and could spend one day a week at Leiden. Soon after I had settled down at Eindhoven, the situation at Leiden came to a head and most of the staff sent in their resignations. Kramers informed me by telephone about developments. He suspected the telephone line might be tapped, so he spoke in Danish. Not a very adequate protection, I am afraid. "Nu gaar det hele i stykker" (Now everything is going to pieces) were his opening words. So I also sent in my resignation. The government reacted by

---

*According to a story told by Herodotus, Cleobis and Biton were the sons of Cydippe, priestess of Hera at Argos. One day, when the white oxen that had to draw her chariot had strayed far into the fields and could not be caught in time, the two boys themselves drew the chariot all the way to the temple, a distance of forty-five stadia, well over five miles. The mother prayed to the goddess to grant her sons the greatest happiness imaginable. That night they died in their sleep.

arresting a number of professors and lecturers and confining them in a prison camp as hostages. I barely escaped; when I came back by train from a Sunday outing with my wife and children, Verwey was waiting for me at the railway station: my neighbors had reported that some uniformed people had called at my home; evidently they wanted to arrest me. So for a few days I went into hiding. But somehow Philips convinced the authorities I was useful and my name was struck from the list.

In those days I read a story *The Lost Ship.* * It tells about the radioman on a tugboat who misses his boat and loses his job because of his entanglement in a dubious love affair. Somewhat later the tugboat is shipwrecked while trying to save another ship, and most of the crew, including the new radio operator, are drowned. The jobless radioman is desperate, feels he ought to have been that radio operator, but his father tells him he should be grateful to God and that this despair is a lack of humility, is "hoovaardij" (an old-fashioned Dutch word that expresses just this kind of pride and for which I have found no satisfactory translation). I felt somewhat like that, but when I tried to explain this to my father he reprimanded me for my "hoovaardij": he used exactly the same word. Perhaps he was right.

I shall come back in a later chapter to the history and the structure of the Philips Company and its research laboratories. Here I want only to tell something about my wartime experiences. The Philips works were under German supervision, and the small group of Dutchmen in charge—Holst, the director of the Research Laboratory, was one of them—had a difficult time of it. Anton Philips, the younger—and greater—of the two brothers who created the Philips Company, had escaped just in time to the United States. His son, Frits Philips, had remained behind. He showed great courage throughout the occupation; an account of his recollections of that period can be found in his autobiography.[10] However, for someone like myself, who did not yet have anything to do with top management, there was surprisingly little German influence and interference. I believe there was something like an official program of research that had to be approved, but it was couched in general terms and, anyway, there was no control on its being observed. The explanation for this situation, which might appear almost too good to be true, may have been

*Willem de Geus, *'t Verloren Schip*. Amsterdam, 1941. The author, a former sailor, later in life sold welding rods for Philips. His real name was Willem Spruit.

roughly as follows. The Germans must have realized that it was useless to try to have the Dutch people do military work. Sabotage was to be expected—and in research you do not even need active sabotage to get no results; moreover, secrets would almost certainly leak out. On the other hand, as long as the Germans expected to win the war, there was some point in preparing for future innovations. For this the German laboratories had no time, so why not give the Dutch laboratories a free hand? Their results might come in handy afterwards.

A young German engineer had been appointed for the day-to-day supervision of the laboratories, and there we were lucky. Did he want to avoid making enemies because he counted on a German defeat? Was he at heart an anti-Nazi? I do not know, but he behaved very decently. He always gave a few hours' warning when he came to inspect a laboratory, so that anything outside the approved pro-gram—for instance, the manufacture of small clandestine radios to listen to the BBC—could be removed. He was interested, in a rather amateurish way, in modern physics and I spent several hours explain-ing a few things to him about atoms and quanta.

In the course of 1944, however, signs of German impatience began to penetrate even the sheltered regions of the Philips Laboratories. The liberation in September 1944 came just in time.

I had some contacts with Germans from the Müller factory of X-ray equipment (which had been a branch of Philips for many years) and later with one or two physicists that were interested in nuclear instrumentation, but, on the whole, such contacts were perfunctory and superficial. Still, I remember one remarkable conversation with F. Kirchner, a German experimental physicist who came to Eind-hoven looking for some equipment. I asked him whether he still believed in a German victory—it was well after Stalingrad—and he admitted sadly that this was unlikely. "Unless," he said—and sud-denly his face showed a glimmer of hope—"unless Germany and Russia could still come to terms, for that would be an unassailable bloc. And, after all, our ideologies are essentially the same." Since I abhorred German National Socialism and had no great love for the Russian kind of so-called Communism either, I could almost accept that last remark, but it was a strange remark to come from a German: the National Socialists had always advertised themselves as the de-fenders of Europe and European traditions against the threat of

Communism, and the fighting between Germans and Russians had been most embittered.

My real colleagues in Germany I did not meet save for three exceptions. These three occasions impressed me deeply, each in its own way, and I feel confident that I can render the gist and even almost the exact wording of the essential parts of our conversations.

The first one to come to Holland under some arrangement for scientific exchange was Richard Becker, who in 1937 had been forced against his will to leave Berlin and to accept Born's chair at Göttingen. (When the war came to its end he must have been glad to find himself in Göttingen and not in Berlin.) This was at the time when Germany was already facing the possibility of a defeat at Stalingrad. He inquired first of all about the fate of a Jewish student of his, Dr. G. Heller. Unfortunately, Philips had not been able to keep him out of a concentration camp, and although at first they could offer a certain protection, he finally died. Then he asked about our life, about conditions in the Netherlands. "Yes," he said, "I understand you must hate us, and I know that conditions elsewhere are even worse. And if you consider that I am also in a way responsible for that, I cannot deny it. I have already suffered some losses in my circle, and I may well suffer more. *Die Rechnung wird schon präsentiert werden* [the bill will be presented, all right], and I don't want to deny there is justice in that. Still, you must understand that I am German. I do not want to see our troops annihilated at Stalingrad, and if I am called upon to assist the war effort of my country, I shall feel obliged to do so. Perhaps it is illogical, but that is my position."

Heisenberg came to Holland in 1943. I believe he made use of the same arrangement for cultural exchange to visit his old friends. Although on principle we did not favor such exchanges, we were glad to see Heisenberg. I am convinced that he really cared how physicists in Holland were making out and might have tried to help in special cases (but he considered it was quite beyond his powers to do anything at all for Jewish physicists or their families). Kramers arranged a program for his visit. I remember that I met Heisenberg and that we went for a brisk walk. I cannot retrace the itinerary, but in my mind it is associated with a curving road and a viaduct. I even have a feeling that I might recognize the spot today—if it were still there unchanged, which is improbable considering the extent of roadbuilding that has been going on in my country since the Second World

War. It was during that walk that Heisenberg began to lecture on history and world politics. He explained that it had always been the historic mission of Germany to defend the West and its culture against the onslaught of eastern hordes and that the present conflict was one more example. Neither France nor England would have been sufficiently determined and sufficiently strong to play a leading role in such a defense, and his conclusion was—and now I repeat in German the exact words he used—"da wäre vielleicht doch ein Europa unter deutscher Führung das kleinere Uebel" (and so, perhaps, a Europe under German leadership might be the lesser evil). Of course, I objected that the many inequities of the Nazi regime, and especially their mad and cruel anti-Semitism, made this unacceptable. Heisenberg did not attempt to deny, still less to defend, these things; but he said one should expect a change for the better once the war was over. And one had to recognize that they were a consequence of the great power of the leader that was also part of the German tradition. I cannot deny that I was somewhat impressed: I had always admired Heisenberg, not only as a physicist. For me he represented much of what was valuable in German culture. He was a good musician and a good sportsman, and his knowledge of classical languages was far superior to mine. But when later in the evening I thought over what he had said, I realized with a shock that his ideas came uncomfortably close to the usual German propaganda. In later years I have often wondered why Heisenberg spoke the way he did. Did he still believe Germany was going to win the war? Then it would be understandable that he tried to console me by pointing out that this might not be altogether bad. And perhaps he might already have been thinking of the possibility of finding in me, in the long run, someone who could be of some help in rebuilding continental European research. After all, he had more or less offered me a position way back in 1934. The trouble with this explanation is that it no longer did look as if Germany were going to win and that Heisenberg had anyway been convinced from the beginning that Germany was going to lose. About this there can be little doubt: in this respect, his widow's recollections are in complete agreement with what I heard from former students of his, to whom he had said: "Once the United States gets into the war, as it is bound to do, it will be like a game of chess in which you start out with only one rook against your opponent's two." I was also told that one of the reasons he gave just prior to the war for not wanting to stay in the U.S.A. was that he felt that after the unavoidable defeat of Germany he would be able to do more for its cultural resurrection if he remained on the spot.

Maybe, however, he still cherished the hope that Germany might be able to negotiate an honorable peace. Maybe he felt that pleading the German cause—even pleading it to an unimportant person like myself—might contribute something to reaching that goal. Be that as it may, Heisenberg's approach showed little understanding for the feelings of people in an occupied country.

The third visitor was Hans Kopfermann. We knew each other fairly well. He was an experimental physicist, but he belonged to the Copenhagen clan: he had worked on spectra at Bohr's Institute, and was a specialist on hyperfine structures, a field in which I had done theoretical research. He came to Philips on some business connected with the purchase of scientific apparatus, and very discreetly, without showing that we knew each other, asked me whether he could see me in private. I went to see him at his hotel in the evening and proposed that we should go to my home for a quiet talk. While we were walking along the dark streets of Eindhoven, he said, "Won't you get into difficulties with your colleagues if they should find out that you are receiving a German at home?" I had a ready answer: "Now that you have said that, I don't care."

> Becker: Die Rechnung wird schon präsentiert werden (The bill will be presented).
>
> Heisenberg: Vielleicht wäre ein Europa unter deutscher Führung das kleinere Uebel (Perhaps a Europe under German leadership might be the lesser evil).
>
> Kopfermann: Kommen Sie nicht in Schwierigkeiten mit Ihren Kollegen (Won't you get into difficulties with your colleagues)?

These three short sentences epitomize the attitudes of three physicists who were certainly all three men of integrity, who were all three anti-Nazis, but who were also in some way German patriots who did not want to forsake their country the way Einstein had done —and had been forced to do—and who therefore were willing to accept a measure of compromise. Of these, Kopfermann had the greatest sensitivity, Becker perhaps the greatest sense of justice, and Heisenberg, who was by far the greatest physicist, the least understanding of the situation. This same lack of understanding may have played a role in the fateful wartime discussions between Heisenberg and Bohr on the possibility of nuclear weapons. Mrs. Bohr told me shortly after the war that Heisenberg had at one time proposed to organize a meeting on theoretical physics at Copenhagen, "just as in the old days," and that he thought he would be able to arrange for the necessary travel permits. But the meeting would then have to take place at the Deutsches Haus, an institution that was regarded by the Danes as a most objectionable center of Nazi propaganda. And Mrs. Bohr said that it had made Niels very, very sad that Heisenberg, whom he admired and who had in many ways been a close friend, should have so little understanding of the feelings of the Danes. If there is truth in Heisenberg's statement that Bohr mistook his intentions, this may partly have been the result of this disappointment.

A genius is someone who can create things that are initially beyond his own understanding. In that sense Heisenberg was certainly a genius, and this goes but rarely together with a special gift for understanding the feelings and the ways of thinking of others. Heisenberg did not have that gift. Perhaps his greatest shortcoming was that he was unable to grasp the full measure of depravity of what was then the ruling group in Germany. A Europe under German leadership would have meant a Europe under Nazi leadership, and it would have been a horrible thing. That became increasingly clear thereafter. At the time of our meeting the persecution and annihilation of the Jews had not yet reached its gruesome climax.

Liberation and the First Years After

When I look back at the first months, the first years even, after our liberation, I am overwhelmed by a wealth of recollections, confused, confusing, and many of them not very important. I shall not try to write a systematic account of that period, but just jot down a few points.

We had been lucky. The German air raid the day after the arrival of the British Second Army might have been followed by a tank battle in the streets of Eindhoven, the city might have been reconquered, or—almost worse—it might have been no-man's-land for some time. But it did not happen that way. Local tradition asserts that the Germans had to turn back because a bridge on which they had counted to cross the Dommel, the small river running through Eindhoven, was too weak for their tanks. For many years that bridge carried a signboard—now it has been removed—with a little poem that said in essence that we had been saved by this bridge. "Her weakness was her force. . . ." Anyway, although we lived for some time in a narrow corridor, we did not see the German troops again and in the course of the autumn and winter the corridor was gradually widened until the whole of the southern part of the Netherlands was free. There were no more German air attacks on the city either, but of course you never felt quite sure about that. And later we could hear V-1s passing over. They were not aimed at us but a few came down prematurely in or near Eindhoven, and they were unpleasant to listen to, although you hoped the noise would not suddenly stop. One had to learn to live with such fears.

One of the first days after our liberation, I was standing in the street, looking at the military traffic, when an air-raid alarm was given. It turned out to be a false one, but far away some airplanes could be seen or heard. People started to run in panic. A little British hospital orderly was standing at a streetcorner, watching the scene disapprovingly. He beckoned to me and said, "That's no good; people should not panic." "We had a nasty air raid the other day," I explained, "so people are afraid of bombs." He looked at me in surprise. "Of course they are. Everyone is afraid of bombs. I am afraid myself. But that is no reason for running. If bombs really are being dropped you may look for shelter; if there is none you just have 'to take it.' Do not panic, take it." Had he said that one should not be afraid it would have been of no use to me; you cannot eliminate fear by a simple statement like that. But his candid admission that he was afraid, himself, helped me a lot. I felt that if he could behave calmly even though afraid, I should be able to do the same. I am still profoundly grateful to that little man. I am sure that he did his duty as a hospital orderly, even under fire and air attack. I hope he came through the war unscathed.

Material conditions of daily life were at first rather meager. There were food shortages: of course, military transports had priority. Once most of the province had been cleared of Germans the situation improved. The farms of the area could supply potatoes, rye, vegetables, milk, and some meat. On Saturdays I often went on bicycle into the country to get some extra food for the family and our guests (about whom more later). I had found a farm at no more than twenty miles distance where they had children somewhat younger than our oldest ones and the farmer's wife was quite willing to barter food for clothes. The accepted basis for such transactions was prewar value, but that left plenty of scope for sharp bargaining, which she obviously enjoyed. But she was a kindhearted woman after all, and when we had reached agreement I was often invited to share the noon meal, usually a stew of potatoes and carrots with a generous piece of sausage or bacon. At another farm I was able to get a goose for Christmas. To ride a bicycle carrying a live goose, its head just sticking out of a rucksack, was for me a new experience—I rather liked these rural excursions.

For electricity the city depended for quite some time on an emergency plant at the Philips works. Maximum power for each household was fifty watts and if I remember rightly the total daily consumption was limited to 0.15 kwh. Highschool boys came and read the meters every day; they were proud of their authority and quite severe against transgressors. A local broadcasting station operated at the Philips works and the only official station in the liberated part of the country—it had the proud name "Vrij Nederland" (Free Netherlands)—transmitted the latest news. Sometimes it would announce, "Will all listeners please turn off the light and listen in the dark; the power plant is overloaded." Fuel was scarce too, but there are quite a few woods around Eindhoven and one could get some firewood. My wife and I always managed to heat one room with a little stove on which we also did much of our cooking. Gas was supplied only a few hours a day. Clothing remained rationed for several years, and in the beginning one could not buy anything—but what does it matter if your clothes are old and worn if everyone else's are too? Repairs were essential and for that you needed thread and needles, and needles especially were hard to come by. On one of my first visits to England I asked the wife of one of my friends, Mrs. Nemet, to buy me some needles. "How far can I go?" she asked, and I, completely ignorant about the price of needles, said that I thought a pound's worth would do nicely. It turned out to be an impressive collection that served my wife for years, and she made many friends happy with one or two.

All these little shortages were no real hardship, and in a way I enjoyed them. They made daily living a bit of an art. They taught you the relative importance and unimportance of various things. And they enhanced the value of simple pleasures. A visit to friends means more if you have to walk an hour to get there. A good meal is more enjoyable when it takes some effort to get the ingredients. I wrote about the disadvantages of fast travel in Chapter 4. Someday we shall have to reduce our energy consumption and our consumption in general, and then we shall remember these lessons of the past. But no amount of reasoning will induce us to cut down our spending before it is really necessary. Then we shall find that it is a thing sooner done than said.

Our house—a rather small one—was always full those days. We put up some "line-crossers," people who had managed to escape from the occupied part of the country. Two girls from Amsterdam, medical students active in the underground, had come as messengers. One of them found a job as an assistant nurse and stayed with us; the other, after delivering her message (and, probably, after having received new instructions), went back. It was a great relief to see her again several months later, but she had grim stories to tell about the situation in the big cities of Holland, where people really were starving.

One day a young Jewish physicist turned up who had been hiding in the east of the country. I found him a job at Philips, but he felt that, being one of the few Jews who had escaped, he should do his share in the war effort. So he enlisted and served as a meteorologist with the air force. Later he became a professor at Eindhoven.

From time to time there were also officers billeted at our place, who spent a few days at a training center at Eindhoven. I believe the instruction they received had something to do with civil defense but I am not sure about that. Our visitors did not take these lessons very seriously, but regarded them as a welcome outing. We had joyous meals together, consisting of K-Rations, potatoes and vegetables, and oysters. Oysters were the one and only luxury we could offer. The oyster beds some seventy or eighty miles to the west had remained intact, and Eindhoven was one of the few places where their produce could be sold.

Among these officers with their different backgrounds—a waiter from New York (he taught my wife to fry oysters), a saxophone player from St. Louis, a bank director from East Anglia, and so on—there was one who impressed me most of all. On Chinese prints you may see curious, very steep mountains and wonder whether they really exist outside the imagination of the painter. But when my wife and I visited China in the spring of 1977 we were taken on a boat trip near Guilin, and such mountains were all around us. Our guest was like a Chinese mountain. He was the type of Englishman about whom you read in detective stories and other entertainment literature and of whose real existence you were never quite sure. But he was real enough. He was a young lieutenant in the Welsh Guards. He began his few days of relaxation by going shooting (probably quite near the front) and came home with a couple of ducks, which he presented to us. So we invited him to dinner. Did he have any job in peacetime? my wife asked. No, he had left Eton not long ago and had not made up his mind what to do when the war was over. He felt he had not learned very much at Eton; he had been captain of the cricket team; that and a few other things had kept him rather busy. He had just passed the examinations. I noticed he was quite familiar with French. Yes, he had usually spent his summer holidays with an uncle who had a house on the Côte d'Azur. In the winter he often went shooting in the highlands of Scotland. He told us how he had had to fight a bigger German tank. "I thought my last hour had come, but I kept driving around it as fast as I could and kept firing. To my great surprise they ceased fire and four Germans came out, their hands above their heads. They were nice prisoners to have; they did the washing up and they sang quartets after dinner. You are supposed to hand over any prisoners you take within one or two days, but I kept them a week, mainly for their beautiful singing." Later the talk turned to army rations. "I have in my outfit one lorry labeled 'high explosives.' There we keep pigs. You know, they calculate rations with calories and all that, but in a war you get pretty scared sometimes, and then calories are no good. You need decent food. Then we kill a pig." I suggested that he might well add some geese. I felt certain they would thrive in a lorry. "Anyway," he said, "they never give us enough material, so from time to time you report having lost a tank, and ask for a replacement. I am at present three tanks up." Needless to say, he accepted with complete ease the rather primitive conditions of our establishment.

Before I return to science, technology, and the Philips Laboratories, I have to mention one more activity in which I played a modest role: De Tijdelijke Academie (Temporary Academy). There were many young people in the southern part of the country who wanted to begin academic studies, or who had had to interrupt their studies and wanted to take them up again. Already before the final capitulation of the Germans on 5 May 1945, their numbers were increased by young men who had been forced to work in Germany and now returned. So plans were made to create an emergency university. C.J. Bakker was one of the driving forces, the great Van der Pol was a wise and prudent president, and I was assigned the role of rector. In continental universities the "rector magnificus"—addressed in Germany as "Magnifizens"—was the president of the academic senate, and also had some jurisdiction over the students. I was definitely a rector non-magnificus, for our enterprise was singularly lacking in splendor. We had no buildings of our own, lectures were given in pubs and in churches, in shops that had gone out of business, and so on. I registered all students myself, sitting on a threadbare, red overstuffed chair in a room above a pub called "Het Rozenknopje" (The Rosebud). We got a pretty good teaching staff together—mainly from Philips, but we also found some outstanding secondary schoolteachers. Some of them later became full professors at one of the official universities. So did our chief medical man, an outstanding pediatrician. The elements of calculus were taught by Dr. L.M. van Rees, a Jewish mathematician who had been teaching at a secondary school in Rotterdam. When the anti-Jewish measures became more and more stringent he had managed to get out of the country with his wife and two sons and had reached the Swiss border. There he had asked for asylum, but the Swiss frontier guards sent him back and told him that if he should ever try again he would be handed over to the Germans right away. They were dead tired by then and had hoped to find at last a safe lodging for the night, but they had to go. They got back to Belgium and there—I believe with the help of a Catholic monastery—they got reasonably good forged papers and survived. We learned about his existence through the office of the Dutch government in Brussels and we soon came to terms. We even found a house for him and his family. Later that year a third child, a girl, was born. B.G. Escher had found it wise to leave Leiden and was living in Breda, thirty miles from Eindhoven. So I pedaled to Breda and found him quite willing to come to us and to teach crystallography and the elements of geology. It is not a great distance but my bicycle was old and rickety and on the way back I must have looked rather tired. In any case, I was stopped by some British soldiers who were having a meal at the roadside. "Here, have this," they said and handed me half a tin of corned beef. It was delicious.

Our "academy" functioned until the beginning of 1946. Of course, it was far from a complete university. We offered mainly first- and second-year courses in the sciences and in engineering, and also a first-year premedical course. For older students we were usually able to find ad hoc solutions. I examined for instance a few older students in theoretical physics. My opinion was later accepted without question by my colleagues at the real universities. Students worked hard, and I am convinced that our initiative not only saved them one or two years but that we also helped many, who had become unsettled during the war years, to refind their bearings.

One ticklish question was that of admission. During the occupation Leiden University had remained closed but the other academic institutions continued in a way. However, in 1943 the authorities imposed a rule that all students had to sign a declaration of loyalty. It was argued by some that the best policy would be to sign: one should not feel morally tied by such a signature enforced by a power that did not itself show any respect for the rights of man nor for any written or unwritten law. But the large majority was against that. From then on anybody who signed was regarded as a traitor. Should such "traitors" now be admitted? A special committee, of which I as the rector was a member, was formed to decide such questions. In that committee I experienced for the first time in my life how difficult it is to pronounce judgment when there is neither written law nor precedent and how disappointingly inadequate our "natural sense of justice" can be. We decided at once that someone who had been active in a National Socialist organization, or who had served with the German army, would not be admitted. That was simple— and I do not know even whether there were any such cases: people in that position did not want to expose themselves.

We further distinguished four categories. There were the students who had not signed and who had gone into hiding. They were regarded as the best. The next group had not signed, but for one reason or another they had not found it possible to hide and they had been sent to Germany as forced labor. They too were at once accepted. Then came those who had signed, but who had not continued their studies, had possibly worked in the underground and so on. There we had to judge case by case. And finally there were those who had signed and continued their studies, the "profiteers." They would have to wait until the universities in the north reopened. I believe our rulings corresponded to the general feeling among the young people of those days. And I do not think it caused great and unjust hardships. Being excluded from the privilege of getting a head start at the Tijdelijke Academie was not a very heavy punishment. During the spring of 1945 I was invited by the University of Louvain to give a lecture on our academy. My talk was received with great enthusiasm but afterwards one Belgian said to me, "You Dutchmen are completely crazy. Of course your category two is the worst of all. They were collaborators; they worked for German industry. And it would have been easy for them to avoid that: they had only to sign a valueless piece of paper. Your category one is slightly better, but they also should be blamed: there were not so many good hiding places, and these should have been reserved for people who were really in danger of their lives, for Jews and for people sought after for political reasons. To occupy such places for the quixotic reason of not wanting to sign was certainly reprehensible. Students who signed were all right. If they did not continue their studies because of their solidarity with non-signers, that is understandable though unreasonable! The best course to take was to sign and to prepare oneself for the reconstruction after the war." There is much to be said for this practical Belgian approach. Yet I do not think that our judgment was wrong; once a majority had decided on a certain course of action, solidarity had to prevail.

A few days after our liberation, Goudsmit turned up in Eindhoven, in military uniform and accompanied by Colonel Pash. We later learned that this visit was part of the Alsos mission that had to find out as much as possible about German work on nuclear explosives. He asked me to try to remember every little detail about German interests in nuclear physics and its instrumentation. "Believe me, it is very important," he said. "I cannot tell you why." For Goudsmit himself there was no joy in coming back to his country; he knew already that his parents had perished. Not much later a British team came to find out what the Research Laboratories had been doing in wartime. One of the members of the team was G. Bennet Lewis, whom I had known at Cambridge. There were one or two items that were of immediate interest to them, and measures were at once taken to ensure transfer of the know-how involved. They asked what we had done about radar and were amused when they discovered that we did not even know the word, although we had heard vague rumors. The Germans referred occasionally to the "Rotterdam-Gerät" because they had found fairly intact radar equipment in an airplane shot down near Rotterdam.

Gradually, our contacts with the outside world increased, especially after the German capitulation on 5 May, and the Japanese capitulation on 8 August 1945. I received letters and parcels from friends and colleagues, some of them having a box at the post office at Santa Fe as the sender's address—which puzzled me until I learned about Los Alamos.

I should also mention the visits of G.P. Kuiper, the Dutch astronomer, who had settled in the United States in the thirties. I do not know exactly what his status was in the U.S. Army, but apparently it gave him a great deal of freedom. He had always been a man of almost unbounded energy and enterprise. Now with his jeep he performed chivalrous actions about which he told with great gusto. One of his exploits was to find Max Planck and his wife, who were living under very primitive circumstances in a farmhand's cottage that was already overcrowded without them. He brought them to Göttingen, where a relative could put them up. The story is also told in Hermann's short biography of Planck.[11] It was Richard Pohl who had suggested to Kuiper that he should undertake this mission and —in an answer to Hermann's enquiry—Pohl described how Kuiper, after having delivered the Plancks at their relatives', came to report, tired but happy. And then Pohl adds, "I got my best bottle of Rhine wine out of its hiding place and we talked deep into the night." (And here I should add that Pohl took great and justified pride in his cellar. His best bottle must certainly have been worth drinking.)

Another of Kuiper's stories. Driving between Amsterdam and Rotterdam he saw a young woman with a bicycle sitting at the roadside and weeping. He stopped and asked what was the matter. An American officer getting out of a jeep and addressing her in Dutch must have appeared to that girl almost like an angel from heaven. She was supposed to get married at the town hall in Rotterdam that morning, but some essential papers had been missing; she had cycled to Amsterdam to get them, but now, on the way back, she was exhausted and unable to make it in time. So Kuiper loaded her and her bicycle into his jeep and got her to her wedding on time. "I did not even kiss the bride," he added virtuously.

I emphasized already that we in the south of the country suffered no real hardships and that our material situation rapidly improved. Yet, when in the course of 1946 I went with a group of Philips people to Switzerland in a bus to see the Baseler Messe (Basel Fair) and to resume contacts, life in Switzerland and especially the shop windows in the Bahnhofstrasse in Zürich seemed almost ridiculously luxurious. I had a similar impression in Stockholm; on that visit, however, I also enjoyed the simple and heartwarming hospitality of Oscar Klein and his family in their cottage in Dalecarlia, where I spent the days around midsummer.

Mail services soon became fairly reliable. The *Reviews of Modern Physics* published a special issue dedicated to Niels Bohr on the occasion of his sixtieth birthday on 7 October 1945. My contribution —it contained my work on Onsager's relations—arrived in time, but if I remember rightly there was no time for proofs. The article contains a few minor errors I might have corrected. In connection with the fiftieth birthday of Röntgen's discovery of X-rays, Dr. Oosterkamp and I wrote a review article on modern X-ray tubes for the periodical *Experientia.* We wrote it in our best German, again there was no time for proofs. When we finally saw the thing in print it was in English: the editors had wanted to give that issue an international character and had had it translated. I am happy to say that I did not notice any serious errors in the translation. All the same, it was a curious experience: I did not recognize our article. The English was fluent and presumably more idiomatic than my own, but if we had written in English it would have been a different article. However, no real harm was done, and we have seen in Chapter 5 that misunderstandings about proofs could even happen in complete peacetime.

In the years that followed I paid many visits to England, where I was well received by my old low-temperature friends. I went also to the United States several times and on three occasions was visiting lecturer there (at Johns Hopkins in the spring of 1947, at Ann Arbor in the summer of 1948, and at Princeton in the spring of 1951). I also attended conferences in France and Italy, but the first time I again took part in a scientific meeting in Germany was in the early summer of 1951. It was a conference at Heidelberg in honor of the sixtieth birthday of Walther Bothe, and many physicists from outside Germany had decided that this was a good occasion to reestablish scientific relations. Pauli also attended. In a later chapter I shall come back to the discussions I had with him on that occasion.

I had come to Heidelberg by car, and after the meeting I drove Maria Goeppert Mayer* and Houtermans in leisurely stages to Copenhagen.

I mentioned Houtermans earlier. Let me now tell a bit more about him. He was a man with pronounced left-wing sympathies, and when Hitler came to power he promptly left the country and went to England. He was not happy there, so he and his family went to Kharkov, then a flourishing center of physics. During the Stalin purge he was arrested and accused of being a German agent. His wife managed to escape to Finland with their two children, received some assistance from Niels Bohr, and went on to the United States, where she became a physics teacher at Sarah Lawrence College. Houtermans was put under such pressure to confess that he finally admitted to having been in contact with two German spies, called Scharnhorst and Gneisenau. These are the well-known names of two armored cruisers that were sunk in the battle of the Falkland Islands in 1914 and of two larger battleships that played a role in World War II. According to the story he later told me, his confession was, at first, accepted; however, when his case came to court, the judge at once recognized the names and Houtermans was able to convince him that he had made his confession under duress. During the brief German-Russian alliance he was, against his urgent request, handed over to the Gestapo, but somehow—I do not know the details but believe he rendered some service to the German forces in occupied Russian territory—he was released. He then worked for a time at Berlin with Manfred von Ardenne, a well-known technical physicist. He was an inveterate chain-smoker, so he devised a research project involving irradiation of tobacco. It was accepted as "Kriegswichtig" (important for the war effort) and he got a whole bag of tobacco-dust. When that had been smoked he managed by hook or by crook to get another consignment, but this time he went a bit too far and he found it safer to leave Berlin for Göttingen, where he arrived with the corpus delicti. I have been told that dried bramble leaves sprinkled with corpus delicti made an acceptable smoke. In the meantime he divorced Charlotte, his first wife—without her knowledge—and remarried.

*Maria Goeppert Mayer, 1906–1972, daughter of a professor of pediatrics at Göttingen, got her PhD in her home town in 1930, then married the physical chemist Joseph C. Mayer and moved to the United States. She received the Nobel Prize in 1963.

Houtermans was an entertaining passenger. He would, for instance, explain that the Roman Empire had left permanent traces and that the former *limes* of the Imperium could be inferred from the way in which potatoes were prepared: if they were just boiled with salt you were certainly outside that borderline. "Deep in the land of salted potatoes," he sighed after a mediocre meal in a countryside inn. He also pointed out that before World War I a coachman in Vienna would address a fare from whom he expected a generous tip as "Herr Baron." In the twenties and early thirties intellectuals were appreciated, and passengers became "Herr Doktor." But since World War II we live in an executive age, and consequently cabdrivers now use "Herr Direktor" as the preferred form of address. I wonder whether he was in league with Pauli to pull my leg? Anyway, he continued: "Words originally designating members of the aristocracy, like *Messieurs* in French or *Gentlemen* in English, in the long run find their way to lavatory doors. How pleasant it would be to live in a world where every loo is inscribed *Directors!*" Once when the Autobahn crossed a tiny brook, he made a sweeping gesture and proclaimed: "The Efze, Hessen's stream, not Hessen's border." He was parodying a favorite theme of former days: "The Rhine, Germany's stream, not Germany's border."

Houtermans was also an unofficially accredited Hilbert-story teller. David Hilbert (1862–1943) was one of the greatest of the many great Göttingen mathematicians. He was reputed to be extremely absentminded, and rather one-sided in his abilities and interests, and these two characteristics gave rise to innumerable stories, some of them obviously not strictly in agreement with the fortuities of reality. About his somewhat feebleminded son, who looked rather like himself, he once said, "Poor boy, everything he has inherited from me, everything; only mathematics he inherited from his mother."

After the conference in Copenhagen, Houtermans asked me whether I could make a small detour on my way back to Holland and drop him at Bielefeld, where his first wife was visiting her mother and sister. And so I became, willy-nilly, a match(re)maker.

It would have been out of character if this curious mission had not been accompanied by some bizarre incident. We had left the Danish islands and were driving south along the east coast of Jutland. Two girls were asking for a ride, we stopped, they got on board and to our surprise asked whether we could take them along to Holland. They were on their way to Paris, where they wanted to celebrate the fourteenth of July. During the summer, long-distance hitchhikers are quite common in Europe, but they invariably carry big rucksacks. These girls wore light summer dresses and had no luggage besides a medium-sized handbag. They were Swedish, knew some English, and they had a small-scale map of Europe. No, they did not know anyone in Paris—not yet, that is. I mentioned the absence of luggage. "We have what we need," was the answer. "After all, this is just a little trip. Last summer we toured the whole of Italy, down to Sicily." They said they were just factory workers, but liked a bit of a holiday. Perhaps they were, but of all the dozens of hitchhikers I have transported in the course of the years, they were the most astounding. Houtermans, far more experienced in the ways of the world than I, could not place them either. They looked frivolous in a rather cheap way, but behaved with exemplary modesty. At the frontier they suggested they should leave the car somewhat before and walk through independently, so that they would not cause any inconvenience. Obviously their papers were in order and they knew the ropes: they had already been waiting quite some time when I had finally cleared the car. When Houtermans and I stopped for the night they did not give the slightest indication that they wished to emulate Mary the Egyptian (as I explained, they did not look saintly anyway). They simply suggested they might stay in the car, but there are limits to my generosity (and to my confidence in flashy girls). They had no German money, so I lent them some, and did not expect to see them, or it, again. Next morning they were waiting beside the car. When we stopped for coffee and later for a meal they refused an invitation to join us: they could do without.

In the afternoon we arrived at Bielefeld, Houtermans, nervous and excited, asked me to come in with him. I think that to begin with both Charlotte and he were happy that there was a neutral person present (Charlotte's sister sneaked out to have a look at our cargo). After a while, when the first uneasiness had disappeared, I felt that I was no longer wanted and continued my trip.

We arrived in Eindhoven rather late at night, and I was rather at a loss when the girls asked for a cheap inn. You usually do not know hotels in your hometown anyway, and the kind of place they were looking for would hardly be on the list for Philips visitors. So I took them home. My wife was only mildly surprised: it was not the first time I had come home with somewhat unlikely hitchhikers. She put them into a spare bedroom, but before they retired they paid her, in Dutch currency, what they had borrowed. Quite a tactful gesture! Next morning they were gone before we came down. They left a note thanking us (and—of course—without pinching anything).

Somewhat later Houtermans divorced his second wife and remarried Charlotte. But it did not work out; they had grown too much apart. They were again divorced and she returned to the States, a sadder and a wiser woman. When Houtermans married again—not with number two—one of his colleagues sent him a telegram that became famous: "Mit den üblichen Glückwünschen" (with the usual congratulations).

Houtermans had by then become professor at Bern. For all I know he did a wonderful job, modernizing the laboratory and giving new life to research in physics. "If you want to see an authentic early-twentieth-century laboratory, come and visit me," he wrote me after his appointment. "But you have to come soon, for I am going to change all that."

A little tower with a winding staircase was part of the physics laboratory. The heavy central column bore on its top the zeropoint of the Swiss geodetic survey. It was also supposed to provide a vibrationless support but, according to Houtermans, the stability left something to be desired. Hence his remark: "When I stand in my institute and push I can waggle the whole of Switzerland."

However, the heavy strains of his eventful life must have been too much for him. He might have entered a serene old age, financially rather well taken care of, appreciated by his students and co-workers and, to the end, a sound experimental physicist. It did not happen that way; he died an alcoholic.

8

Holst and the Philips Research Laboratory

The Philips Company

When I joined the Philips Laboratory in 1942, it was still the only place in the Netherlands outside the universities where research in physics at an academic level was going on. How did it come into being? One must be careful not to ascribe the birth of any great organization exclusively to the efforts of one man. The times must be ripe, the general social and economic trends must be taken into account. Yet the course of events within the general socioeconomic framework does depend on the actions of a few outstanding people. In this sense it was Gilles Holst (1886–1968) who created the Philips Natuurkundig Laboratorium,* although he could only achieve what he did achieve because he worked for a fast-growing and prosperous industrial enterprise. The rise of the Philips Company itself formed part of the stupendous development of electrotechnical industries that had set in towards the end of the nineteenth century and that was spreading all over Europe and the United States. However, the company would not have grown in the same way without the existence of the remarkable Philips family. So let me try to sketch its origins.†

*In Dutch this means simply Philips Physical Laboratory. The usual abbreviation inside the Company is Nat. Lab., and this sometimes gave rise to confusion. From time to time I received letters addressed to Philips National Laboratory—almost a *contradictio in terminis*.

†There is a biography of Anton Philips by P.J. Bouman. It is very readable and re-creates a general atmosphere, but it is not entirely reliable. The first part of a very thorough study of the history of the Philips Company and its founders by A. Heerding appeared in November 1980. My own sketch is mainly based on these books, combined with what I gathered from casual conversations with old-timers.

Zaltbommel was—and in many ways still is—a quiet provincial market town on the southern bank of the Waal, the main branch of the Rhine before it splits up into the intricate network of its delta. The motor road from Utrecht to the south and the railway cross the river slightly to the east. The bridges command a view of the town and of its Gothic church and tower, but the ancient town and its waterfront are left undisturbed. In the fifteenth century Zaltbommel was a Hansa town. Then, and also later, most trade went by water. Seen from the river, Zaltbommel is less striking than some of the other former Hansa towns along our main rivers. It turns inwards rather than facing outwards, but within the confines of its former ramparts it contains many impressive buildings and stately private homes. The railway and its bridge were built in the late eighteen-sixties, the road bridge as late as 1932. Before that time there was only a car ferry, and I remember that the opening of the bridge was quite an event. The Dutch poet Martinus Nijhoff (1894–1953)—grandson of the founder of the well-known publishing house that published the collected papers of H.A. Lorentz and the even more impressive *Oeuvres Complètes* of Christiaan Huygens—even mentions it in a sonnet:

> Ik ging naar Bommel om de brug te zien
> Ik zag de nieuwe brug. . . .

> [I went to Bommel to see the bridge
> I saw the new bridge. . . .]

It was there that Benjamin Philips, a Jewish businessman, who continued his father's trade as a wholesale dealer in tobacco, met his future wife, married, settled down, and prospered.

In 1820 Lion, his oldest son (1794–1866) married Sophie Pres(s)-burg, a highly cultured and well-educated woman, who came from a long and distinguished line of rabbis. Her sister Henriette married a German lawyer, Heinrich Marx (1782–1838), and became the mother of Karl Marx (1818–1883). In 1826 Benjamin Philips and his wife and children renounced the Jewish faith and entered the Dutch Reformed Church. From then on the Philipses were counted among the leading citizens. It is interesting to note that Heinrich Marx entered the Lutheran church in 1817. One of Karl's biographers remarks:

Although there was no Jewish education or tradition in the upbringing of their children—indeed, the home was deliberately separated from family connections—Jewish self-consciousness was to some extent unavoidable.[1]

The same may have been true for Lion and his family. Not only was Lion a successful businessman, he was also a very enlightened man who was interested in many subjects, including the sciences. This appears clearly from the correspondence with his nephew Karl Marx, who held him in high esteem. In one letter Karl speaks for instance in a laudatory way about the book of Grove. William Robert Grove (later Sir William, 1811–1896) did important work in electro-chemistry, developed an improved voltaic cell that was at one time very popular, and was the first to make a hydrogen-oxygen fuel cell —he called it a gas-battery—an idea that went into near-oblivion until it was revived after the Second World War. The book Marx referred to, *On the Correlation of Physical Forces*—it is Grove's only book—appeared in 1846 (sixth edition 1874). It contains an early, and essentially independent, formulation of the law of conservation of energy. Readers who relish curious coincidences will be amused by the fact that Grove was one of the first persons ever to make an incandescent electric lamp with a metal filament. He put a platinum helix into a glass vessel and heated it by an electric current (1845). Rather surprising is the remark he made at the jubilee meeting of the Chemical Society* in 1891: "For my part I must say that science to me generally ceases to be interesting as it becomes useful." One would hardly expect this almost Pauli-like attitude from an eighty-year-old research man who had made important contributions to technology, and it was certainly not an attitude that would have been approved by Lion's grandsons. Frederik Philips (1830–1900), the seventh child of Lion and Sophie, married a gentile and continued to live at Zaltbommel and to deal successfully in tobacco and coffee, together with a partner, G.R. Peletier. Although he lived in a small town, his business gave him many contacts with the outside world. He was anything but a narrowminded provincial, and he was well aware of the technical-industrial expansion that was beginning to change even the Dutch economy. He had his finger in many pies. For a brief period he even owned the local gasworks; they gave him much trouble and little profit. Having been more or less forced to

*He was one of the original members. Cf. the article by E.L. Scott in *DSB*.

accept against his wishes the supervision of an agency of the official Nederlandsche Bank, he gradually developed it with one of his sons into a full-fledged banking house. In the early nineteen-seventies it was taken over by a larger banking concern that, in turn, became part of a still larger one.*

The Philips Company owes its existence to Frederik, to his oldest son Gerard (1858–1942), and to his youngest son Anton (1874–1951). Gerard began his schooling in Zaltbommel, but since the local secondary school offered only a three-year course, he had to go to Arnhem for two years in order to prepare for the final examination. He went to the same school Lorentz had attended, and from what Mrs. de Haas-Lorentz tells about the youth of her father, we can infer that it had some highly competent teachers of science and mathematics. Lorentz himself was at that time living in Arnhem and teaching at an evening school. Since Lorentz also replaced a temporarily disabled teacher at the HBS and gave talks to the society "Wessel Knoops," an organization of the type mentioned earlier, and since he was already a local celebrity, it is not improbable that Gerard was aware of his existence. Bouman in his book retells the story that a lecture by Lorentz actually stimulated Gerard's interest in science, but not only is that story fictitious, it is even out of character. Gerard's interest in science came much later: to begin with, he was a down-to-earth engineer. He passed his examination in 1876 and went to Delft. He first studied civil engineering but in 1883 he got his degree in mechanical engineering. He next went into shipbuilding, first in Vlissingen (Flushing) in the Netherlands and then in Glasgow, in those days a main center in that field. It was in Glasgow that he found his true vocation. In those days shipbuilders were beginning to install electric lighting with incandescent lamps, and this novel development fascinated Gerard far more than the ships themselves. He realized that his training in physics was inadequate, enrolled as a student at Glasgow University, and worked in William Thomson's (Lord Kelvin's) laboratory (and also with Professor Jamison at the College of Science and Arts). Some laboratory reports have been conserved. They deal, among other things, with the accurate calibration of electrical resistances. Gerard held a variety of jobs; he was technical representative of the British Brush Company in Germany

*The firm in question was Mees & Hope, which became part of the ABN (Algemene Bank Nederland).

and of German firms in London, and in 1889 he represented the AEG (Allgemeine Elektrizitäts Gesellschaft) in connection with one particular tender in Amsterdam. These occupations provided him with many contacts and with valuable insight into the structure of industry. I shall not relate the various steps and proposals that preceded the founding of the Philips Company; they can be found in Heerding's book. My short and possibly inexact summary should suffice to make it clear that when in 1891 Gerard Philips, financially backed and morally encouraged by his far-sighted father, began to manufacture incandescent lamps in a former buckskin factory in Eindhoven, he was well prepared for the job. He had the necessary technical competence and he knew his way around the world of industry. Moreover, during his years in England he had stumbled upon a process for making carbon filaments invented there, which he had tried out on a small scale with a friend of his in Amsterdam and which appeared to him to be suitable for mass production. Mass production was what he was aiming at, and he went about his work with remarkable singleness of purpose. Of course, there were the usual teething troubles, but by the end of 1894 manufacture was running more or less smoothly and beginning to make a profit. It was then that Anton entered the firm.

Anton Philips, sociable, sporting, outspoken, and generous, was in many ways the opposite of his reticent brother. Intelligent he certainly was, but his scholastic performance was not outstanding and he did not even finish the course at the commercial college in Amsterdam where he enrolled in 1891. In 1893 he started to work in a firm of stockbrokers in Amsterdam and in 1894 he moved to London. It is said that he was impressed both by the thrills of the London Stock Market and by the solid reliability of the firm he was working for, a well-established firm in the City. In January 1895 he began to sell incandescent lamps. It soon became clear that he was a superb salesman. Stories handed down by word of mouth usually suffer one of two fates. They are either forgotten or they are canonized in an embellished form. Some of the yarns told by Anton himself and by others about his adventures and achievements in those early days may have followed this pattern, but nobody will question his talents.

He was enterprising and untiring and willing to get up in the small hours of the morning and to return late at night (and later expected the same from his staff); he was good at establishing friendly relations and he aimed at creating permanent business rather than at grasping a quick penny. In the course of time it became evident that he was also a capable and farsighted negotiator, and later still that he was a real enterpreneur with a feeling for innovation and for technological trends. By contrast, Gerard, the hardworking and competent manufacturer, might in later years appear almost narrow-minded; for him electric lighting was the end of the journey, not the beginning. True, he was sixteen years older than Anton, but it was also a question of character. Of course, I did not know Gerard and I met Anton only occasionally, and then only after the Second World War when he was past his prime. My description is mainly based on what I gathered from Holst. He respected Gerard, who had hired him, but he venerated Anton.

Gerard's preoccupation with the manufacture of incandescent lamps was well known, even outside the firm. I remember that during a visit Mervin Kelly, the president of Bell Laboratories, paid to Philips, I showed him a rather amusing demonstration involving alkali halides irradiated by polarized light. Dr. Tromp, vice-president in charge of engineering, was also present and asked how this effect might be applied. He may have been slightly shocked by my answer: "The beauty of this phenomenon is that I see no possibility at all that it will ever be applied." Kelly laughed and said: "It seems things have changed. I have been told that in the old days when Gerard Philips was present you had always to say, 'We believe that in the long run this will lead to a reduction in the cost of production of incandescent lamps.' "

It was not to be expected that the relations between two persons as different as Gerard and Anton would always be entirely harmonious. They were not, but together the two brothers created a great company.

On the eve of World War I, Philips had become one of the major European lamp manufacturers and its sales were by no means limited to the Netherlands and its colonies; Anton had seen to that. However, from the point of view of technology its position was vulnerable. So far the company had followed its competitors. It had not contributed important new ideas of its own, but it had been skillful at rapidly absorbing and adapting procedures invented elsewhere. This practice had been made easier by the fact that until 1910 the Netherlands had no patent law. But in 1910 a patent law came into force, and moreover, Gerard had been deeply impressed by the work at the General Electric Laboratories in the United States, work that had led to tungsten filaments and to gas-filled lamps. He saw that he needed a research laboratory as well.

It took some time before the right man had been found to head this new venture.[2] W.J. de Haas liked to tell that he had been invited to come to Eindhoven and could have taken the job, but he had not felt attracted by the prospect and had declined. I am not certain that Gerard Philips would have endorsed this story: it is quite possible that he had had a look at De Haas and had decided that he was not the man he was looking for. C. Dorsman also declined, so it seems, and there may have been others, who were either considered unsuitable or did not like the idea of working in industry. In any case, on Thursday, 23 October 1913, there appeared in the *Nieuwe Rotterdamse Courant,* the leading newspaper, an advertisement: "Wanted: a capable, young doctor of physics, particularly also a good experimenter." On Saturday the twenty-fifth Gilles Holst sent in his application, or rather, asked for an interview. He was promptly invited, Gerard Philips and he soon came to terms, and on 2 January 1914, Holst started work. His official appointment also began on that day, not on the first of January—that was a holiday, and Gerard Philips was not the man to begin working relations by paying one day's salary for nothing!

Gilles Holst

Gilles Holst was born on 20 March 1886 in Haarlem, where his father was manager of a shipyard. After leaving the HBS he worked for half a year in his father's shipyard and half a year in another industrial enterprise. I think this period convinced him already that he did not want to become a mechanical engineer. "I did not like cast iron," was one of his favorite sayings later on, and that short sentence summarizes much of Holst's approach to science and technology. The point is that in mechanical engineering one can carry out elaborate calculations, but at the last moment one has to introduce large safety factors, because the properties of the material might show a large spread. For Holst this was unacceptable. As a matter of fact the wish to understand and to control materials was later to become one of the guiding principles of the research he directed. He was convinced that real technical progress could be based only on a better understanding of processes and materials. He was a physicist, not an engineer, but a physicist who was willing to concentrate on problems that held technological promise as well. In short, he was just the man Gerard Philips needed. But I am running ahead of my story. After this practical year Holst went to the famous ETH (Eidgenössische Technische Hochschule) in Zürich and began to study electrotechnical engineering, but that did not satisfy him either. Perhaps there was still too much cast iron around. In any case after two and a half years, and after having passed the intermediate examination, he changed to physics and mathematics, and in 1908 he obtained the degree of "Geprüfter Fachlehrer" (certified special-subject teacher). It was the same degree Einstein obtained in August 1900.

He then worked for one year as assistant to Professor H.F. Weber. This was the same Weber who had encouraged Einstein not to give up after he had failed the entrance examination for the ETH; had during Einstein's student days exclaimed in exasperation, "You are a clever fellow, but you have one fault. You won't let anyone tell you a thing"; and who, after Einstein had obtained his degree, had refused to appoint him as assistant, and had, according to some utterances of Einstein, even intrigued against him. Holst returned to Holland and became assistant to Kamerlingh Onnes. Curiously enough, in 1901 Einstein had written to Kamerlingh Onnes asking whether he had a job for him. There is no evidence that Kamerlingh Onnes ever answered—it is unlikely that he would have taken an application on an open postcard very seriously.*

There is a lesson to be drawn from this story of purely accidental coincidences. A great genius like Einstein may have more difficulty in finding his place in the world than a man like Holst, who was clever and competent, and who was receptive to revolutionary new ideas but would not create them himself, even though he took part in one of the most astonishing discoveries in solid-state physics, the discovery of superconductivity. I have given some indication of his role in that discovery in Chapter 6. At Leiden he also assisted Madame Curie, who wanted to examine whether radioactivity is influenced by temperature; no effect was in fact found. His main work

*The original of this remarkable postcard—with prepaid return addressed to Einstein but never mailed—is in the Boerhaave Museum in Leiden. Cf. *Nederl. Tijdschrift Natuurk.* A 45 (1949):131.

was in the best of the old Leiden traditions. It dealt with the equation of state and the thermodynamic properties of ammonia and methyl chloride. It was a solid piece of work on the basis of which he got his doctor's degree at Zürich in the summer of 1914. Yet it was hardly in line with Holst's personal preference: he had no special liking for precision measurements and even less for least-square fitting of semi-empirical formulae to experimental data.

I soon discovered that in the Philips Laboratory precision measurements were the exception rather than the rule, and what has been called the physics of the next decimal place was not held in high esteem. There were, however, some notable exceptions. During the occupation one team had found time quietly to push up the precision of measurement of small-phase angles of electrical condensors (capacitors) to one part in ten million. Holst's reaction was characteristic: "An angle of one in ten million!" he said. "That is one millimeter at ten kilometers' distance. No, that does not interest me." All the same, the measuring apparatus designed for that work later proved useful in several other investigations. I also felt a sense of pride when, during a visit I paid to the National Bureau of Standards, the two elderly staff members who were responsible for measurements on capacitors admitted that for very small loss angles the Eindhoven method was superior: their own measurements did not go much below one in a million, which, at that time, was adequate for all questions coming in.

Holst, who was soon joined by Ekko Oosterhuis, a less brilliant but extremely solid and commonsensical physicist, tackled his new job with great energy and gusto. At first he did much of the work himself and published an impressive number of papers. He assisted manufacture in many ways. He established reliable "photometry," measurements of the light output of lamps. He studied the properties of tungsten, the sputtering of metals, and the properties of thin layers. He was one of the pioneers of image transformers. Teves, co-author of the first paper on infrared tubes (tubes of that type became known during the Second World War as sniperscopes) later applied the same idea in his image intensifier. (Image intensifiers will be discussed in Chapter 9.) Together with Oosterhuis, Holst initiated work in the field of gas discharges, and this became one of the major fields of activity of the Philips Laboratory. In a way it was a field with a venerable past. Already before 1900 it had led to investigations on cathode rays, to the discovery of X-rays and of the electron. Holst took it up, because he foresaw possibilities for important new light sources. There was also another, more fundamental reason for his choice. He felt that Bohr's theory had for the first time made it possible to arrive at a real understanding of many of the features of gas discharges and especially of the light emission by the gas in a discharge tube. That was typical of Holst: he followed what was happening in basic research and when at a certain moment he felt a subject was approaching a state where it might possibly be applied for practical purposes, he would tackle it.

The Philips Company grew and moved into new fields, especially in the twenties. The research laboratory grew too, and more often than not it paved the way for new ventures. No wonder that Holst's personal publications became of lesser importance, but all of us knew that he was the driving force behind much of the work. In the matter of publications he certainly did not walk in Kamerlingh Onnes's footsteps. He would never insist on being co-author (let alone main or only author) of a paper when the actual work had been done by someone else, although he may have at times been disappointed by the fact that few of his co-workers proposed themselves that he should be co-author. I know he was grateful to the few who did. His attitude was a noble reaction to what must have been a great disappointment of his younger years.

Also, in other ways, Holst in later years sacrificed his personal ambitions as a physicist to his work as a leader—a leader rather than a manager—of a rapidly growing research establishment. Such a sacrifice never comes easily. Balthasar van der Pol, for instance, who was one of the most widely known members of the research staff, was never willing to make it. "In my younger years I was a pretty good experimental physicist," Holst once said to me rather wistfully. Of course it was his own choice, and there were certain compensations, financially and otherwise, but I do not think Holst enjoyed having power as such. The following little story, which circulated in the laboratory, is true in spirit, although I cannot vouch for the facts. A famous character in the laboratory was Nolleke, an elderly man of limited abilities who carried out some menial tasks. "I have always been slow of understanding and short of memory," he used to say, "and that has served me well." One morning Holst arrived somewhat late when Nolleke was sweeping the floor in front of his office. They exchanged greetings and Nolleke added: "You have a good life, Mr. Holst. You come when you like, you go when you like, and nobody to ask any questions." "Well," said Holst, "it isn't always easy, you know. There are many problems, you have to deal with many people . . ." "Maybe," muttered Nolleke and went on with his sweeping. "But I suppose the wage is also proportionate."

Holst's freedom of coming and going turned him on one occasion into an accessory to the stealing of radio sets. He lived in a pleasant country house some five miles to the south of Eindhoven. One day a technician came to him with the following story. He had to do some testing of radio sets at night and he lived close to Holst. Could he occasionally put a radio in the trunk of Holst's car? He could then easily pick it up in the evening. Holst, not suspecting any mischief and not being a stickler for strict observance of formal rules—in principle, nothing should go out of the gates before a number of forms had been filled out, approved, signed, stamped, and handed over to the porter—readily gave his consent. Later it turned out that the radio sets were stolen.

I have already made some remarks about the Philips Laboratory in wartime. I do not feel competent to write a detailed history. Let me only say that Holst managed to keep the laboratory going, to keep the program virtually free from German interference, and to keep a number of people out of the hands of the Germans. He must have taken considerable risks in doing so. About those he never talked.

He retired in 1946 and was succeeded by a triumvirate: Herre Rinia, engineer and inventor; Evert Verwey, physical chemist; and myself. By then the laboratory was again in full swing. We had to try to continue according to the principles that had been laid down by Holst. Let me now say something about these principles. Holst himself was not the man to write extensive papers on research management (or on any other general subject). He was primarily a man of action, not a philosopher. Not that he was reticent about his ideas, but they are to be found scattered in lectures and addresses and in aphoristic remarks remembered and treasured by his co-workers. I mentioned above: "I didn't like cast iron." About examinations, he said: "Difficult examinations make for a stupid nation." About studies: "Clever people may study a long time; stupid ones should finish their studies rapidly." (Unfortunately in Holland it is usually the other way around.) On the fact that employees that are less than honest in their dealings with third parties cannot be trusted in their relations inside the company either: "There exist no scoundrels with built-in rectifiers."

When Holst came to Philips he had no special experience or knowledge concerning light and lamps, but, as we have seen, he was a well-educated, versatile physicist. That was the type of man Holst himself later tried to find. Usually he did not look for specific skills: he might ask a PhD in organic chemistry to tackle work on oxidic semiconductors or a PhD in low-temperature physics to dive into TV circuitry; in that way he could bring a variety of competences and abilities into the attack on specific problems.

He rarely gave his staff accurately defined tasks. He tried to make people enthusiastic about the things he was enthusiastic about—and usually succeeded. He knew that one can order a man to sweep floors, and, perhaps, to perform routine measurements and routine calculations, but that one can hardly expect really original work from a man who does not believe in the project he is ordered to undertake.

Anyway, he did not believe in a strict hierarchic structure. I quote from his inaugural address at Leiden, where he was an "extraordinary professor" from 1929 to 1939:

> If organization is carried too far, it leads to a narrowing rather than a widening of our vision. Another serious disadvantage is that one cannot take into account the personal characteristics of the individual researchers, of whom especially the most original ones have difficulty adapting themselves to an organization.

In those cases where a job was too much for one man, so that a team had to be created, Holst tried to create multidisciplinary teams. He definitely refused to divide his laboratory into chemistry, physics, engineering, etc. A team working on, say, ferrites, would include physicists, but also chemists, crystallographers, and electrical engineers (although these might be on loan from other groups of the laboratory).

He was convinced that one should publish; this was made possible by the existence of patents:

> Publication of results has another great advantage, the possibility of having first-class scientists working for you. I am convinced that many of the very best physicists that are now working in industry would have changed position long ago if they had had to keep their results secret.

So much for his relations with his staff. In his dealings with product divisions he insisted on a high degree of independence for the research laboratory: it reported to the top management of the company, not to the divisions. Of course, divisions could formulate problems, and in really urgent cases Holst was always willing to muster a team to come to the rescue, but the decision rested in practice with Holst and his staff. Only in very rare cases would top management —of which Holst was part—give definite directions. The final stages of product development and the day-to-day troubleshooting had to be done by the factory laboratories, and Holst encouraged the transfer of suitable staff from research to development. Holst had to fight pretty hard for his freedom but got the support of Anton Philips. For his successors life was far easier; by then top management was convinced that in a highly diversified firm like Philips such independence is the only viable solution. I served under three company presidents: P.F.S. Otten, Anton Philips's oldest son-in-law; F.J. Philips, Anton's only son; and, my final year under H.A.C. van Riemsdijk, Anton's youngest son-in-law (Gerard had no children). All three of them accepted and if necessary defended the independent position of the research laboratory. Of course, we were occasionally taken to task for having neglected this or that subject, we were sometimes instructed to do more—rarely to do less—for this or that industrial division, but the basic principles of our organization were never questioned.

Holst was strongly opposed to detailed budgeting by project and even more against regarding expenditure on the research that led to a specific product as initial costs for that product, but he always carefully considered whether it was justifiable to have so-and-so many men working on a given subject.

All these principles would, however, have been of no avail without Holst's extraordinary gift for entering a field of research at the right moment. I pointed out already that he believed real progress could be made only on the basis of real understanding. That led him to tackle gas discharges, and later to initiate a program in solid-state physics. He also had a keen eye for technological trends, and he early realized the enormous potential of radio and electronics.

His choice of specific development projects was not always as fortunate as that of fields of research. His feeling for immediate markets was sometimes at fault and he underestimated the problems of development and engineering. To give one example: he was aware, earlier than many others, of the advantages, of the necessity even, of micro-documentation. Then, instead of studying the whole problem with all its technical, optical, psychological, and commercial implications, he concentrated the research program on the study of a grainless photographic process. This work had some success, but it was irrelevant for the later development of the subject, in which Philips played no part.

It is fashionable these days to distinguish between "phenomena-oriented" and "mission-oriented" research. In the case of Holst I wonder whether the distinction is much to the point. Holst was definitely interested in phenomena and their theoretical explanation, but at the same time he was always thinking about applications. He became interested in phenomena when he began to see a possibility of harnessing them for industrial purposes.

I have tried to summarize Holst's principles in the form of ten commandments:

1. Engage competent scientists, if possible young ones, yet with academic research experience.
2. Do not pay too much attention to the details of their previous experience.
3. Give them a good deal of freedom and give a good deal of leeway to their particular preferences.
4. Let them publish and take part in international scientific activities.
5. Steer a middle course between individualism and strict regimentation; base authority on real competence; in case of doubt prefer anarchy.
6. Do not split up a laboratory according to different disciplines, but create multi-disciplinary teams.
7. Give the research laboratory independence in choice of subjects, but see to it that leaders and staff are thoroughly aware of their responsibility for the future of the company.
8. Do not run the research laboratory on budgets per project and never allow product divisions budgetary control over research projects.
9. Encourage transfer of competent senior people from the research laboratory to the development laboratories of product divisions.
10. In choosing research projects, be guided not only by market possibilities but also by the state of development of academic science.

I trust that in summarizing in this way I have not unduly distorted Holst's views. I do realize however that I have not really re-created the image of his personality, of his persuasive enthusiasm, of his quick and witty repartee (and his lack of formal oratorical skill), of his way of approaching people—essentially kind but wavering between shyness and bluntness. Holst himself used to say that he was more interested in things than in people. I wonder whether this was true. I believe he realized that setting objective goals was the best way to encourage people and even to help them overcome personal difficulties.

As for myself, I am grateful for his confidence and his guidance, even though he may have made me forsake my true calling.

Why I Stayed with Philips

Once the war was over I might have gone back into academic life. Many of my colleagues were surprised and some still regret that I did not. They may even surmise that there must have been a definite turning point, a crisis, when after much deliberation I made a clearcut decision that set a new course for the years to come. That was not really the case. Things just "happened to happen."* Perhaps it is one of my weaknesses that I never firmly mapped out a plan, either for my work or for my private life. Horace, in the same stanza that contains the proud adage of the Royal Society of London, "Nullius in Verba," claims that he tries "to bend the world to himself, not himself to the world."[3] That is a thing I have never really tried to do. I always more or less drifted along . . . and somehow winds and currents were more often favorable than otherwise. Let me sketch the sequence of events that gradually led to a new course and to a career I would never have envisaged in my younger years.

I had every reason to be glad I went to Philips in 1942. As I explained already, together with my wife and children I came through the occupation in relative comfort and safety. Not only that: I could continue to work under conditions far more favorable than those prevailing in the Dutch universities during the occupation. On the one hand I could do some fairly fundamental theoretical work, on the other hand I learned something about subjects I had never

*Dr. Seuss, *The 500 Hats of Bartholomew Cubbins*, final sentence. During a visit to Minneapolis in 1980, thanks to the kind offices of Dr. and Mrs. Stuewer, my wife and I could witness a remarkable performance based on this remarkable book at the Children's Theater.

studied before and that I found quite interesting, such as medical radiography.

Also, during the first years after the war the Philips Laboratory had far more to offer than the universities. The number of competent senior physicists working at Philips was much larger: it would be several years before an expansion of Dutch academic research set in. We received back numbers of scientific journals quite a bit earlier— the Philips organization in the United States had seen to that. I could travel freely—to England, to Scandinavia, to the United States— resume old contacts, establish new ones. If someone should blame me for having indulged in a kind of scientific joyriding in those years, there is some truth in such criticism. I also enjoyed being able to dispense some gifts here and there. Radio receivers may seem a trivial matter nowadays, but just after the war they were hard to come by and they provided much-needed contact with the outside world. My wife remembers how she delivered a radio to Kramers at Leiden; for me the one I gave to Coster at Groningen was even more significant.

Dirk Coster, born in 1889, had been professor of physics at Groningen since 1924. He is one more example of a scientist who began his career as a schoolteacher. He then obtained a stipend that enabled him to study at Leiden and also to get a degree in electrical engineering at Delft. From 1920 to 1922 he worked in Manne Siegbahn's laboratory at Lund in Sweden on X-ray spectroscopy and on the basis of that work got his doctor's degree at Leiden in 1922. He next worked for a year at Bohr's Institute at Copenhagen, where together with Hevesy he discovered a "missing element," hafnium. At Groningen he and his pupils continued work in X-rays, but he also worked in optical spectroscopy and was one of the first Dutch physicists to start work in nuclear physics. Considering the limited resources he had, a surprising amount of work was done under his direction. He was also a courageous man. I mentioned his role in arranging the escape of Lise Meitner, and he tried—unsuccessfully —to intercede on behalf of the parents of Goudsmit. His later years —he died in 1950—were rendered difficult by a progressive paralysis. I like to think that the radio I brought to his home provided some little distraction.

Of course, the essential thing was the work itself, and that was fascinating and challenging in many ways. Although industry only rarely offers really profound problems, it makes up for that by offering a great variety. If you are a fairly competent and all-round theoretician who likes to jump from problem to problem, an industrial research laboratory is not a bad place to be. And occasionally industry does lead to fairly fundamental problems. I still believe that the best work I ever did was my work on Van der Waals forces, and that work was prompted by a suggestion of Overbeek—then at Philips, later at the University of Utrecht—who had been studying suspensions of quartz powder used in manufacture. I shall come back to that theory later on.

There was also a very simple practical reason for staying at Philips: there was no suitable academic position available in the Netherlands just after the war. I did get several offers from the United States, especially a rather attractive one from Rochester, where Lee Dubridge tried to hire me as a successor to Weisskopf (who had been my successor at Zürich). These offers I declined: I did not want to become part of the brain drain. That was perhaps partly a matter of laziness: it is quite a job to emigrate with a young family, but there also entered a question of national pride, and even of more parochial Philips pride. Today I have reservations about the capitalist system, about the rat race of competition, about the consumer society to which a company like Philips caters, and about the commercial and industrial practices that are unavoidable if one wants to stay in the game. Just after the war we had no such misgivings. Holland had to show that it could prosper without its colonies, it had to reconstruct and expand industry, and we in the research laboratory had to show that we could face our American colleagues on equal terms, even though we were living in an impoverished and badly mangled country.

One particularly tempting offer came from England. The Plummer Chair of Theoretical Physics at Cambridge had been vacant since the death of R.H. Fowler (1889–1944), and some of my younger friends there had suggested that I might be a suitable candidate. However, the senior faculty members decided that if they had to get someone from Holland anyway, they should first try to get Kramers. Kramers told me in confidence that he was going to accept. It was for me a slight disappointment, but I had to admit Kramers was the better man. Moreover, he was not entirely happy at Leiden in those days. For him a change might be a good thing, whereas I, for the time being, was quite happy in my job. Just at that time Holst asked me to become one of the triumvirate that would succeed him on his retirement. Was it a wise decision of Philips to offer me that position? I sometimes wonder whether I would not have been more valuable as a general theoretical advisor. Should I have refused? That is a thing not easily done, especially since I did not see a plausible successor to Holst under whom I should care to work. And after the Cambridge episode I did not expect a comparable possibility outside Philips in the near future. So I accepted. A short time later Kramers decided after all to decline the offer and the position was then offered to me. But I felt that I had made my decision and could not go back on that.

After having become a co-director of the research laboratory, and even more after having become a member of the top executive committee of the company in 1956, I found it increasingly difficult to disentangle myself from Philips. There were some decisions to make later on. Kramers died in 1952 and I might have become his successor. But when I went to have a look at the Institute of Theoretical Physics at Leiden I felt depressed. The old rooms, the old library, were almost exactly as they used to be, without new life, without an indication of growth. There was hardly any administrative assistance, there were only very limited funds for traveling and for inviting lecturers from abroad. It was returning to a past I had liked but that should be changed, and I had been too much part of that past to be able to effect the change. Still later I was on one or two occasions sounded out in connection with the directorate of CERN, the important European research center at Geneva. I know—and knew then —that I was unsuitable for that job.

I must admit that in later years financial considerations also played a role. I do not think that my wife and I are particularly greedy, but when you are accustomed to a very generous salary it is not easy to go back to a lower level. Sometimes, when traveling in a chauffeur-driven company car—usually a Cadillac or a Chrysler— or crossing the Atlantic as a first-class passenger and possibly receiving some extra VIP treatment at airports, or flying in Europe on a company plane, I remembered with regret the contempt with which we looked down on that kind of traveler in the old days, when we might spend the night on the wooden benches of a third- or even fourth-class carriage on a German railroad, or cover considerable distances on our bicycles. I felt I had sunk rather low on my erstwhile scale of values and had really come down in the world. However, one does not easily go back to the simple life of earlier days—until one is forced to do so, as we were during the occupation. Then you find it is quite possible, even enjoyable.

Let me repeat once more that I really liked my position, especially the freedom that was generously given for my own work and the intellectual challenge of technical problems. The management side I did not like, but initially I took that rather lightheartedly, as I had done at Leiden. I have never been able to identify myself with an industry the way Holst did; throughout the years the world of academic research remained my spiritual homeland, and my modest contributions to physics and mathematics have given me more satisfaction than the fact of having a top position in one of the biggest companies in its field. This may have had the advantage of enabling me to look at the affairs of the company and its laboratories with a measure of impartial detachment.

Mervin Kelly, the president of Bell Labs, once said about his job: "I don't know that I really like the work, but that is what I am being paid for, and what one is paid for, one should do." I was shocked by that remark. I felt that one should first decide what one wanted to do and then look around to see whether someone was willing to pay for it. Of course, one might have to accept a compromise and I am even willing to admit that one must do what one is accepting payment for. The difference between "what one is accepting payment for" and "what one is being paid for" may seem slight, but I think it is important.

All the same, I had high regard for Kelly. He was heading the world's greatest research organization and he was a man of real vision. During a visit a few years after World War II he suggested that Philips should put a major effort into electronic computers. He predicted that this was going to be an activity at least as important as telecommunications, but whereas in telecommunications positions were already well established and one had moreover to deal with governments, in computers the field was still wide open. IBM had a powerful sales and service organization, but in electronics they were still babes in the wood. If we were quick we should be able to cash in on our superior knowledge in that field. I wonder what would have happened had we listened to that advice. What did happen was that by the time Philips—like RCA, like General Electric—tried to get really active, IBM had long since remedied its shortcomings.

During that same visit, at a dinner party at the home of one of our vice-presidents, he insisted on the importance of birth control for the third world and on the need for more research in that field. My wife and I were thoroughly amused by the reactions of those present; several of the men and women looked embarrassed. In the rather provincial ambiance of Eindhoven high life, women were apparently not supposed to have an intelligent opinion on such things, still less to talk about them. I am afraid that the ridiculousness of the situation was lost on Kelly. A sense of humor was not one of his main virtues.

That became apparent on another occasion when he was staying at the home of Frits Philips. The dog of the family was also named Kelly, so during this visit Mr. and Mrs. Philips carefully avoided calling it by its true name. I was present when coffee was served after dinner and the dog, rather wet and muddy, came in to beg for a biscuit. "Kelly, don't put your dirty paws on my clean tablecloth," Mrs. Philips burst out, and, seeing Dr. Kelly's surprise, she had to explain. He did not really enjoy this incident.

Many years later, I showed Sir Arnold Weinstock, the energetic managing director of GEC (the British General Electric) our laboratory. He asked me about the influence of the industrial divisions on our program and I explained that they could—and of course did— voice their wishes but that they had no control; and in a flippant mood I added, "Anyway, the commercial people themselves usually don't know what is good for them." The big boss was really shocked. "If anyone in my organization said that, his head would fall," was his reaction. "Well, I seem to get away with it," I said and left it at that. We parted on friendly terms. I could have said more. I could have explained that it would not at all be a question of a head falling, but that there were certain conditions under which I was willing to work in industry. If the management thought such conditions unacceptable that was their decision. They would have to look for another man and I would have no difficulty in finding another job.

There were other even more essential conditions, such as the right to publish, the insistence on scientific integrity. Most of them were part of the Holst code and, as I said before, I never had any real difficulty in defending them.

I have sketched the circumstances and events that made me an industrial research director. I like to think that in later years it was also a sense of responsibility that made me stick to that job—responsibility not in the first place with regard to the Philips Company and its shareholders but with regard to several hundred university graduates and a total staff of well over two thousand. I believe I have contributed to providing and defending satisfactory working conditions and have, on the whole, been able to maintain an atmosphere of friendly cooperation. However, to be a director in that way, to keep well informed about the work in the laboratories, and to be a theoretical physicist in my own right, all at once, was more than I was able to do. Gradually I came to realize that being a director of research really meant goodbye to physics. About that I shall write in my next section.

Goodbye to Physics

"A slow sort of country!" said the Queen. "Now, *here*, you see, it takes all the running *you* can do to keep in the same place. If you want to get somewhere else, you must run at least twice as fast as that!"[4]

Did Charles Lutwidge Dodgson have his own none too prominent position in mathematics in mind when he wrote those lines? Probably not, but in our present era of rapidly expanding science they are singularly appropriate for any aging scientist, and they certainly were in my case. At some point in my career I felt that I had more or less mastered the most advanced branches of physical theory and that in some areas of specialization I might occasionally even be slightly ahead of others. But I did not do all the running I might have been able to do, and after I had become a co-director of the Philips Laboratory I began to lose ground rapidly.

Pauli was well aware of this. Whenever I met him in later years he always addressed me as "Herr Direktor." He even told other people "If you go to Holland and should happen to meet Casimir, give him my regards and call him Herr Direktor. *Das ärgert ihn nämlich*" (for that vexes him). I believe that Pauli on the one hand, was amused that a pupil of his should be able to hold his own in an industrial environment, but that, on the other hand, he was both puzzled and disappointed by the fact that someone who was in his opinion "nicht ganz dumm" (not entirely stupid) should of his own free will decide to sink down to what he considered the lower realms of human achievement and activity.

He never missed an opportunity to rub this in. Once, when I paid a visit to Columbia University, it so happened that Pauli was to give a lecture there later in the afternoon and I was asked to introduce him. I gave a short survey of his many contributions to physics and finally mentioned the Pauli effect. Pauli thanked me for my introduction and added that he left it to the audience to decide whether it should be considered a Pauli effect when a theoretical physicist turned into a Herr Direktor.

Not very long before his death in 1958 Pauli paid one more visit to the Netherlands. It was after the death of Kramers, but I think Uhlenbeck was there as a visiting professor. In any case, one evening we were sitting together with a small group of friends, discussing old times. Someone asked me whether I had not had a hard time of it as Pauli's assistant, way back in 1932–1933. Pauli looked at me expectantly, wondering what I would say, so I had to make up some kind of story. "Not really," I said. "Pauli had managed to get a driving license and there was a kind of silent understanding between us that as long as I would not make any critical remarks about his driving he would refrain from caustic comments about my physics. Now I do not want to brag about my physics, but I do believe that in those days it was slightly better than Pauli's driving." Everybody laughed, but Pauli had the final word: "Yes, it may have been like that. But I don't drive anymore. And you, Herr Direktor, aren't doing physics anymore. *Die Sache stimmt noch immer*" (Things still match).

Of course, Pauli's standards were very high. I did publish a few papers, even in later years, but they were not of great importance. They dealt either with the elucidation of some specific point (Nernst's theorem, Bose-Einstein statistics) or with some mathematical problem in classical electrodynamics.

I mentioned already how Pauli reacted to a technological talk in the Zürich colloquium. Even more characteristic was a letter he wrote me on 11 October 1945. He was then still at Princeton, where he had spent the war years, and I had written him on 12 September asking him what were, in his opinion, the really important things that had happened in the United States and England during the war years. We would soon receive all the back numbers of the various journals, and it was important to know what to read.

In his answer Pauli explained that much work had been done on technical problems (microwaves, radar) and on heavy nuclei. Such work consisted mainly in the application of known principles and the intense preoccupation with these wartime problems had left little opportunity for really fundamental work, both in the United States and in England. So we should not worry too much about having "to catch up again": there were only a few things one should read and study. He himself was most grateful to the Institute of Advanced Studies, which had offered him hospitality and where he had quietly continued his theoretical work, but he felt he had not achieved very much. An outstanding event of the war years was, in his opinion, Onsager's solution of the order-disorder problem in a two-dimensional lattice, "a masterpiece of mathematical analysis."

Evidently it did not occur to Pauli that I might be interested in some of the details of radar; unlike many other theoreticians (Slater, Bethe, Schwinger, for instance) he was himself not in the least attracted by the intellectual challenge of technological problems.

Should one call this a blind spot in Pauli's vision? Was there also a slight affectation in his ostentatious disdain of technology? For instance, when he was later asked whether he would take part in the Geneva Conference on Peaceful Applications of Nuclear Energy, a conference of which Bhabha was president, he answered, "I'm not going to attend this uranium-peddler's convention" (Uranhändlertagung). Maybe, but let us be glad there existed and exist people like Pauli, who rate a profound and beautiful theory higher than any practical application. To put Onsager's theory of order-disorder in a two-dimensional lattice far above radar and nuclear chain-reactions is an extreme and most refreshing example of this attitude.

However, during my first years at Philips I did some work that even Pauli regarded as physics. It dealt with two entirely different subjects that had in common that they had no connection with the work I was supposed to do, although they were related to work done elsewhere in the laboratory. To explain the essence of these contributions without going into mathematical detail would be rather hard and I do not think they are sufficiently important to justify the effort. So a very short description will have to suffice. A first group of my papers—already begun at Leiden—dealt with application, extension, and clarification of Onsager's theory of irreversible phenomena, the same theory I mentioned already. The second group of papers—the work was partly done together with D. Polder—dealt with so-called Van der Waals forces, attractive forces between electrically neutral atoms or molecules, and was connected with work of Verwey and Overbeek on the stability of colloids. One of the results of this work, work which was later extended by Lifschitz and by many others, was the prediction of a universal attractive force between two metal plates at small distances—of the order of one-thousandth of a millimeter. Universal force—that is, independent of the properties of the metals as long as they are good conductors. This effect has been found experimentally and its generalizations have turned out to be of some theoretical significance. (It is sometimes referred to as the Casimir effect.)

I mentioned already that in 1951 both Pauli and I attended the Bothe Conference in Heidelberg. During an excursion by boat on the Neckar I explained to him my results on the Van der Waals forces and their relation to field fluctuations in empty space. He began by bluntly telling me it was all nonsense, but was obviously amused when I did not give in. Finally, after I had countered all his arguments, he agreed, and called me a real *Stehaufmanderl*—a toy known as a "tumbler" in English. (The translation is unsatisfactory: the word *tumbler* can mean some fourteen different things, *Stehaufmännchen* [or -*manderl* in Viennese vernacular] only one, defined in my dictionary as "a little doll with a sphere of lead instead of legs, which always gets back to the vertical position.") It is a pleasant memory. The beautiful scenery of the Neckar valley slid quietly past, while we were sitting on deck discussing physics, and Pauli repeated over and over "a true *Stehaufmanderl.*" It was the last time he did not call me "Herr Direktor."

My Later Years at Philips

I became co-director of the research laboratory in 1946 and member of the board of management, the highest executive body of the company in 1956. I stayed with Philips until my retirement in 1972, so I spent more than half of my professional life in an executive position in industry. I tried to maintain an intelligent interest in what was going on in the research laboratories and in the world of physics in general, but, as I explained above, my personal contributions to physics were no longer significant. On the other hand I came into contact with a far wider range of activities and a far more varied assortment of people than I had known before, and perhaps my experiences in industry and my views on the operations of big companies might interest some readers. I shall have to disappoint them, for I have decided not to write about my Philips years, at least not in the way I have been writing about my earlier years. For this I have a number of reasons.

First, the thirty-year delay imposed on the declassification of many official documents is also a reasonable one to observe when dealing with industry. I am not thinking so much in terms of real technical secrets. Technology develops and spreads fast enough to make procedures either obsolete or generally known within fifteen or twenty years. Moreover, the value of industrial secrecy is often grossly overrated. A chief engineer of Citroën, the French motorcar manufacturers, once explained to me the relative unimportance of industrial secrets in his branch. "But," I objected, "you would not like to have your competitors know what your next model will look like." "Ah," he said smiling, "that is different. That is not a technical secret; it is 'le secret du couturier.' " It is a useful distinction. The secret of the couturier is, of course, a typically French notion. For an amusing description I refer the reader to a book by Elisabeth Hawes.[5] One of the great couturières she mentions is Madame Vionnet. Her establishment on the Avenue Montaigne was later bought by Philips and became their Paris head office. According to information received from a French lady I met in the lobby of a hotel in Cape Town and who was representing Dior—almost a neighbor on the Avenue Montaigne—it was Madame Vionnet who introduced the bias-cut skirt into haute couture. It is a gratifying thought that the present Philips office was once the cradle of this daring innovation.

There exists a more important reason for a thirty-year delay. Such a delay amounts to about one generation. Angry young men will have turned into sedate citizens, young professors and directors will have retired, and aged tyrants will have died. One can tell stories without giving too much offense, without really hurting and, above all, without implying that things still are like that.

But even thirty years is not always enough. When you have been in a position of authority you should never write critically or facetiously about your colleagues and even less about those who were in some way your subordinates. You would betray their confidence if you did.

This even means that I should find it difficult to write the history of the scientific and technical work of the Philips Laboratory, for in order to do that well I should have to report failures as well as successes, opportunities grasped as well as missed chances, errors of judgment as well as wise decisions, and not only my own errors but those of my co-workers as well.

In a way it is a pity, for there would be plenty to tell. The Philips Research Laboratory was and still is a remarkable organization staffed by remarkable and talented people. Talented in their special branches of science or technology, talented also in many other directions. That became evident on special occasions, when the laboratory could rise to a high level of playful creativity that would have put to shame our earlier Copenhagen efforts. The performance on the occasion of my retirement was an impressive—and for me a deeply moving—example. But, once more, I cannot re-create the special ambiance of the research laboratory without going into details that I do not feel entitled to put on paper. So I have to confine myself to the rather flat statement that I am proud to have been at the head of this organization and that I am grateful to all the people with whom I have worked.

Yes, there would be plenty to tell. In a big and well-organized outfit like Philips you have to confront many ridiculous situations and you often have to deal with pretty eccentric people. And there were and still are with Philips quite a few men of sufficient stature to deserve being laughed at. Captains of industry and big businessmen often have a curious tendency to take themselves seriously and they even expect to be taken seriously by others. It might be healthy for them to be shown the relativity of that notion. That poking fun at someone does not necessarily mean a complete lack of respect is shown by the following conversation overheard in the cafeteria of the Nat. Lab. One of the senior physicists told that he had spent some time the day before in the restaurant of the railway station in Utrecht. It had been rush hour and the waiters were very busy. A funny thought had struck him: how would the top management of Philips bear themselves if they were waiters? He could see Mr. A walking faster and carrying more plates of soup at a time than any of the others; Mr. B, immensely correct but slightly supercilious; Mr. C with a kind word to little children, perhaps accompanied by a pat on the head, and so on (and here old Philips hands will certainly know who were the men he had in mind). But his story was interrupted by one of his colleagues. "You would have to be quick if you wanted to see them," he said. "All of them would soon have restaurants of their own. Then you might become a waiter there."

One of the advantages of a more facetious treatment might also be that it would help to dispel the idea that a large company proceeds with precise and cold calculation towards one goal: the maximization of profit. I do not deny the importance of the profit motive, but the policy of a firm is the result of innumerable small decisions, influences, accidents. Tolstoy's ideas on war apply in my opinion equally well, or rather better, to the behavior of a large company.

A man who would certainly be a rewarding subject for a candid biography was O.M.E. Loupart, senior vice-president. It was he more than anybody else who, after the Second World War, built up the worldwide network of the Philips sales organizations and who established a modus vivendi within the complicated matrix structure of national organizations and international industrial divisions. He came originally from the northeastern corner of Belgium, a region with a rather complicated political history, where Dutch, German, and French are contending languages. Loupart's Dutch had a slightly outlandish flavor and it certainly did not lack vigor. He liked to use metaphors in which he most effectively blended classical mythology and the simple facts of life. If others tried to emulate his style the result was displeasing; with Loupart even pretty coarse similes seemed somehow appropriate. In his younger years—later he became more mellow—he did not shun violent invective in his dealings with people inside and outside the Philips Company. I was told the following story.

During the thirties there was some collaboration, relating to manufacture and sales of medical apparatus, between Philips and a British firm. It did not go too smoothly: each partner felt the other side was not keeping its part of the bargain. A discussion took place at Eindhoven in which Loupart had to confront a polite, soft-spoken, public-school type of Englishman who was very noncommittal. When his reactions to concrete proposals were only "I am afraid I will have to refer that to our board of directors," Loupart finally got impatient and said, "And if I tell you that I think your board of directors is just one bloody bunch of scoundrels and impostors, what would you say then?" The Englishman was unruffled: "Well, Mr. Loupart, eh, if you would really say that, eh, I might answer almost anything."

I was myself present at a meeting where the first discussions with a Japanese firm were taking place. The idea—which was later carried out—was to start a joint enterprise, and Loupart in his usual flowery way explained that it would be like a marriage where one cannot say who is the boss, but "together we will make beautiful children." This was duly translated into Japanese. The Japanese company president looked very serious, thought for quite some time, and when, finally, his answer was translated it ran as follows: "He [the president] says that in Japan it is usually the husband who is the boss, but he understands your good intentions all the same."

Loupart, in meetings of the board of management and on other occasions, could also embark on lengthy expositions on matters of company organization, of structure of the market, etc. Judged by the standards of pure reason, such harangues were often either completely trivial or quite incoherent. I remember, for instance, a discussion on telecommunications in which one of our senior research people asked him a question. He embarked upon an "answer" that bore no relation whatsoever to what had been asked. "I suppose that answers your question" were the final words of his long and rambling talk. My colleague's reaction was admirable: "Well, in a way it does, and in some ways it does not; we'd better let it go at that." All the same, Loupart was a great man. He made me think of De Haas, whose physics had traits similar to Loupart's economics, and I have often wondered how a discussion between these two would have turned out.

The world of research was far removed from Loupart's own activities, but he treated the laboratory with understanding and even with a certain respect. "But, Mr. X, what manner of coat and shirt are you wearing?" he is said to have asked one of his younger staff members, "You look like one of those people from the Nat. Lab. But you know, those people have brains. In your case your appearance will have to do the job." And there was his famous defense of the research laboratory. "The laboratory is the womb from which all really new activities originate. And [warming to the subject and forgetting his previous metaphors] do not imagine that people are just sitting there and smoking big cigars. . . ."

However, I am lapsing into the very kind of anecdotal treatment I want to avoid. It seems to be my fate that I always remember insignificant details—and forget essentials. For instance, when David Sarnoff, the grand old man of RCA, visited Eindhoven shortly after

World War II, he was accompanied by a secretary, a well-bred young man, who never entered the discussions of his own accord and who, when asked for his opinion, would confine himself to saying, "I think the General is right" (Sarnoff had the rank of Brigadier General). So far so good, but why should I remember that this young man wore a green silk tie on which was displayed an arrangement of little white terriers? I shall henceforth try to be more serious, but I cannot promise that there will not be a little white terrier here or there.

My Later Years Outside Philips

Since in the remainder of this book I shall be far less concerned with people, with their offbeat talents and their foibles, than in my earlier chapters, I do not feel inclined to say much about myself either. It is true that the arguments I put forward against telling funny stories about colleagues and assistants would not necessarily apply, but to present myself as the only amusing—or at least from time to time ridiculous—person in the whole setup goes against my sense of balance. There is more to it. During my years at Copenhagen, Zürich, and Leiden, and even during my first years at Eindhoven, physics was an intrinsic part of my whole life. Being a manager and an executive in a large industrial firm leads to a more pronounced separation between work and home, and one's family is far less involved in one's actual work than at a university. Sometimes, sitting in my Philips office, I would wistfully remember the helium days at Leiden, when my wife would drop in, perhaps with one of the children, to bring me something to eat and to ask how things were going. When in 1956 we decided to move into the countryside, where we found—and gradually improved and expanded—a somewhat ramshackle, but pleasantly crazy house in beautiful grounds, this split became even more pronounced. My youngest daughter once put it very nicely: "When I am asked about your profession, I like to say you are a village carpenter. Of course you are not very good at carpenting, but at least I see you do it. I have no idea about what you are doing at Philips."

I mentioned already that my two oldest children were born at Leiden, a girl in 1936, a boy in 1938, and that a second girl was born at Eindhoven in 1943. Two more girls were born after the war, in 1946 and 1950. When we moved into our new home my oldest girl had already left home to study biology at Groningen University and my boy went to Amsterdam to study economics soon after. I am afraid I was not a very satisfactory father. I was frequently away from home, kept long and irregular hours, and was often tired and irritated by the time I came home. When one day I had nothing much to do and came home at 5:00 P.M., this was so exceptional that our maid-of-all-work, a girl from a neighboring farm, eyed me suspiciously and asked, "What's the matter, got the sack?"

However, I tried to make up for my shortcomings during weekends and especially during holidays. Speaking about holidays, during my early years at Eindhoven I once mentioned to Frits Philips—Anton's son, who was then vice-president of the company; later he was for many years president—that I had been very busy and had not got around to going away on a vacation at all. He looked at me sternly and said, "And you even seem to be proud of that. Probably you are doing your work badly, and in any case you will be insupportable at home. Don't let me hear such nonsense again." Mr Philips's comments were often both blunt and helpful. Another example: during a reception, Bernhard, the prince consort, said to my wife, "I suppose such a professor is pretty difficult to handle at home." She replied very sweetly, "Not really, Your Highness; I find things rather easy." At that moment my sleeve was vigorously pulled. Frits Philips had overheard the conversation and whispered in my ear, "Better buy her some nice flowers, for I don't believe a word." I am afraid he was right.

In any case, my children grew up, not without some strains and conflicts, but without major tragedies or irreparable damage, and each of them has found a place—and a companion—in life. It remains a moot question to what extent intelligence is inherited or mainly a matter of early surroundings—I discussed that in Chapter 4. But whether by heredity or by circumstance, each of my children learned easily. In the long run that may be of minor importance, but it made school years considerably less taxing, both for the children themselves and for their parents.

Now I could write at length about my attempts at being a "country gentleman." About my occasional successes and frequent failures. About dogs and Icelandic ponies, about building a greenhouse and a sheepcot and so on. But although such things were and are welcome diversions, I don't think they are worth writing about in the context of this book. The point is that although I like such activities, I am not very good at any of them. As a craftsman, as a handyman, as a horseman, as a forester, I belong decidedly to the lower brackets, my theoretical background notwithstanding. When for instance I tried in vain to get a lawn mower going, my thorough knowledge of electromagnetic theory and my less thorough but yet adequate knowledge of hydro- and aerodynamics proved of no avail. A sixteen-year-old classmate of my daughter's fixed it within ten minutes. Such sobering experiences were presumably a healthy check to my ego. Their details will be of no interest to my readers.

In 1972 I retired from Philips. During the ten years that followed I have been engaged in numerous activities. I was president of the European Physical Society and president of the Royal Netherlands Academy of Arts and Sciences, I was a member of several committees and gave many talks at universities, at international meetings, and so on. But gradually I am getting more time for just staying at home. By now we have nine grandchildren, and when some of them come to stay with us—which they all enjoy doing—I can give them the attention they are entitled to. It is one of the consolations of aging.

One day we were speaking about collecting things, for, like most children, my grandchildren have various collections: stamps, coins, shells, and so on. When they questioned me I shamefacedly had to admit that I had not at any time built up any collection worth mentioning. But one of my granddaughters—she must have been about nine years old—took pity on me. "But you have, grandaddy," she said, and pointing to my fairly bulky skull she continued, "You have collected a lot of curious things in that little head of yours." I hope she was right and that part of my collection has found its way into my book.

9

Industry and Science After
the Second World War

I have explained why I do not want to write the history of the Philips Laboratory. Perhaps someday the work of Mr. Garrett (mentioned on page 347) will be taken up again, completed, and published. For the time being I can only refer readers who are interested in that history to publications in the scientific literature and to the Philips Technical Review,[1] which at an intermediate level surveys results of the laboratory. My aim in this chapter is a different one.

Since the Second World War there has been enormous progress in our knowledge and understanding of nature. There has also been an equally spectacular advance in our technical capabilities. I want to show that this technological advance was mainly based on prewar fundamental knowledge, but that postwar advance in knowledge would not have been possible without postwar technology. Substantiating this thesis requires some technicality and the remarks made in the introduction to Chapter 2 apply also to the present chapter.*

I shall first discuss these relations in a general and necessarily rather superficial manner. After that I shall give a few case histories from the Philips Laboratory that illustrate my main thesis. In a way, my own career is also such an illustration. For whatever my shortcomings as a director of research, they were not related to my admitted failure to keep abreast of later developments in fundamental

*There is some duplication between this chapter and Appendix B, which describes the situation around 1930. That is understandable: some of the trends originating around 1930 continue at unabated speed right up to the present.

physics. As far as fundamental physics was concerned my knowledge was still adequate.

Generalities

For after all, what is man in the material world? A nothing compared with the infinite, an everything compared with nothing, a middle between all and nought.[2]

These words of Pascal are as appropriate today as when they were written, perhaps even more so. Or at least they can be given a more concrete content. In the century of Pascal the range of astronomical observation began to expand: in 1610 Galileo published his *Sidereus Nuncius* describing the wonderful things that can be seen by means of a telescope. Hooke's *Micrographia*, which can be said similarly to have opened up the field of microscopy, was published in 1665.*

The life of Pascal (1623–1662) falls between these two publications that point towards the large and towards the small. Our earth is part of the solar system, the solar system part of our galaxy, myriads of galaxies constitute the universe. The microscope reveals cells and bacteria, the electron microscope surprising details of their structure. Atoms are ten or twenty thousand times smaller than bacteria; I have discussed their reality in Chapter 2. Ten thousand times smaller than the atom is the atomic nucleus. And beyond that we have discovered a whole new world of fields and ephemeral particles.

Different fields of force prevail in the different areas. The motion of planets and satellites is governed by gravity, by the universal gravitational attraction between heavy bodies. This gravitation appears in a new light through Einstein's general theory of relativity, which also accounts for the possibility of curious objects, such as black holes, and for the existence of a finite expanding universe. Electromagnetic forces hold the atom together, bind atom to atom in molecules, and provide cohesion in solids. We, human beings, are living on the intersection of gravity and electromagnetism. Gravity prevents the atmosphere from escaping into outer space and keeps our feet on the ground; the ground itself derives its strength from electromagnetic interaction.

*In the long run Antoni Leeuwenhoek (1632–1723) contributed more to microscopy than Hooke, but the first of his long series of letters to the Royal Society was written in 1673.

Inside the nucleus entirely new forces are at work—the strong nuclear forces, but electric forces are by no means irrelevant. In lighter nuclei the nuclear forces prevail but in uranium the supremacy of nuclear attraction over electric repulsion is marginal. Injection of one neutron is sufficient to make the nuclear bonds snap and electrostatic repulsion drives the fission fragments apart with an enormous energy. There exists another type of nuclear force: the weak nuclear force that is responsible for a non-electromagnetic interaction between electrons and nuclei, such as is revealed in beta decay, and that also plays a role in particle physics. It has been possible to establish a formal theory that unites electromagnetic force and weak nuclear forces, but so far there is no unification with gravity and with strong nuclear forces. From that point of view our picture of the material world is still very incomplete.

On the basis of the foregoing, the following classification should be understandable:

Astrophysics and cosmology.
Physics of gases, liquids, and solids.
Physics of molecules and atoms.
Nuclear physics.
Particle physics.

Let us now have a closer look at theory. I have explained in a former chapter how the notion of quanta of energy, first introduced by Planck in connection with thermal radiation, successfully applied by Einstein to photo effect and to specific heat of solids, was further developed by Bohr and others in connection with the study of atomic structure and atomic spectra. This development resulted finally in the formulation of quantum mechanics. Then new quantum mechanics began to extend its range of applications: towards larger systems, molecules, condensed matter; and towards smaller ones, the atomic nuclei. There are two important features that distinguish "traditional" quantum mechanics—that is, quantum mechanics applied to atoms, molecules, and condensed matter—from quantum mechanics applied to nuclei. First, in traditional quantum mechanics, particles are conserved; electrons stay electrons; and even nuclei can be regarded as stable particles, with a mass, with an electrical charge, possibly not quite symmetrically distributed, and with a magnetic moment; one has not to be concerned with their transformation by radioactive decay. Second, interactions are known: the only field of force that has to be taken into account is the electromagnetic field. It can manifest itself in several forms: as simple electrostatic interaction, as "exchange interaction," as Van der Waals forces. Physicists are convinced—and I think with good reason—that quantum mechanics and electromagnetism contain in principle the whole of chemistry. They also know that even with the largest computers one can calculate the properties of only the simplest molecules with satisfactory precision. Yet a sufficiently arrogant physicist might maintain that experimental chemistry can be regarded as running analog computers for solving complicated many-particle Schrödinger equations. Does a similar statement hold true for biology? Personally, I do not think so, but here opinions differ.

In nuclear physics the situation is different. We do not know the forces a priori, they have to be deduced from comparison of the results of calculations with experimental data. Moreover, particles do not maintain their character and the number of electrons is not constant. A neutron can change into a proton with emission of an electron, a proton can change into a neutron with emission of a positive electron. In these processes there is also an emission of (almost undetectable) neutrinos and antineutrinos. A positive electron will sooner or later unite with a negative electron outside the nucleus and disappear into gamma (that is electromagnetic) rays. And finally, particles in the nucleus are very densely packed, and methods of approximation that were successful in the theory of atoms and molecules are not applicable. In view of all this it is not surprising that, great progress notwithstanding, an accurate calculation of the binding energies and of the energy levels of complicated nuclei is not (yet) possible. So there is an analogy between complex nuclei and complex molecules, but there is a fundamental difference too: for molecules the first principles from which we cannot calculate their properties are far better known. High-energy physics reveals an entirely new world and to tackle it, theory needs new concepts far beyond those of traditional quantum mechanics.

Extremes meet. Whereas much in astronomy and cosmology is determined by gravity, the stars themselves can be regarded as huge fusion reactors, and in speculations about the origin of the universe the latest ideas of particle physics come into play.[3]

More generally, although the subdivision of the whole of physics into areas is meaningful, although few physicists will be specialists in all these areas, and although unification has not been reached of all the different fields of force, one should not overemphasize differences and thereby forget relations and analogies. Not only is the general methodology of physics essentially the same in all areas, there are also many mathematical procedures and results that can be transferred from one area to the other. The mathematics developed for dealing with mechanical (and acoustical) vibrations could be applied in Schrödinger's wave mechanics; the skills of people trained in wave mechanics were applied to wave guides and resonant cavities during the Second World War. Special solutions of nonlinear equations—solitons—may come to play an important role in many different branches of physics, and so on. Analogies between mechanical or hydrodynamical phenomena and phenomena in particle physics may have considerable heuristic or educational value. I give an example of an analogy between certain aspects of solid-state physics and the theory of positrons in Appendix B.

However, if one gets away from basic principles and methodology and gets down to actual details, then with increasingly detailed knowledge the distance between areas may seem to grow. And now I am coming to an essential point of my argument. Technology is interested in precise details. Not only in the basic explanation of ferromagnetism, but in the properties of specific magnetic materials, to give just one example. For the technologist the separation of the different areas of physics is far more obvious and expedient than for the philosopher. The areas can be distinguished by their applicability. Have recent developments in astronomy and cosmology had any direct influence on technology? Certainly not. Even space travel requires nothing from astronomy beyond classical and basically simple knowledge of the movements of planets and their satellites. (These, it must be granted, it has to know with very great precision.) By the way, the name *cosmonaut* for today's space travelers seems to me highly inappropriate. I do not underestimate the skill, the stamina, and the courage of these people, but from an astronomical point of view their excursions are rather in the nature of a roadside picnic. I could, with better justification, call my youngest granddaughter a globetrotter when she has walked to the end of my garden: she has walked then more than one millionth of the circumference of the earth; the moon traveler has covered less than one millionth of a millionth of the galaxy, which itself is only a minute fraction of the cosmos. Anyway, space travel, observation satellites, communication satellites, and so on do not need the results of advanced modern astronomy. There is no indication that this situation is going to change in the foreseeable future.

Something similar can be said about high-energy physics. Opinions may differ about the year of its birth. Should one say 1932, when the positive electron was discovered? Or 1935, when Yukawa predicted a "heavy electron"? In the same year particles roughly one hundred times as heavy as an electron were found in cosmic-ray research, but they were not the particles predicted, for their interaction with nuclei was very weak. So one should perhaps say the real beginning was after the Second World War, when particles answering more closely to Yukawa's ideas were found—in 1947—and when the big accelerators in Berkeley, in Brookhaven, and elsewhere got into operation. Anyway, since that time this field of research has shown enormous progress, but no technical applications are in sight.

Only the physics of atoms, molecules, gases, and condensed matter, and nuclear physics are of direct technological importance.

In Chapter 5 I mentioned the rise of nuclear physics in the thirties, but I did not really take part in that myself. As early as the thirties the Philips Laboratory became interested in nuclear physics: accelerators were built, some new reactions were discovered, and Philips even acquired the European rights and a non-exclusive U.S. license under the Fermi patents, but that happened before I joined Philips.

Application of nuclear physics are of three kinds: nuclear weapons ("atom bombs"), nuclear energy, and radio-isotopes used in medical therapy and diagnostics and as general analytical tools.

The existence of nuclear weapons has become a decisive factor in international politics, though their use has so far been limited to the two bombs that destroyed Hiroshima and Nagasaki. The hope that their use will not go beyond that is the only hope for survival of our Western civilization. The "secret of the atom bomb" has become a secret of Polichinelle as far as general principles are concerned, but details are strictly classified. Therefore the impact of nuclear-weapons technology on other industries has been small.

Nuclear energy has, so far, given rise to discussions, protests, actions, that seem somewhat out of proportion with its modest contribution to the production of electric power. But the opponents may argue that by protesting against an as yet fairly harmless situation they want to prevent a large-scale application they consider undesirable. On the other hand there are quite a few physicists and others who believe that nuclear energy, either from breeder reactors or from fusion reactors, will provide the long-term answer to the energy problems of the world. Opponents consider this approach far too dangerous.

Radioisotopes are useful and interesting. They are not a major factor in industry.

In short, nuclear *technology* is a vitally important factor in society not because of its actual applications, but because of its potentialities. It is a development of World War II and after. Nuclear *physics* began before 1900 and burst into full bloom in the thirties. The essential nuclear processes underlying nuclear technology were established before 1940. Certainly, nuclear technology did need more precise data on these processes, but it uses only a minor fraction of the enormous amount of data on energy levels, on radiations, on decay, of a thousand different nuclei—data that since World War II have been assiduously assembled by experimenters using ever more refined and powerful apparatus and that have been partly explained by theorists, using ever more powerful mathematical concepts and ever larger computers.

All other technological advances depend on phenomena that are within the scope of what I called "traditional" quantum mechanics, and of macroscopic theory, which is its limiting case. As I explained in former chapters, by the early thirties the basic principles of quantum mechanics and its applications were well established. The most fundamental post–World War II extension of theory that is relevant to our "traditional" domain was the creation of quantum electrodynamics—of so-called renormalization theory—in the late forties and early fifties.[4] This theory leads to small, but accurately measurable, corrections to energy levels and to the magnetic moment of the electron, but these corrections are up to the present of no importance for any technical application I can think of. Otherwise, later developments are a further elaboration of principles known before, or the mathematical solution of problems formulated on the basis of existing theory.

Lest I should be considered an old man dreaming about the past and unaware of what happened in later years, let me at once add that I know very well that these later elaborations far surpass the early work in volume, in experimental refinement, and in mathematical sophistication. A few examples:

In 1907 Einstein based his theory of the specific heat of solids on the assumption that all atoms in a crystal vibrate with the same frequency. Debye, in 1911, made a bold guess for a frequency spectrum. Today, thanks to a combination of theoretical and experimental methods, we know these vibration spectra quite well.

Bloch showed that the energy levels of electrons in metals must be arranged in bands and Brillouin established some basic notions about the structure of such bands. Today, again thanks to a combination of theoretical and experimental work, we know the structure of energy bands in considerable detail. We also know the limitations of this approximate theoretical approach.

In the thirties one was convinced that the explanation of superconductivity should be possible on the basis of known principles. That proved to be true, but one had to wait for the solution until 1957; it involved rather new and surprising ways of dealing with many-electron systems. Later this theory suggested a series of new and interesting experiments.

The invention of the laser has led to renewed interest in and considerable refinement of the theory of optical spectra.

The problem of phase transitions—like melting of solids, condensation of vapors—already tackled by Van der Waals in the eighteen-seventies, has taxed the ingenuity of mathematical physicists to its very limits.

What about classical non-quantal theory? Even here foundations have not really changed; even here there have been spectacular advances. It should be emphasized that computers have greatly enhanced the power of mathematics.

Progress in experimental physics was closely coupled to theory. (That is a difference from earlier times when experimental discovery often went far ahead of theory; witness for instance the discovery of superconductivity. To a certain extent, this situation still prevails in astronomy and in particle physics.) The invention of the laser was preceded by its theoretical prediction, a prediction based on ideas formulated by Einstein as early as 1917. Something similar holds for nuclear spin resonance. As to the Josephson effect—or perhaps one should say Josephson effects, the effects connected with the tunneling of a persisting current through a thin layer of material that in bulk is not superconducting—let me quote Josephson himself.

Experimental investigations during the past couple of years into the behaviour of tunnelling supercurrents have resulted in the observation of most of the properties predicted as well as others which were unexpected at the time of discovery.[5]

I should add that Josephson's arguments were at first not universally accepted.

Of all the discoveries in solid-state physics, the observation that positive holes can be injected into n-type germanium and survive there for quite some time had the farthest-reaching technical consequences. It might have been predicted; to the best of my knowledge it was not.*

I am afraid that the foregoing enumeration will not be very helpful to a reader unacquainted with the subjects I mention. I can only hope that even if what I have written is to him no more than a list of vaguely familiar terms, he will yet believe me that there has been and still is enormous research activity.

There is another feature I should like to get across. That is the highly professional character of most of that research. Take whatever topic you like and you will nearly always find that much has already been done, both theoretically and experimentally, and that you have to make a thorough study before you can hope to do something new. Moreover, experiments are getting more and more subtle and the chances that you can do important work with simple apparatus that you can build in a few days with your own hands are becoming slight. In my younger years, when I was fairly active in research, I was in a way an amateur—in the sense that I did things just for the fun of it, but also in the sense that I could move from subject to subject, could "sip every flower and change every hour," as Blackett, quoting the *Beggar's Opera*, once said of me at an anniversary dinner of the Royal Society.[6] That way of non-planning is getting more and more difficult.

Let us now look at the relations of science and technology in a

*The notion that an unoccupied place in an energy band, a "hole," behaves as a positive particle became current around 1930. Compare Appendix B.

different way: let us take technology as our starting point. I find it useful to distinguish four branches of technology, to wit:

1. *Energy technology.* This comprises not only power plants and prime movers but all engines and mechanical contraptions where power is an essential factor: motorcars, locomotives, airplanes and rockets, harvesters, dragline excavators, and bulldozers.
2. *Information technology.* Telecommunications, radio, and television; microdocumentation, phonograph records, tape-recording; calculators and computers.
3. *Technology of materials and chemicals.* New metallic alloys, ceramics, plastics, and artificial fibers; medical drugs, agricultural chemicals, paints, etc.
4. *Medical and biological technology.* New procedures in diagnostics and therapeutics; genetics applied in plant breeding and in animal husbandry, genetic "manipulation"; biological methods of pest control.

Of course, these "branches" are not mutually exclusive. Take agriculture. Tractors, harvesters, and other machinery are clearly energy technology, but artificial fertilizers, pesticides, and herbicides belong in group 3. There is even an information aspect: farmers have to be instructed and to be warned so that they can anticipate frosts, invasions of parasites, and so on. But the essence of farming is biological technology. Airplanes belong to energy technology, but in building an aircraft, choice of materials and rigorous control of their properties is essential. Modern aviation would be impossible without radio and radar, and the extensive instrumentation also belongs to information technology. Even medical technology is important: the psychological and physiological reactions of crew and passengers to the curious conditions created by modern aviation have to be taken into account.

It is obvious that in our century, and especially since the Second World War, there have been striking advances in every branch of technology. I shall not enter into that at present. Here we are concerned with the question of to what extent this progress was connected with research. I have discussed the relation between nuclear physics and nuclear energy. All other branches of energy technology do not involve recent advances in the principles of physical research.

They have to use mechanics, hydro- and aerodynamics, and thermo-dynamics, disciplines whose fundamental principles were established quite some time ago. This does not at all imply that work in this sector is therefore less "scientific" and certainly not that it is less difficult than work in other fields. On the contrary, it is for instance well known that aerodynamics is a very difficult subject. And the existence of large computers notwithstanding, a complete a priori calculation of the behavior of an aircraft is still impossible: one does need wind tunnels.

Until the arrival of the transistor, information technology was mainly based on classical electrodynamics and on classical mechanics; that was sufficient to describe the behavior of electrical circuits and the motion of electrons in electron tubes. Modern ideas on the structure of matter did not have a decisive influence. But again it is clear that the "branches" I defined cannot be strictly separated; this technology needed special materials—insulators, for instance, and magnetic material for transformers or dielectrics for capacitors.

Vacuum tubes put special demands on materials. A 1936 book by Espe and Knoll gives an impression of the wide range of materials that were then in use in the manufacture of vacuum tubes.[7] However, most of the impressive amount of data presented is purely empirical and uninfluenced by the modern theory of solids. This situation began to change in the late thirties and early forties. Work on magnetic materials did at least in part take notice of theoretical developments. The invention of the transistor was even more closely linked to the quantum-mechanical theory of the solid state.* It was then that quantum mechanics, created two decades earlier, began to play a decisive role in technology. The further development led to integrated circuits. In that technique a number of transistors and other components, like rectifiers, capacitors, resistors, and their con-

*Here a digression on nomenclature may be useful. I use *semiconductor* for "a substance whose electric conductivity is between that of a metal and an insulator, especially germanium and silicon," and *transistor* for "a small electronic semi-conductor device performing functions similar to those of a vacuum tube." I take these excellent definitions from the Random House Dictionary. Unfortunately the word *transistor* is also used as a *pars pro toto*—synecdochically if you prefer Greek to Latin—for "a small radio that uses transistors," and this deplorable usage has led quite a few people, and especially the commercial people at Philips, to call a transistor a semiconductor, a most objectionable habit. I also disapprove of the term *chip* for an integrated circuit. The same dictionary defines, "a tiny square piece of thin semiconducting material on which an integrated circuit is formed." All right, but chip has many meanings, and I do not like to call a beautiful painting a canvas either.

nections, are made simultaneously on one small thin slice of silicon. This has inaugurated a new era of electronics. It can be characterized —with only a little exaggeration—by saying that any electronic circuit, however complicated, that can be designed can also be made reliably and cheaply provided it is manufactured and sold in sufficient numbers to recover the cost of design and of making the masks required for its manufacture. We see the consequences of this all around us.

About materials and chemical compounds I have little to say. Chemistry certainly uses methods of preparation and analytical tools that stem from physics. Of course it could not proceed without knowledge of chemical elements, of valences, of chemical thermodynamics, of reaction kinetics, of physical chemistry in general; by now the attitude of the director of a well-known Dutch ceramics factory who is rumored to have said to a colleague he met at a convention of a chemical society that he did not really believe in atoms is almost a thing of the past. But the reduction of chemistry to quantum mechanics that is possible in principle has so far had little influence on the practice of chemistry. Neither is much known about the relation between the structure of complicated molecules and their biological action. The search for drugs with specific biological action is still largely an experimental craft. So is the search for materials with specified properties, although—as pointed out above—in looking for materials meeting the requirements of electro-technical industries modern theoretical considerations do supplement the time-honored method of trial and error. About medical and biological technology I cannot speak with any authority.

Let us again change our point of view and look at the technology of research. There are different levels of experimentation. There is the string-and-sealing-wax approach. An experimenter can improvise apparatus, using simple, easily obtainable materials and components, he can make gold-leaf electrometers—much of the early work on radioactivity was done with those—with gold leaf obtained from a bookbinder's and perhaps using a tobacco tin as a container. He may hunt through a five-and-ten looking for suitable material. Otto Frisch relates that he bought womens' black underwear to line the walls of a cloud chamber; I remember reading about someone using pie pans as electrodes for a high-voltage generator; and once I bought myself a couple of wooden hoops I found in a toyshop and which I used to make a coil for compensating the magnetic field of the earth. Most childrens' games have survived rather well, but propelling a hoop with a stick has fallen victim to modern traffic, at least in my country. Of course plastic hoops knew a brief revival as hula hoops. I was, for many years, a member of the board of trustees of the Technical University in Delft. One day the prince consort, Bernhard, paid us a visit and it so happened that I had to accompany him in the car that was taking him around. The hula-hoop craze was then at its height, and some girls were practicing on the sidewalk, rapidly swinging the hoop around their waist, letting it go down to their knees, bringing it up again, and so on. Prince Bernhard eyed them wistfully, then asked me, "Professor, can you do that?" I answered truthfully "No, Your Highness, I have tried, but I cannot." "I cannot do it either, but my daughters seem to have no difficulty at all." "Yes, Your Highness, same thing with my daughters." "Fortunately my wife cannot do it either." I am not a convinced royalist, but I have the greatest respect and sympathy for our three successive queens, and the idea of Queen Juliana trying her luck with a hula hoop almost smacked of lese majesty. I might have added, "My wife is quite good at it," for she was, but I kept my silence; which goes to show that I am not entirely without talents as a courtier!

Even in as well-equipped laboratory as the Nat. Lab., toys sometimes came in handy. We once had to irradiate a number of samples with a high-voltage electron beam. The cheapest and by far the quickest way of arranging the experiment was to buy an electric model freight train. It entered and left the high-voltage room through real tunnels—that is, through holes in the wall. The samples were put on numbered open freight cars and the train—controlled from behind an observation window—made a number of stops so that each sample was irradiated for an appointed time, after which it returned to the station where it was unloaded.

The next level is that of the skilled artisan, of the glassblower and the mechanic. Some experimental physicists were highly expert themselves, and this still holds true today, but there have been many experimenters who were outstanding at devising experiments, at designing apparatus, and at interpreting results but who were clumsy with their hands. It is said, for instance, that when J.J. Thomson's technician came to the critical stage in the assembly of a piece of apparatus, one of his colleagues would try to draw the professor away, lest he should try to carry out the operation himself.

Heinrich Hertz, on the other hand, seems to have been a good handyman. According to his biographers, he spent quite some time as a boy at cabinetmaking and at turning. When the craftsman who had taught him learned about his appointment to a professorial chair, he exclaimed, "What a pity! That boy had the makings of a first-class woodturner."

In any case, laboratory workshops and their staffs have played an important role in the development of physics. Closely related to this are the many small firms, staffed by craftsmen, who made and still make instruments for the market.

The third level is that of industrially produced equipment that requires production facilities surpassing those of the artisan's workshop. Scientific instrumentation has become a big business, and computers have also become indispensable research tools.

One can add a fourth level, where the experimenter requires a complicated system, the parts of which are far too big for him to make in his own shops, but where the system as a whole is so large and so unique that no commercial firm would be willing to deliver it as a turnkey project. The installations of Kamerlingh Onnes—he bought pumps and compressors—were an early example. Today large particle accelerators and astronomical observatories are certainly in this class.

A reasonably complete list of scientific instruments now in use would fill a book of many pages. What follows is only an indication of the various price categories.

Particle (high-energy) physicists and astronomers, especially radio astronomers—again they meet—use the most expensive apparatus. Large particle accelerators may cost up to a thousand million dollars, and large aerials for radio astronomy cost at least a hundred million dollars. Astronomical satellites are not cheap either—even if one discounts the costs of the launcher. The development of these rockets has of course required huge amounts, but they were not paid out of science budgets.

Cyclotrons and large Van de Graaffs are in the ten-to-hundred-million-dollar range. So are large computer systems. It is curious how terminology changes. When Cockcroft and Walton accelerated protons with roughly half a million volts, it was certainly considered high voltage. Today someone working with particles accelerated to ten or

twenty million volts is considered to be engaged in low-energy nuclear physics. In other walks of life words are often devalued—compare Houtermans's ideas on "gentlemen"—but in physics and technology they are continually upgraded. A pressure of one millionth of
a millimeter of mercury was fifty years ago regarded as very high
vacuum: one would not call it that today. One degree absolute is no
longer a very low temperature, one part in ten thousand is no longer
high precision in electrical measurements.

In the hundred-thousand-dollar range we find electron microscopes, mass spectrometers, elaborate recording spectrometers, and
so on. There are many instruments between ten and one hundred
thousand dollars: X-ray equipment, spectrometers, and so on. Simpler electronic instruments such as oscilloscopes are between one
and ten thousand dollars. Below that there is a vast choice of routine
instruments, made in large numbers for technical applications, ser-
vicing, testing. Small programmable computers and calculators may
cost a hundred dollars or less. And finally we get down to the indispensable tools of the theoretician: paper, fountain pen, pencil, and
ballpoint. Although I consider ballpoints unsuitable for serious writing, they have the unmistakable advantage that you can use them to
write and draw on paper napkins in restaurants.

I am afraid that my very sketchy treatment will be insufficient to
convince all my readers that modern research really needs modern
technology. I shall add a few details later on, but only a few.

For I feel that visiting one or two well-equipped research
laboratories, looking at their equipment, and obtaining a rough indication of its costs will be far more convincing than any amount of
writing I could do.

Case Histories

I shall discuss the following subjects:

1. Thermionic cathodes.
2. Magnetic materials.
3. Image transformers and camera tubes.
4. Scientific instruments.

These case histories are typical for an industrial research laboratory
and illustrate the general principles I want to set forth.

In all four cases the work was a technical success and led to new industrial products. From the point of view of scientific progress the results were of an average level. Our work on magnetic materials certainly contributed to a better understanding of what happens in such materials, but I could have chosen other investigations, for instance on the stability of colloids, that contributed far more to basic science, although they had a smaller technical impact.

The cases I discuss are also average in another respect. The work could not have been done without the knowledge provided by academic science, but this had to be supplemented by trial-and-error methods. I could have given examples where theoretical prediction played a more dominant role.

I shall not describe extensively where and how this work depended on preceding scientific results. Let me only point out that one cannot think about electron tubes without knowing that electrons exist, that one needs vacuum, that one cannot start designing image intensifiers without the basic notions about light quanta and photoelectricity. I leave it to the reader to trace such lines in more detail.

I want to make it clear that I am not describing my own work and therefore I have to mention a few names. But I had many co-workers of equal merit I don't mention by name. H.J. Lipkin, in a useful textbook,[8] states in the introduction to his bibliography that, in the text, he has not given any references to original publications and that he has tried not to mention any names. In some cases, however—for instance in the case of Casimir operators—it was inconvenient not to use the name of an author associated with a frequently occurring mathematical notion and he apologizes to those authors. Perhaps I should follow his example and apologize to the few I do mention rather than to the many I do not.

CATHODES

In an electronic tube, whether rectifier ("valve"), amplifier, transmitting tube, or cathode-ray tube, electrons move in vacuum under the influence of electric fields originating from their fellow electrons (space charge) and from electrodes, or of magnetic fields originating from magnets or coils. But how does one get these electrons? Inside a metal, electrons are more or less free to move about but they are held in the metal by an attractive potential. How can we get them out? There are a number of ways. One can apply a very strong external electric field and pull them out. This is really possible and the phenomenon is known as cold emission; it has been applied in a few special tubes but it is of limited applicability. In other cases, cold emission is rather a nuisance: it may lead to electric discharges in high-voltages tubes (for instance, X-ray tubes) in places where we do not want any electric discharge.

We can also shine light on a metal and if the wavelength is sufficiently short (for every metal there exists an upper limit), electrons will come out. The mechanism is roughly that an electron absorbs a light quantum and thereby gains sufficient kinetic energy to jump out of the metal. This photoelectric effect is important for converting light into an electric current and is applied in many measuring and control devices, but it is not a practical way of getting electrons into a radio tube.

If fast electrons impinge on a metal, they will impart energy to the electrons inside that metal and some of these will be thrown out. This is known as secondary emission. In favorable cases one impinging electron will lead to more than one emitted electron. That is known as electron multiplication and some important measuring devices are based on this. Again, it is not a practical method for populating the vacuum of a radio tube with electrons.

Electrons are also emitted when an ionized atom—an atom robbed of one electron, for instance—impinges on the metal. That is what happens in gas discharges and how electrons were discovered, but since I am talking about vacuum tubes that does not concern us.

There remains thermionic emission. If we heat the metal, the electrons will get a higher energy; both the average energy and the energy spread will increase and a fraction of electrons will have sufficient energy to leave the metal. That is the method of getting free electrons commonly used in radio tubes.

The fact that a current through vacuum can be drawn from incandescent wires was known to Edison. The phenomenon was investigated in greater detail by O.W. Richardson (1879–1959) and by I. Langmuir (1881–1957), both experimentally and theoretically. New quantum mechanics led to some quantitative changes but the general features of the theory remained the same.

A tungsten wire heated to a very high temperature is a serviceable cathode. It can be used, for example, in electron microscopes, but the high temperature makes for short life and has other disadvantages too. Now the energy an electron needs to get out of a metal is not the same for all metals and can be changed by surface layers. That can qualitatively and even semiquantitatively be understood on the basis of fairly simple theory. This combined with trial and error led in the twenties and thirties to the development of highly efficient oxide cathodes. Such a cathode may have the form of a short length of nickel tube that can be heated by an internal spiral (the field of the heating current is therefore completely screened off and does not influence in any way the behavior of the electrons). A layer of a paste of barium carbonate, usually mixed with strontium carbonate, is applied. While the electron tube is evacuated the cathode is heated and the carbonates decompose. There remains a layer of barium (and strontium) oxide. The oxide is then, during an activation process, partly decomposed by electrolysis, and the final result is nickel covered with a porous fine-grained layer of barium and strontium oxide, the grains being at least partly covered by a thin film of barium. We meet here with a situation that is quite common in industry. Although the basic principles are clear the behavior of the cathode depends on minute details of grain structure, surface structure, quantity of barium on the surface of grains, quantity of barium in the barium oxide crystals—it will contribute to the conductivity of the grains—and so on. Procedures are gradually optimized by trial and error and finally there results a manufacturing discipline that is rigorously observed although it is quite possible that part of it is "superstition." It is not surprising that from time to time there would be a "cathode crisis" in the factory and sometimes the research laboratory was called in to help locate the trouble. On one occasion it was found, for instance, that minute quantities of chlorine can "poison" the cathodes.

An electron of sufficient energy to leave the metal or the oxide grain will hover near the surface and may be trapped again unless it is drawn away by an electric field. The saturation current is reached when all electrons that are leaving the solid are carried away. For traditional radio tubes the current is always far below the saturation current, and if one tries to draw the saturation current, the cathode is rapidly destroyed either by overheating or by further electrolytic decomposition of the oxide. This limitation of currents to be drawn became of particular importance for microwave tubes, the tubes that were specially developed for radar and for radio links, and various methods were found to increase the conductivity and the stability of the oxide layers. Large numbers of reliable tubes were made, but the procedures remained rather "alchemistic." A colleague from one of the leading American firms making special tubes once told me the following amusing story. They had been making a special type of tube successfully for some time when they ran into emission trouble. They decided to seek advice from a consultant. He carefully went through all the manufacturing details, then asked how they degassed their components. When he heard they heated them in high vacuum he felt he had spotted the source of the trouble. He advised them to do the final degassing in pure hydrogen. They followed his advice and the troubles disappeared. A year or so later they ran into trouble again and after having tried in vain to get rid of it they got hold of the same consultant. He again went through all the details, checked that they were still degassing in a hydrogen atmosphere, and this time he could not get them out of their difficulties. So they finally went to another consultant. He soon spotted the source of their troubles: they ought to degas in high vacuum, and certainly not in hydrogen. They followed his advice and the troubles disappeared. "If we ever run into trouble again we know what to do," said my colleague. "We just change consultants once more."

In the Philips Laboratory we did much work in order to understand better the details of what happened in the oxide layer. Progress was made, but an attempt to obtain a simpler situation was even more successful. The initial idea was to get rid of the powdery oxide layer altogether and to use metallic alloys containing some barium as cathode material. A solid solution of barium in copper did have satisfactory emission but it also had a most unsatisfactorily high evaporation of barium, which made it practically useless. It was at that point that Mr. Lemmens enters the story.

Mr. Lemmens (born 13 January 1906) was heading the glassblowing shop and the tubemaking shop of the research laboratory. He was a man with little or no theoretical training, but he had enormous experience, was skilful and inventive, and was also a pretty good manager. I seldom heard any complaints either from his subordinates or from his "clients" in the laboratory (and this is all the more unusual, because when things in research do not go as desired—and they rarely do—the workshops are a favorite scapegoat). He was energetic, outspoken, and temperamental. He was anything but a petty, thrifty petit bourgeois: like many people in the south of the Netherlands he had something of the "grand seigneur" about him. Once, when he had received a sizable bonus for one of his inventions, he promptly invited his whole workshop to an evening party where he regaled them with fried chicken and ample beer. He had forgotten about the income tax, though, and later in the year, we had to supply him with a supplement for dealing with that. A remarkable thing about him was that he was willing to listen to advice from people with greater theoretical knowledge. Not from everyone, of course: he was pretty shrewd at picking his advisers. He was killed on 9 June 1963, when he crossed a major road on his little motorbike just in front of an oncoming car, in a way a double victim of his job. He had lost one eye in a glassblowing accident, so his vision towards the left was impaired—and, as his widow later said, he did not pay attention to traffic; he was always thinking about his inventions.

Now Lemmens was familiar with the existence of porous tungsten. Pieces of tungsten sintered from powder behave like blotting paper, and he had the idea of filling the tiny pores with carbonates. I think that, to his way of thinking, this did not seem much different from an alloy, for he probably had no idea at all about the size of atoms. Also, he always referred to "carbonates" as "carbonades," the latter being the usual Dutch designation of a broiled chop (according to the Oxford Dictionary the word went out of use in England during the seventeenth century; the second [1958] edition of the big Webster still recognizes "carbonado"). Anyway, he tried to impregnate the tungsten with his "carbonades" and then to activate the cathode in the usual way. It did not work: the carbon dioxide, liberated when the carbonates were decomposed, broke up the tungsten. And then Lemmens had a remarkable thought, one that would probably not have occurred to anyone thinking along theoretical lines. If I cannot put the carbonates in, why not put them behind? And he put a pellet of carbonates in a little chamber, the front wall of which was formed by a disk of porous tungsten; the other walls were molybdenum. A heating spiral was arranged behind the rear wall of this chamber. And, indeed, this arrangement, later known as an L-cathode, worked. The carbonates when heated were transformed into oxides and the oxides gave off some barium vapor that would diffuse through the pores and form a surface layer on the outside of the tungsten disk. The tungsten acts in a way like the wick in a good old-fashioned kerosene lamp. One now had a cathode with a solid metallic surface that could be machined to close tolerances, one could draw currents up to the saturation current without damage. It is true that there was also a certain amount of barium evaporation, but in many cases the evaporated barium did not give trouble and the layer was continually renewed. That is why cathodes of this general type are often called dispenser cathodes.

Important contributions to the further development of dispenser cathodes were made by the North American Philips Laboratories. Now here I could try to embark upon an explanation of a rather unique type of corporate structure, but I am no expert in the field and would probably make mistakes. For me, it was sufficient to know that North American Philips was not an affiliate of Philips Eindhoven in the usual sense of the word and that I had no formal right whatever to determine the program of its laboratory, but that arrangements existed through which each company had access to the know-how and patents of the other. Therefore, it was useful to compare programs, in order to avoid duplication and to provide mutual support. One of the main contributions of North American Philips was to make a systematic search for other compounds providing barium, and they found that barium calcium aluminate ($5BaO \cdot 3CaO \cdot 2Al_2O_3$) was the best choice. This can either be used, mixed with tungsten powder, in a separate chamber behind the porous disk as in the original L-cathode or, alternatively, the porous tungsten can be impregnated, as in the first attempts of Lemmens.

A further interesting improvement came again from Eindhoven. If the cathode could be run at a lower temperature, then the evaporation of barium would be further reduced and the life of the cathode would be lengthened. But that would be possible only if the electron would need less energy to get out—or, in technical terms, if the work function was lowered. A simple theoretical argument by one of our research people, Dr. Zalm, made it plausible that the work function of a barium-covered metal would be the lower, the higher the work function of the bare metal. A paradoxical conclusion, but it was confirmed by experiment. With a thin layer of osmium on the tungsten, a dispenser cathode will need a temperature 100° C lower than a plain tungsten dispenser cathode.

These osmium-covered dispenser cathodes are almost ideal heavy-duty cathodes.

The art of making radio tubes for generating very short radio waves had made enormous progress during the war years: such tubes were required for radar. Philips obviously was, to begin with, quite a bit behind its American and British colleagues. The dispenser cathode enabled us to design some tubes that put us at least in some respects again in the front line of development.*

Of special importance were triodes for wavelengths below ten centimeters. In principle, they were conventional triodes, with a cathode, a grid, and an anode, but they had a planar instead of a cylindrical geometry, a very small clearance between cathode and grid, and a grid wound with very fine wire, and they operated with a high current density. It was thanks to these tubes that a French affiliate of Philips obtained important orders for television relay links.

The research laboratory tried pushing this type of tube to its very limits. One attempt, which never came off, proposed to use a grid wound with wire only two and a half micrometers (that is one ten-thousandth of an inch) in diameter. Watching a mechanic winding those grids always reminded me of Andersen's fairy tale about the emperor's new clothes, where impostors pretend to weave beautiful garments that can be seen only by men competent for the position they hold. In our case, the wire could not be seen either, but, unlike the emperor, the grids were not left naked.

MAGNETIC MATERIALS

The phenomenon of magnetism was known already to the ancient Greeks. It is mentioned by Plato in one of his Dialogues, and, of course, by Aristotle. Lucretius, in his *De Rerum Natura*, devotes almost two hundred lines to the subject. The opening words of this section are "In matter of this sort many principles have to be established before you can give a reason for the thing itself."[9] These words are still very appropriate today: a long preamble would be required if I were really to explain the finer details of the theory of magnetism.

*We were, for instance, able to make reflex klystrons for wavelengths of four and of two millimeters and even set up a small-scale manufacture in the research laboratory. There was a time when these tubes were much sought after by research workers in the field of microwave spectroscopy.

This I shall not do, although there is some quite elegant physics involved.

Everyone has, at one time or another, played with magnets. All of us are also familiar with the magnetic compass, which came into use during the Middle Ages (the Chinese knew it earlier). But even well-trained scientists or engineers may not be fully aware of the very wide range of useful applications of magnetic materials.

Since early days magnetism has been associated with magic and occultism, and also with medicine. Serious practitioners believed in the influence of magnetic fields on the human organism; mountebanks found here a rewarding subject. That kind of "application" I shall not discuss: it did not form a part of our research program.

Permanent magnets are used in loudspeakers, in small dynamos (for instance those for bicycle lights), and they provide the magnetic field for magnetrons, the tubes that emit radar signals. They can be used for temporarily fastening things—like drawings or calenders on a steel cabinet. There are also "soft" magnetic materials: they become strongly magnetic in a magnetic field which can be produced either by a permanent magnet or by an electric current but they lose their magnetism when the field is removed. The cores of transformers and the antenna rods in small portable radio sets are in this category and so are the cores of coils used in electric circuits. Such materials can be used in fields where the magnetization is proportional to the strength of the applied field—this is the case for normal transformers—and also in the "non-linear" range of fields. The magnetic material on the tape of a tape recorder (either for sound or for television) is used as a semipermanent material. The electric current that represents the sound or the picture is recorded as magnetization of the grains on the tape. It is a permanent record, but it can be wiped out by a magnetic field. Memory cores in computers are also of this semipermanent type.

Magnetic materials can be of widely different chemical constitution; metallic, or non-metallic compounds of various kinds. The Philips Laboratory has worked on many classes of materials. In the case of metallic alloys it showed in 1937 that the magnetic performance of Ticonal, an alloy of iron with nickel, cobalt aluminium, titanium, and copper (essentially the same material is known as Alnico in the United States), is much improved (by a factor of 3 in the famous figure of merit BH_{max}) when the material is cooled down from a high temperature in a magnetic field. The true reason for this was found after the Second World War, and then it was also found that pieces of Ticonal, in which all the crystallites have roughly the same orientation, react even more favorably to cooling in a magnetic field: the figure of merit can be stepped up by another factor of 1.6.

However, the greatest contribution from the Philips Laboratory was in the area of oxidic compounds of elements of the iron group. Let us look at that in some detail.

Pure iron is good soft-magnetic material. Yet a transformer with a solid core of pure iron would have very poor performance at mains frequency and at higher frequencies it would be completely useless. That is because of eddy currents induced by the changing magnetic field. In power engineering this difficulty is obviated by using an iron alloy with higher resistivity, and, even more important, by laminating the core. At higher frequencies such measures are no longer adequate and one has to use iron powder embedded in, for instance, a plastic matrix. It would be far more advantageous, however, if one could find an insulating material with favorable magnetic properties. Then the material could be used in bulk; lamination or powdering would become superfluous. That was the target clearly proposed by Holst and taken up by J.L. Snoek and others.

There was a starting point. The oldest permanent magnet known was the lodestone. It is magnetite, a mineral with the formula Fe_3O_4. Now this material itself is unsuitable; it is not a magnetically soft material and, moreover, it is a rather good conductor. But its formula can also be written as $FeO \cdot Fe_2O_3$. This shows clearly that one-third of the iron atoms are two-valued, two-thirds are three-valued. Now one can look for other ferrites, compounds of the type $MeO \cdot Fe_2O_3$ where Me is any two-valued metal or, possibly, combination of metals. It is among such compounds that suitable materials were found, to begin with mainly by trial and error. Later the experiments were supported by better insight into the relations between magnetic properties and the arrangement of atoms in the crystals, carefully determined by X-ray diffraction. The basic theory is due to the French theoretician and Nobel Prize–winner Louis Néel. The materials were prepared by a sintering technique. The materials are of fine-grained crystalline structure, and the name "black ceramic" is fitting, because of the appearance of the material and because of the method of preparation. The name "ferroxcube"—a name invented by Snoek—was introduced as a proprietary name but has almost become a generic term. Snoek was definitely not amused when one morning he found on his desk a meat cube bearing the label Veroxcube, suggesting that it was prepared from veritable ox-meat. Probably it was not (whereas ferroxcube really does contain iron oxide).

These ferrites have found widespread application, and in the course of time it became more or less possible to provide tailor-made ferrites for specific purposes. A special kind of ferrite is, for instance, used in magnetic memory cores of computers.* A little ring of ferrite can be magnetized in either of the two tangential directions and the magnetization can be switched from one direction to the other by a current pulse through wires passing through the ring. One ring can store one bit of information. (Magnetization going around clockwise can, for instance, be counted as "yes" or as "one," magnetization going around counterclockwise as "no" or "zero.") In large computers thousands and thousands of tiny rings are combined. A factory of memory stores, for instance one run by North American Philips, always struck me as a remarkable outfit. On one side of the factory a better kind of rust and some other equally mundane substances were ground, mixed, pressed into shape, and fired in large ovens. Apparently a somewhat messy—although in reality well-controlled —operation. On the other side, ladies with manicured hands were threading colored wires through little rings and weaving them into mats to the accompaniment of soft music—a kind of glorified kindergarten activity. These small rings found a counterpart in very large rings (about half a meter outer diameter) manufactured for the accelerators at Brookhaven and at Geneva.

A curious incident happened during the deliveries to Brookhaven. The factory had guaranteed certain characteristics and according to their measurements the material was just within the specifications. Brookhaven's measurements gave more favorable values. We sent one of our best specialists on measurements to investigate. He found that the Brookhaven values were correct and that an error had crept into the Philips measurements, an error for which the research laboratory was responsible. The Brookhaven people were pleased. It is always nice to be right and if you get, moreover, a better

---

*The idea of using ferrite rings as memory devices did not originate at Philips. This type of memory is now being ousted by semiconducting devices.

material than you have bargained for, you are, of course, doubly pleased. I hate to think what their opinion would have been if our error had been in the other direction. Now they could only conclude we were dumb but honest.

While writing this book I have, time and time again, been confronted with the problem of translation: rendering one pithy sentence is often more difficult than translating a whole passage. Problems of translation can also play a role in business. A curious example arose in connection with ferrites. Philips was suing an American firm for infringement of one of our ferrite patents. The whole question had been carefully prepared and one of our research chemists was going to accompany the patent lawyers as an expert. As a final preparation he was reading an Ellery Queen story in order to get some local atmosphere and improve his colloquial American. Suddenly he was worried: Ellery and his father were living in a *brownstone*. Where had he recently seen that word? It must have been in one of the Philips patents. He investigated and here is what he found. In Dutch, manganese dioxide is called *bruinsteen*, in German *Braunstein*. In the original Dutch patent it was stated that manganese should be introduced into the mixture preferably in the form of very pure "bruinsteen." In the American version this was translated as ... *brownstone*, and not as *pyrolusite*, the correct translation. Brownstone, a reddish-brown sandstone, whose color is presumably due to iron oxide, is certainly not a source of pure manganese! He then looked at the major Dutch-English dictionaries and found that all of them had the translation *bruinsteen = brownstone*. Hectic activity followed. A number of dictionaries were sealed and affidavits were sworn that these were the dictionaries in regular office use in the Netherlands. But nobody ever noticed the error and the matter was settled out of court. However, the dictionaries might defend themselves. In the Oxford dictionary there occurs one quotation from a chemical text from the year 1873, "Hyperoxide of maganese or brownstone," and a German-English dictionary from 1859 also lists *Braunstein = brown-stone*. So it seems that English chemists, much influenced as they were by German chemistry, have during a certain period used *brownstone* as a translation of *Braunstein*. This nomenclature disappeared in England, and in the United States brownstone has an entirely different meaning, but the tradition survived in Dutch dictionaries.

There exists an entirely different group of ferrites, the so-called hexaferrites, compounds of the type $MeO \cdot 6Fe_2O_3$ that also happen to have a hexagonal crystal structure. Some of them make pretty good permanent magnets, especially $BaO \cdot 6Fe_2O_3$, known as ferroxdure in Europe and as magnadure in the United States. The first sample was obtained by accident. A technician had been ordered to make a preparation of such a composition that it would almost certainly have been a kind of ferroxcube. He made an error when turning the chemical formula into grams of raw materials, and the result was a permanent magnet. Of course, not at once of optimal composition. Then phase diagrams were ascertained, crystal structures were determined, and so on. Still the fact remains: although scientific methods were required to arrive at an optimum, the first step was a weighing error. Research people shamefacedly have to admit this is so.

Ferroxdure is not as powerful magnetically as Ticonal, but it is cheap, containing no rare elements. Also, though it has a rather low saturation magnetization it has a very high coercive force, which makes it particularly advantageous for some purposes. It is for instance very suitable for magnetic toys. My colleague G. Rathenau, who is a very conscientious man, therefore considered it important to ascertain that it was not poisonous. Large quantities were fed to rabbits, but it did not do any harm. Some ferrite has occasionally been fed to human beings. It was mixed into the contrast medium used in radiography of the gastrointestinal tract. One can then exert some force on the bowel by an external magnet. I do not think the method has found widespread application, although it seems more reasonable than a cure of rupture described by Ambroise Paré:

A certaine Chirurgion who deserveth credit, hath told me that he hath cured many children as thus: He beates a loadstone into fine powder, and gives it in pappe, and then hee annointes with hony the Groine, by which the gut came out, and then strewed it over with fine filings of iron. He administered this kind of remedy for ten or twelve dayes. . . .[10]

What I have told about magnetic materials is very sketchy and peripheral. I am afraid it does not in any way convey an impression of all the work that went on, of the large number of preparations that were made and tested, of all the measurements over a wide range of frequencies, of the X-ray work and the calculation of crystal structures, of theoretical models and their comparison with experiments, but I have to leave it at that. The reader who really wants to know about ferrites is referred to the scientific literature.[11]

IMAGE TRANSFORMERS AND CAMERA TUBES

I have mentioned already that Holst together with J.H. de Boer and M.C. Teves was a pioneer in the field of image transformers. Let us look more closely at the principles involved. Infrared radiation cannot be seen by the human eye. However, an infrared light quantum is able to eject an electron from certain metals, for instance from cesium-covered silver. Preparing efficient photocathodes is another somewhat alchemistic process; the theoretical background, though adequate for a general understanding, is insufficient for prescribing the details of manufacturing procedures. Now an electron, once out of the cathode, can be accelerated by an electric field. If it then is made to hit a layer of fluorescent powder, a number of light quanta will be emitted. If the arrangement of the electric field is such that the electrons impinging on the fluorescent screen form an image of the photocathode, then a "picture" in infrared projected on the photocathode will be transformed into a visible picture on the fluorescent screen. Satisfactory tubes of this kind were made in the Philips Laboratory, but so far as I know—I was not yet with Philips in those days—no attempts were made at manufacture. During World War II, tubes based on that principle were made in large numbers, mainly in the United States. Together with suitable optics they became known as sniperscopes.

After the war Philips obtained military orders for such tubes, and a new electrode system was designed. The matter was, to begin with, regarded as very secret; even the fact that such tubes were made was not supposed to leak out. This was, in a way, inconvenient. Tubes of this kind were useful for some experiments on semiconductors, but in the research lab we could not use Philips tubes without prohibitive security measures. Fortunately, sniperscopes could be bought freely on dumps, so we used those, or said we did, and everyone was happy.

In those days, the number of motor cars in Holland was still small, and since rail traffic was also not yet back to normal, there were many hitchhikers, and car owners had been officially asked to help enlisted soldiers on weekend leave. I have transported many and found it most instructive. I remember, for instance, one young man who traveled from Amsterdam to his native village, Middelbeers, a very small village indeed in very quiet rural surroundings. "I shall be glad to get home," he said, "for what can a man possibly do in Amsterdam over the weekend?" I was glad to agree, that, compared to Amsterdam, Middelbeers offered far greater opportunities for a pleasant weekend. But back to image transformers. One of my passengers explained to me that they had great fun in the dunes: they would stake off an area with red flags and then go rabbit shooting. But now they had something really new. They had rifles with some special kind of telescope, made by Philips, and some searchlights radiating invisible light—he thought they called it infrared—and now they could shoot rabbits in the dark. Most exciting sport. I never told our security people about this conversation.

I mentioned M.C. Teves. He was not connected with the military project, but he turned his experience to good account in another way: he developed an X-ray intensifier. There were two methods of X-ray diagnostics: X-ray photography and fluoroscopy. The disadvantage of photography is that one has to wait for films to be developed —especially during surgery that can be awkward—and that you cannot study motion. The disadvantage of fluoroscopy (that is, the observation of a fluorescent screen) is that the X-ray dose received by the patient is rather high, even when the intensity of the X-rays, and therefore the brightness of the fluorescent image, is reduced to such a low level that the observer needs a long time of dark adaptation before he can properly observe. The X-ray intensifier increases the brightness of the fluorescent image in the following way. (I am describing the original process that was used in the Philips Laboratory and later in the factory for many years but that is now obsolete.) A fluorescent screen is prepared by embedding the appropriate powder in silicone plastic. This is put into a vacuum envelope and degassed. Silicones can stand a sufficiently high temperature for this. A transparent photoelectric layer is then deposited on this screen. An X-ray quantum absorbed in the fluorescent screen will give rise to a number of light quanta. About 10 percent of the light quanta will yield a free electron. These electrons are now accelerated to 25 kV and imaged on a second fluorescent screen. This leads to a total light emission that is about fifteen times as large as that from the first screen. A much greater increase in brightness is due to the fact that the area of the second screen is about eighty times smaller than that of the first screen. The total gain in brightness is about 1200. It is no exaggeration to say that image intensifiers helped to inaugurate a new era in X-ray diagnostics. For instance, the catheterization of the heart, the possibility of which had been shown already in 1929 by Forssman and which had been developed into a clinical method by Cournand and Richards, can be safely and conveniently carried out only with the help of X-ray intensifiers.

Philips was not the only firm in this field. As a matter of fact, although our program was started independently, Westinghouse was a few months ahead of us in bringing out an intensifier, designed by J.W. Coltman. I believe that our equipment was better adapted to the needs of the radiologist; there had been close contacts with the medical profession, and the optics group in the research laboratory had given valuable assistance. Yet I found it surprising that with their head start Westinghouse was not able to keep up with us. I learned the true reason many years later during a scientific meeting on the Greek island of Crete, where I met a physicist who, at the time, had been working at Westinghouse. He explained to me that just then Westinghouse was getting an order for the first nuclear-powered submarine and the laboratories were combed for scientists and technicians who knew something about radiation. As a result Coltman was left to deal with X-ray intensification almost singlehandedly. It is an instructive story: it shows that it is sometimes an advantage not to have government contracts.

Television camera tubes are quite different from image transformers and intensifiers; they fulfill a different function and operate on a different principle. The two types of tube have in common that their functioning depends on thin layers of special materials with uniform and narrowly prescribed bulk and surface properties. Therefore, they require even more rigorously observed manufacturing processes than normal tubes. A rather successful project of the fifties and the early sixties was the development of the "plumbikon," a new camera tube.

Here I am again meeting the difficulty mentioned by Lucretius. A long introduction would be necessary for a real explanation. I restrict myself to a short one, probably not understandable to the layman and superfluous for the expert. The television pictures we see at home are formed by an electron beam that rapidly and line by line

scans the surface (625 × 25 lines a second in most European countries, 525 × 30 lines a second in the United States). The intensity of
the beam is at every moment determined by the incoming television
signal. In a camera tube the optical image is first transformed into an
electric image: high surface charge corresponds to high light level.
An electron beam scans and wipes out this electric image twenty-five
times a second and while doing this provides the information required for the transmission. The scanning beams in the individual
receivers must be synchronized with the scanning beam in the camera tube. That is achieved by means of special synchronization
pulses. All that is pretty complicated electronics and things get even
more complicated in color television. From the physicist's point of
view, however, the essential question is: how do we transform the
optical image into an electric image, and what are the properties of
this electric image?

For many years RCA had been undisputed leader in this field. The
tradition in this field of special tubes was established by Zworykin.
Vladimir Kosma Zworykin was born in 1889, studied in St. Petersburg (Leningrad) and with Langevin in Paris, and came to the United
States in 1919, after the Russian revolution. He worked first with
Westinghouse, but when he demonstrated his electronic television
system at a meeting of the Institute of Radio Engineers in Rochester
in 1929, David Sarnoff at once realized that here was the man he
needed for one more new venture, and Zworykin became a leading
figure in RCA research.

When I first met Zworykin, shortly after the war, he was no longer
a young man, but he was still full of ideas. His English resembled that
of Gamow—I believe they knew each other rather well—so I had not
the slightest difficulty in understanding him. He later remarried but,
at the time, he was living by himself with a male servant. One of his
hobbies was to reconstruct Russian cooking without a cookbook. He
remembered what things should taste like and had some rather
vague ideas of ingredients, and his cook kept changing recipes until
Zworykin thought it tasted all right.

He also told me about his lack of success with an electronic dog
trainer. The dog was wearing a collar with a little receiver that would
transform a signal from a transmitter into an electric shock. You
ordered the dog to lie down or to come and, if he did not obey, you
gave it a shock. The dog learnt rapidly to avoid the shocks and soon

the transmitter was no longer required: the dog would follow commands perfectly, but only as long as it was wearing this special collar. Without a collar it would just run away and do as it pleased; such dogs are known as stupid dogs.

The best of the stories I remember had to do with a reading aid for the blind he had been developing. He did not try to synthesize real speech sounds on the basis of optically presented text. His idea was simpler: if I remember rightly letters were just scanned and black or white was acoustically translated into high or low. In any case what was presented to the blind reader was an acoustical signal that was easy to obtain but in which letters were encoded in a complicated way. Zworykin convinced himself that it was not too difficult to learn that code, but would it really be possible to read comfortably —a thing that is achieved by Braille? A psychology department undertook to investigate this. They found a young lady who was willing to make the effort; she was not blind, but she was always carefully blindfolded during the experiments. Soon she read fairly fluently, she gave many demonstrations, and her performance kept improving. So much so that Zworykin became suspicious, for she read faster than he considered possible on the basis of the theory of information. But his remarks were brushed aside by the psychologists: they felt that human capabilities are above simple rules about amounts of information that can be processed. Then there was suddenly complete silence. When he had not heard anything for several months he rang up the man heading that project. "I'd better come to see you," was the faltering answer. So Zworykin wrote a few words on a sheet of paper, put it on his desk, and waited for his visitor. "I do not know where to start," the latter began. "I suppose it is this," said Zworykin and handed him the piece of paper on which he had written "Miss so-and-so cheated you." "How did you know?" asked the man. "It was pretty obvious," said Zworykin. "Nobody could read as fast as that using this system." What had happened was this. The young lady had discovered that the blindfold left a tiny gap through which, after some practice, she could just manage to read. One day, someone attending one of her demonstrations put a sheet of black paper between her and the text. She burst out weeping, disappeared into the ladies' room, and was not heard of again.

Back to pickup tubes. One of the tubes developed at RCA was the vidicon, in which the electric image is obtained by photoconductivity. Some substances have the property that they become conductive when illuminated. If one side of a thin layer of such a substance is covered by a transparent but conducting layer (tin oxide on glass is a favorite choice) that is maintained at a positive potential, and if the other side has been brought to zero potential (that is the potential of the cathode) by a scanning beam of electrons, then if an optical image is projected on the layer, currents will flow at the illuminated spots, the amount of charge transported being proportional to the intensity of light and the duration of the illumination.

Now the RCA vidicons worked well, they were comparatively easy to make and simple to handle, but the quality of the pictures obtained was insufficient for studio work. So it was decided to look for a more suitable substance than the antimony sulfide and arsenic sulfide used by RCA. Lead oxide, PbO, seemed for several reasons a promising choice. And then, for many years, a group of people was struggling to get a layer of lead oxide two centimeters in diameter and fifteen one-thousandths of a millimeter thick that had all the properties required to get high-quality pictures. The well-known saying in *Tristram Shandy,* "Tis known by the name of perseverance in a good cause,—and of obstinacy in a bad one," certainly applied here. We were often blamed for obstinacy, especially by the commercial department selling tubes. They could not believe that we should be able to beat RCA at its own game. But it turned out to be perseverance after all. Again the final solution was found by combining trial-and-error with theoretical analysis and prediction. Accurate electrical measurements, chemical analysis, and X-ray studies all played their part; and the irrepressible optimism of Dr. De Haan, the leader of the team under the capable supervision of my deputy director, Dr. Bruining, played a decisive role. But I am beginning to mention names and that I should not do where my staff members are concerned; there are so many of comparable merits.

Once we knew how to make perfect tubes our troubles were not yet over: we had to convince the factory that they should make these tubes. And not only that, we had to convince them that to begin with they had to use exactly our procedures. Even that hurdle was taken, and after a while the factory made plumbicons without much difficulty and of a more reproducible quality than we had ever been able to attain in the laboratory.

I believe that even today the vast majority of color broadcasts in the world are made with plumbicons made either by Philips or by our licensees. Not everyone is convinced of the necessity nor even of the desirability of television broadcasts, I am well aware of that. But once you accept television, you have to admit that color adds greatly to the information content. That is especially true for educational programs (like, for instance, some extremely beautiful German broadcasts on animal behavior). And I am convinced that the Philips Laboratory made an important contribution to the excellent quality of color television pictures that now prevails, especially in Europe.

Let me finish by stressing that my own contribution to this success is limited to having picked the right people for the job and to having defended it against adverse criticism. I was never thoroughly familiar with all the details and would never have been able to do this kind of work myself.

SCIENTIFIC INSTRUMENTS

I mentioned that some of the tubes and materials we developed in the Philips Laboratory came to play a role in academic research projects. We also built more comprehensive equipment. Holst, in the years before the Second World War, was not a great believer in this. "An incandescent lamp is about the only thing you can profitably sell to an academic laboratory" he once said to me. That was of course an exaggeration, but as always there was, at the time, a good deal of truth in his statement. The firms that did make some profit in this line were, as I pointed out, of the artisan type, and that type of production does not thrive in a big company like Philips with its large overheads and with an organization that is too cumbersome and inflexible for dealing with short runs of sophisticated, not *very* expensive apparatus. There had been one notable exception, the Cockcroft-Walton generators. They were developed in the X-ray laboratory, a small outfit, housed in the Nat. Lab., but until 1942 with a separate budget and independent management.

Dr. A. Bouwers was the originator and manager of this group. He was born in 1893, started out by qualifying as a schoolmaster—one more example of this route to academic excellence—then studied at Amsterdam. He joined Philips in 1920. Bouwers was a man of vision and of considerable inventiveness. He had also the energy, and the push and the skill—and on occasion the ruthlessness—required to realize his many ideas. Anton Philips had a weak spot for X-ray activities, although for many years they were unprofitable. Maybe he foresaw a brighter future, but I think it was also the humanitarian aspect that appealed to him. In any case, he was willing to support Bouwers and to give him a free hand, even though Bouwers's inventions, always ingenious and usually basically sound, did not always lead to industrial and commercial success. There can be no doubt, however, that the X-ray Division of Philips would not have developed the way it did without Bouwers.* And the influence of Bouwers on the trends of development of X-ray equipment extended far beyond Philips.

Through X-rays Bouwers became interested in high voltages; a book that appeared in 1939 shows the extent of his interest.[12] And so Bouwers, together with his co-workers, began to build cascade generators. One of them was installed at Cambridge, as mentioned earlier. One rather powerful installation remained at Eindhoven. It was installed in a hall generally known as the Circus and was used for experiments in nuclear physics and also for producing radioactive phosphorus for medical and biological research.

In 1942 Bouwers left Philips and became managing director of De Oude Delft, a firm that rose to some prominence in the years after World War II in the field of professional optics. For many years its bread-and-butter line was the famous Odelca camera, an instrument for so-called indirect X-ray photography: the X-ray image produced on a fluorescent screen is photographed on a smaller film. To do this efficiently an optical system with good definition and very high aperture is needed. The Odelca camera has been of particular value for mass chest surveys, where full-size films would be far too expensive and unwieldy. So Bouwers's transition to the Oude Delft did not mean that he entirely left the field of X-rays, and his knowledge and renown in that field stood him in good stead.

After Bouwers's departure the X-ray laboratory ceased to exist as a separate unit. I was more or less put in charge of the remaining group. More or less, for under the Holst regime positions were not very clearly defined. The man who later continued the tradition in scientific apparatus was A.C. van Dorsten; W.J. Oosterkamp con-

---

*The division later also dealt in other types of medical apparatus and eventually changed its name to Medical Systems. Against my vain protest: I hate this insipid use of the usually meaningless word *systems*.

tinued work in X-ray diagnostics. I cannot claim to have made any important contributions to X-ray diagnostics or therapy, to high-voltage engineering, or to scientific instrumentation, but I learned enough to be able to see the importance of image intensifiers, and I also encouraged work on scientific instruments. Or perhaps "encourage" is not the right word: the people concerned needed little encouragement, but they did need some support.

In the years after World War II a number of cascade generators —sometimes complete with discharge tubes—were sold to university laboratories, mainly in the Netherlands itself and in Great Britain. However, in the long run they could not compete with Van de Graaff generators, which are in principle simpler and have a number of technical advantages.

Already during the war, C.J. Bakker and F. Heyn began to design a cyclotron, and they even began to order components, although the German supervisors had not given permission for building such a machine. I had little to do with this activity. I once had to receive the chief supervisor and some of his staff members and I had to show them the Circus. Fortunately, there had been time to wrap up the various components for the cyclotron, which were also stored in that hall. Leaning against one of the packages, I expressed my regret that we were not allowed to work on a cyclotron. After the war the original design was changed: the machine became a synchrocyclotron. Arrangements were made to install the cyclotron in Amsterdam; Bakker had accepted a chair of physics in the university there. A small Philips group headed by Heyn also moved to Amsterdam.*

The fact that our first cyclotron was now nicely accommodated at Amsterdam did not mean the end of cyclotrons in Eindhoven. Several cyclotrons were manufactured at the factory in close collaboration with the research laboratory. To begin with, they were synchrocyclotrons, essentially copies of the Amsterdam machine; later, far more sophisticated "isochronous" cyclotrons. Van Dorsten's team at the Nat. Lab. became highly proficient in designing such machines and in calculating exactly the required electric and magnetic fields. But there was no permanent business in that area, the team was dissolved, two of its members became professors at the Technical

*Bakker, born 1904, was later appointed as director general of CERN in Geneva. In 1960 he died in an air crash. Heyn became a professor at Delft. The story of the cyclotron is told in some detail in *Philips Technical Review*.

University in Eindhoven and the last prototype was donated to that same institution.

This work on accelerators provided contacts with leading scientists and leading centers of research. On the whole such contacts were useful and pleasant but there could be awkward moments. For instance, during a visit to Oxford, I was taken to task by Lord Cherwell because of the poor quality of some of the components of a cascade generator. The complaints were justified, but during the rapid buildup after the war it was difficult to avoid such mistakes entirely. Anyway, there was little I could do about it, apart from speaking or writing to the man in charge of the factory.

A cyclotron was built for Irène Joliot-Curie. She came to Eindhoven to discuss details and I had dinner with her on that occasion. Unlike Lise Meitner, who was full of energy at seventy, she struck me as rather worn out. She died soon afterwards, on 17 March 1956, before the machine was completed.

Her husband took over; he survived his wife by no more than two years. He was only fifty-eight when he died, but I had the feeling that during the last years of his life he lived already very much in the past. He spoke mainly about the thirties, the heroic days, when Irène and he did the work that gave them a Nobel Prize, the days when the clan of active physicists was so much smaller.

It was not without a feeling of regret that I saw Philips leave this field. I had not done much work in nuclear physics myself, but as a young theorist I had kept in touch with the developments of the thirties and I knew personally many of the leading figures of that exciting era. Joliot greeted me as an old friend: although we had not met very frequently, he recognized me as one of the clan. Now the links with such people and their laboratories were severed. My regret was not only a question of personal vanity and nostalgia. In the thirties Philips, on the basis of its technical competence, had been able to improve considerably on the original equipment of Cockcroft and Walton. Later our isochronous cyclotrons were at least as good and in some ways better than the best university-built machines. But times were changing: in the field of accelerators Philips's competence was no longer superior to, and not even on a par with, the competence of the most advanced centers of "big research." And we could not afford to build up such competence. The cyclotrons had still been profitable but that was partly due to our way of accounting.

Every industrial division had to pay a fixed percentage of cost price for research, but the amount of research assistance given was at the discretion of the management of the research laboratory and was not rigorously determined by the amount paid. For cyclotrons the cost of the research effort involved was far in excess of the average percentage. That was acceptable as long as it was a question of starting a new activity; it could not be accepted as a permanent state of affairs. The "small" cyclotron at CERN, a synchrocyclotron for 700 million volts, was already above our possibilities: we could not have afforded to construct a prototype of such a machine at our own risk, or for our own purposes. But at least we provided the high-frequency system. With the big accelerators we should have been utterly unable to assume a leading role. In this field scientific apparatus building had overtaken us and left us far behind.

Electron microscopes became a more permanent activity. I related in Chapter 5 how much I had been impressed by the lectures on electron microscopy I attended at Berlin in the summer of 1932. Before the war there was slow but steady progress, especially at Siemens. After the war several industries and many research laboratories entered the field. In the Netherlands work in electron microscopy began at the Technical University in Delft, where a young graduate, J.B. Le Poole, built excellent instruments embodying many ingenious new features. Shortly after I joined Philips, Holst arranged a collaboration between Delft and Philips in this field. The first electron microscopes Philips produced industrially were entirely designed in our research laboratory but incorporated many of Le Poole's ideas. Later the development was taken over by the factory.

Electron microscopes have become indispensable research tools, especially in biological research. Optical microscopes improved the resolving power of the human eye by a factor of several hundred; electron microscopy gives another factor of a thousand. It opens up the whole range between the wavelength of visible light and the dimensions of the atom and shows that living organisms contain highly complicated structures in that very range. It should be pointed out that at present the patterns of microelectronics do not usually go below the optical limit (although boundaries may be quite a bit sharper than a wavelength), but that will only be a question of time.

Of course, what I described so far is not at all a complete list of scientific instruments made by Philips. I mention in passing the X-ray diffraction apparatus, originally developed at North American Philips, and the low-temperature equipment based on the so-called Stirling cycle. Liquid nitrogen has many applications and even biological ones: quite a number of our low-temperature installations have been sold to artificial-insemination centers and are used for the storage of bull sperm. I made some remarks on artificial insemination in Chapter 4, and I trust that my readers will be impressed by the way in which my seemingly extravagant digressions fit into my story. I can even establish a link with magnetism: the sister of one of our leading researchers in magnetism was a sculptress specializing in portraits of animals. Most of her clients wanted sculptures of their favorite domestic pets, but once she got a commission to make a statue of Tadema IV, a famous pedigree bull, to be put in front of the artificial-insemination station where the animal had performed. She was quite nervous when the statue was going to be unveiled. She was not afraid of connoisseurs and art critics, but here her work had to be judged by people who did not have any theories about art but who knew everything about bulls. But when after the necessary speeches the statue was uncovered, there was an approving murmur: "That is him!"

10

The Science-Technology Spiral

In later years I was often invited to speak on themes like science and technology, physics and industry, industrial and academic research, and so on. Sometimes such lectures were perfunctory and uninspired; no more than mediocre PR work. Sometimes—as, for instance, in my talk at the First General Conference of the European Physical Society in Florence in 1969—I would speak briefly and somewhat flippantly about the general issues and then proceed to treat in detail some scientific and technical problems. And, now and then, I spoke with real conviction, especially when I was facing an audience that I expected would disagree with—or better still, be irritated by—some of my statements.

However that may be, my ideas about the relations of science and technology became gradually more distinct. I do not want to say that I developed a new philosophy. Let me rather say that I worked out a form of presentation which, I believe, enables one to see the different aspects of these questions in their right perspective. Such clarification made me more keenly aware of the enormous influence of science-based technology on the affairs of the world. As a scientist and industrial research manager I have good reason to be proud of the achievements of science and technology, but as a citizen of my country, as a citizen of the world, as a human being, I am profoundly worried rather than pleased.

The closing chapter of my book will deal with these questions. It is not a straightforward continuation of what I have written so far, but what I have written so far provides background and illustration for my more general considerations.

As an introduction to my discussion of the relations between science and technology I present three caricatures. Or perhaps, using a much abused modern term, I might call them models. They are caricature models because they exaggerate one specific feature.

First model. Science and technology are really independent of each other. Any relations that seem to exist are incidental and not really important. Many down-to-earth engineers will feel that way; so will highbrow mathematicians.

A specialist in the theory of the distribution of prime numbers and an accountant are both dealing with numbers, but that is about all they have in common. A theoretical physicist working on quantum electrodynamics and a designer of electrical appliances will hardly feel that the Greek root *electr–* occurring in the names of both professions brings them very close together. Edison was certainly not proficient in Maxwell's mathematical theory of electromagnetic fields. That did not prevent him making many inventions that had to do with electricity.

Especially in older industries, like the ceramic industry, like many branches of metallurgy, like the textile industry, the practical engineers feel they owe next to nothing to science, whereas scientists will certainly make no claim that they have created those industries.

Yet it is clear that this picture does not correspond to reality. Engineers construct electric motors and dynamos, but they only started doing this after Ørsted and Ampère had discovered the force between electric currents and magnets and after Faraday had discovered electromagnetic induction. Maxwell predicted and Hertz discovered electromagnetic waves; it was only then that Marconi began to apply them for telecommunication purposes. Vacuum electronics was preceded by J.J. Thomson's discovery of the electron, solid-state electronics by the quantum theory of electrons in metals and semiconductors. Such considerations bring me to my second model.

Second model. Scientists do not really care for technology. They are feasting at the high table of the gods, but from time to time, perhaps just by accident, or perhaps out of pity and benevolent generosity, they drop a few crumbs that are then gingerly picked up by lesser people, like engineers and captains of industry and generals and statesmen.

That model will not be popular with engineers and captains of industry, but let me be honest, it is more or less the way I felt as a young man; many theoreticians must have felt that way, feel that way today. And especially pure mathematicians may be inclined to think along this line. G.H. Hardy, for instance, writes in his *A Mathematician's Apology:* "[I will say only] that very little of mathematics is useful practically, and that little is comparatively dull."

But the "crumbs from the table" are often quite large chunks, and moreover, the table might well be empty without the crumb-gatherers. (This last remark is more appropriate for experimental physicists than for mathematicians.)

Third model. The course of technology and of science is dictated by capitalist enterprise. By influencing the appointments of university professors, by subsidizing certain types of work, by making it clear that scientists will have a hard time finding a job unless they specialize in the directions industry prescribes, capitalism keeps a firm control over scientific programs. If it seemingly allows a certain amount of "free" research, that is part of a policy of "repressive tolerance." That pseudo-freedom serves to weaken the spiritual freedom rather than to strengthen it.

This model is not any closer to reality than the first two caricatures, although there was a period—I believe in this respect the worst is over—when it was put forward with almost paranoiac zeal by so-called progressive students. For one thing, it overrates the acumen of big business. No one foresaw—neither the physicists nor the engineers and least of all the leaders of industry—that quantum mechanics would lead to a theory of semiconductors that would open up an entirely new era of electronics and of data handling. Electromagnetic waves were not discovered at the request of postmaster generals who found post coaches and even trains too slow. The discovery of the electron was not ordered by industrialists who were already envisaging enormous sales of television sets. Nor did the Curies and Ernest Rutherford start their work on radioactivity because boilermakers told them to look for a new source of heat.

The model also underrates the stubborn independence of the academic research worker.

After having dealt summarily with these three inadequate models I shall now treat in greater detail my own favorite model, which I like to call the science-technology spiral.

Let us first look at science as an independent stream. Seen from a distance it might appear to proceed steadily and systematically, but in reality it has a complicated, motley structure. Lines of research often run in parallel: the same results are obtained almost simultaneously in different places. Sometimes apparently divergent lines later converge and a synthesis is reached. There are truly blind alleys, but a seemingly blind alley may also have an issue leading to new vista. Some indication of all this can be found in earlier chapters of this book. Only a few scientists are working at the very front, are creating entirely new concepts or discovering entirely new phenomena. Most research workers are doing work of extension and consolidation. They are tying up loose ends, are refining or clarifying mathematical theories, are perfecting measurements or looking for further examples of essentially the same phenomenon.

The development of technology presents a similar pattern. But the stream is even broader, the structure even more complicated. There are more simultaneous developments in parallel, more methods and products that are discarded as obsolete.

What I have described so far is just an elaboration of my first model but now there comes a further step. Technology today *always* draws upon earlier scientific results. Occasionally studies appear in which this is denied; they are foolish. The fact of the matter is simply that for some branches of technology fairly old science is sufficient. For some mechanical engineers a few notions of Newton's mechanics, and Euler's and Bernoulli's theory of bending beams, may suffice; even so, these engineers are depending on earlier scientific results. And they will use at least some mathematics, if only negative numbers, decimal fractions, and the number zero. For some electricians Ohm's law may be about all they require, but Ohm's law is not very old; in Chapter 2 I mentioned the celebration of its 150th birthday. An electrical engineer dealing with electromagnetic waves will have to use the complete Maxwell theory. Chemists will, at the very least, have to use the notions of atoms, of chemical elements, of valence. For organic chemists Van't Hoff's ideas on stereochemistry, that is, on the spatial arrangement of atoms in molecules, are indispensable.

In many cases technology has to apply more recent scientific results. But technology never uses the most recent and hardly ever the most profound results and notions of academic research. There is always a time lag of ten to twenty years. During the last hundred years this time lag was remarkably constant; it is now showing a tendency to increase. This I consider a very important point, and

since in popular writings one may frequently encounter the erroneous statement that the transition from basic science to technology
is taking place more and more rapidly, I shall illustrate it by a number
of examples.

X-rays were discovered in 1895; within ten years there came into
being considerable industrial activity in the manufacture of diagnos-
tic X-ray equipment. The existence of the electron was ascertained
around 1900; within ten years the first triodes were manufactured.
In 1888 Hertz discovered electromagnetic waves; in 1897 after successful experiments Marconi founded the Wireless Signal Company
—again a time lag of less than ten years. On the other hand the notion
of positive holes in semiconductors appeared in the late 1920s. The
transistor was invented in 1948, and that was the first instance of an
application of quantum mechanics. It is true that the first "atom
bomb" was exploded within seven years after the discovery of nuclear fission, but that was an all-out effort, with the whole power of
science and American technology, driven by a sense of extreme
urgency, behind it. And one should not forget that the discovery of
fission was preceded by almost half a century of nuclear physics.

The whole field of particle physics (or of high-energy physics, if
one prefers that name) that began in the thirties and that is now the
most spectacular "growth point" of physics has so far not found any
direct application.

So much about time lag. What about my second statement,
namely that the most profound notions are rarely applied? Einstein's
general theory of relativity is a grandiose conception. I do not see any
applications at all. The refinements of quantum electrodynamics are
not quite in the same class, but they are certainly an intellectual tour
de force; they have no impact on technological problems. Whether
the refinements of statistical mechanics and especially the mathematical theory of phase transitions will ever be applied, is, in my
opinion, rather doubtful too. As to mathematics, one does not have
to share Hardy's opinion that all applicable mathematics is rather
dull, but it is undeniable that there exists a whole body of beautiful
and profound mathematics that is not directly "useful."

In short, there is some truth in our crumbs-from-the-table model.
Perhaps we should slightly modify the metaphor though. It is not
crumbs but substantial helpings of solid food that are served, but
those that are seated at the high table reserve the choicest and most

refined dishes for themselves; they would not be appreciated by the *profanum vulgus* anyway. Which reminds me of the following story.

My wife and I once had to receive a rather large number of guests, and a caterer had supplied an elaborate cold buffet supper. Next day, we had to go out and we told the children there were plenty of leftovers that would make an ample meal. When we later asked them how they had fared they said it was all right, only there was quite a quantity of curious little black balls. "Those we did not dare to touch, so we fed them to the dog." (It was *not* a little white terrier but a Great Pyrenees). Similarly, mathematical caviar will not be palatable to practically minded technologists.

The progress of science on the other hand depends on technology. I have described that in some detail in Chapter 9. This is obviously true for experimental work, but since the arrival of computers also part of theory has become technology-dependent. And here there is no time lag. On the contrary, science likes to use the very latest and the most advanced technology. Astronomers are particularly good at that. Hardly had the first artificial satellite been launched before they began to make plans to send measuring equipment into outer space. Radio astronomers have used the very latest in low-noise detecting devices and so on.

This completes my science-technology spiral: technology uses scientific results but with a time lag; science uses technology without a time lag.

We can now add a few additional touches. Results of academic research that are potentially applicable are often not sufficiently precise or do not relate to the most useful materials or structures. Industries will then try to exert some pressure on academic research to supply the missing data. They will give research contracts, create special positions, endow new laboratories. To that extent there is some truth in our third model, but only a grain of truth. The arrangements I described may lead to some influence of industry on academic research—which I do not consider undesirable. They do not at all provide a firm grip, let alone complete control. And, in many cases, they are insufficient to provide the information industry needs. It is the primary task of industrial research laboratories to complement results of academic research that are applicable in principle so that they become applicable in practice. Their way of working is not different from that of the majority of scientists I indicated above.

They too are tying up loose ends, are refining or clarifying mathematical theories, are perfecting measurements or looking for further examples of essentially the same phenomenon. The difference lies in the choice of subjects and, possibly, in the motivation. Here it is important to distinguish between the motivation of the research worker and the motivation of the enterprise by which he is paid. A man working on the electron optics of color cathode-ray tubes may not be much interested in television and not at all in manufacture and sales of TV receivers. For him the problems are almost problems in pure research: difficult mathematical problems have to be solved, ingenious computer programs have to be worked out, and so on. The managers of the enterprise on the other hand are not primarily interested in the esoteric beauty of mathematics or the elegance of programs. For them, it is the practical results that count. The well-known saying that beauty is in the eye of the beholder can be used, mutatis mutandis, in connection with the difference between pure and applied research. I mentioned the intellectual challenge of technological problems before. It is an essential prerequisite for the productivity of industrial research laboratories. It is also a cause for concern. To that I shall have to come back later on.

Occasionally industrial research laboratories do also contribute to basic principles, to fundamental knowledge. That does help to raise the level of all their work, but it is not their main task. (I mentioned two examples from my own work in an earlier chapter.)

Such fundamental work in industry has its counterpart in technological work done in academic research establishments. A simple example is the Dewar flask (thermos or vacuum flask) invented in 1892 by Sir James Dewar as a research tool but now an almost ubiquitous domestic utensil. It is very much in evidence in China and in a lecture the author Han Suyin gave at Geneva, which I happened to attend, it figured as one of the blessings the Chinese occupation—she called it differently—of Tibet had brought to that country. Astronomers have contributed to the development of optical instruments; some of the data-handling methods developed by high-energy physicists may find application in medical diagnostics although the results of high energy physics do not. And so on.

Let us now look at the influence of wars. Pauli, in the letter I mentioned before, wrote that because of the war effort the development of physics had almost come to a standstill. Others have emphasized the stimulating role of military requirements. The apparent contradiction is easily resolved. In times of war there is stagnation at the very front of fundamental research but the time lag between existing basic knowledge and technical application is shortened. The development of radar technology, of tubes generating centimetric waves, of wave guides, and so on did not invoke new physical principles, but no one had yet got around to designing such things. After the war much fundamental work—leading to several Nobel Prizes—was done with this type of equipment. As another example, I mentioned already the short time required for developing the first nuclear bombs.

Although I expect that most scientists and technologists will more or less agree with my exposition, it will certainly be criticized by historians and social scientists. They will object that I have only spoken about science and technology as if they were the only things in today's world that count, that I have given no explanation whatever why an applicable form of science was born in our Western world, that I have entirely neglected the influence of economic and political factors on technology and on science.

Such objections are not unfounded; I am well aware of the limitations of my description and of my competence. That is the very reason that I have spoken about a presentation and not about a philosophy. I believe my model gives a fairly correct picture of the way knowledge and technological skill are acquired; it does not deal with socioeconomic aspects nor with deeper reasons.

The influence of society on trends of development is clearest in technology. In the nineteen-thirties, radio broadcasting conquered the world. Therefore, radio tubes became cheap and reliable, and this had an important influence on the development of nuclear physics. The course of physics and of other sciences might have been different if commerce and finance had not asked for powerful computers after World War II. (An IBM director once said to me: "We don't claim to have technical skills superior to those available inside Philips, Siemens, and a few others. But you people have never felt the real need for flamboyant thrusts towards large-scale data-handling.")

Furthermore, there will be no science without scientists, and there will be very few scientists if no one is willing to pay them a salary and to pay for their equipment. Governments have to decide how much they want to spend, so they must have some form of science policy, but the influence of science policy on the content of science has been slight: it does influence the size of the scientific effort, not its nature.

All the same I agree that one can speak about a technology-economy spiral, or feedback loop, that is working in parallel with the science-technology spiral, and this interaction deserves a less superficial discussion than I have given above. The fact remains that I could get as far as I did without taking them into account. Therefore I like to say—using an admittedly poorly defined and rather vague notion —that the science-technology spiral is an almost autonomous mechanism.

Let me try to develop this idea somewhat further. In Chapter 5 I mentioned the lecture in which Schrödinger tried to establish parallels between the "Zeitgeist," the spirit of the age, and the trends of physics; I also mentioned that his ideas did not impress his colleagues. If such parallels really existed one would expect to find systematic differences between physics in Soviet Russia and in the countries of the West, for Russia has a different economic and social order. I do not see any systematic difference. Maybe prefaces and introductions occasionally contain some reference to the official religion of the country—that is, to dialectical materialism—but even if the reverence shown should be genuine, such declarations of faith do not influence the scientific content.

Modern science in China is still in its early stages, but it is developing rapidly. Some physicists there told me that the teaching of Mao had greatly influenced their work. But what did this mean? That they had taken to heart Mao's admonitions concerning perseverance and self-reliance. So they had worked very hard and had not given up when a problem seemed almost insoluble. I am not poking fun at this; if a leader can inculcate in his people a willingness to work very hard and to face difficulties, that can be a good thing; again this did not change the physics.

Of course, there have been unfortunate examples where a government really sought to influence science. In Chapter 7 I briefly discussed the attempts of some Nazis to introduce a special brand of German physics. They did not succeed. In Russia at one time a special kind of genetics—or rather non-genetics—was officially imposed. That must have done great damage to Russian agriculture, but in the long run the course of genetics in Russia was not changed.

If we look at differences between individuals, we find a similar situation. Max Planck was a somewhat conservative and deeply religious man. Langevin was a communist and a rationalist. Both were receptive to new ideas, like Einstein's theory of relativity, but Planck's own work was more revolutionary than that of Langevin.

One would expect technology to show greater differences, since it is more closely connected with the needs and wishes of the people, as I mentioned before. But when, after my retirement from Philips, I began to visit Eastern European countries, I was almost disappointed. One is told so often that in our capitalist world the customer is manipulated by the industries to buy products that do not really correspond to his needs, that one begins to wonder whether in a communist country all the utensils of daily life should not be quite different. They are not. Cutlery and crockery, chairs, tables and beds, clothing, and bicycles are not in any fundamental way different from ours. If there are differences they are either the result of traditional, pre-revolutionary folklore—for instance, embroidery in Rumania— or they simply show that the technology is somewhat behind ours.

In China one eats with chopsticks and wears what we call a Mao suit. But I have carefully examined a Chinese bicycle. It is amazingly like ours. Apparently our much-blamed market mechanism has led to a product that is considered useful and desirable also in communist surroundings; the same is true for sewing machines.

As to the interaction between science and technology, so far I have not been able to find striking differences between the East and the West. I once had an interesting discussion with a professor teaching Marxist-Leninist philosophy at a university in the DDR. He had given a lecture in which he emphasized among other things how important it was that students during their first year should get a thorough course in Marxism-Leninism, "So that they acquire a scientific basis for all their later studies." I told him that he had not convinced me, to which he answered, smiling, that he had not expected he would. But when I began to ask him concrete questions about the relations between universities and industry, and especially about the transfer of academic research results into industry, he had to admit that Marxism did not provide helpful answers and before long he was asking my advice. When the conversation turned to certain electronic aids for handicapped people, I remarked that in our part of the world it often happened that later sales would be insufficient to recover the development costs and that no one was willing to foot the bill. I supposed that in his country the situation would be different in that respect. He sighed and said, "I'm afraid not. If with you a project is economically not viable, it is usually not socially justifiable with us."

If I am right in my observation that, so far as science and technology and their mutual relations are concerned, there is no great difference between countries with entirely different socioeconomic systems and that there is no great difference either between persons with entirely different ideologies, then the question "Who is controlling the spiral?" can have only one answer: nobody. The progress of the science-technology spiral can be favored or retarded by government measures, by wars, and by revolutions, but such influences do not control the nature of the progress. So perhaps my statement that this spiral is an autonomous force is not quite so vague as it appears at first sight.

I might end my book here. Looking back on half a century of activity in the realm of physics and its applications, I feel that I have played my little part in this powerful spiral. There were the days of my youth when I could watch the unfolding of great new theoretical concepts in an atmosphere of serious play, and could make an occasional contribution myself. There followed years when I supervised how some of these concepts were turned into practical realizations and how some of these realizations could be made to assist further scientific progress. I feel some regret for problems I might have solved in my younger years and did not solve—but most of them have been or will be solved by others. I feel some compunction because of projects I should have forced through at Philips and did not force through—but somehow the company survived my shortcomings, and in any case I helped to create meaningful and interesting work for many scientists and technicians. Sometimes I feel unhappy because I can no longer follow the real pioneer work in physics, let alone contribute to it; neither can I walk as fast and as far as I could formerly. Fortunately, there are still plenty of things I can do and enjoy doing.

One of the things I feel obliged to do is to add a few more pages. For, although I may gradually settle down to a serene old age and a perhaps not wholly undeserved rest, science and technology do not rest at all but are running ahead at unabated and threatening speed. I have related how in my younger years my colleagues and I playfully disregarded possible social impacts of our work. I do not want to commit the same error once more, do not want "to stretch the folly of my youth to be the shame of age." In my final pages I shall try to formulate some of my doubts, some of my apprehensions.

The close alliance of science and technology has changed the pattern of life of so-called developed countries. Or rather, it shapes the material framework of life in those countries. Nearly all production processes—including those of agriculture—are highly mechanized industrial processes. In our homes we are surrounded by innumerable appliances. Our locomotion, our traveling is motorized. Even our recreation—look at a modern winter sports resort—even our sexual morals and practices depend on mass-produced industrial commodities.

Scientists and technologists have good reason to be proud of their achievements. Proud of our ever-deeper understanding of the properties of the material world, proud of our ever-greater technical skills and our ever-greater mastery of the forces of nature. In our Western world much inhuman toil has been eliminated. Life expectancy has risen. Many diseases that formerly caused widespread misery and death have been conquered by the advances of curative medicine and by better diets and better sanitation, based on increased biochemical and microbiological knowledge. Third World countries that want to improve the lot of their people know that it is essential for them to acquire a measure of competence in science and technology. That is admitted even by thinkers who, on the whole, take a dim view of our industrial civilization. Even Marcuse, whose rather obscure writings were at one time regarded by hippies and contesting students as a theoretical background for their behavior, even Marcuse writes in his *Essay on Liberation:* "Electrification indeed, and all technical devices which alleviate and protect life...."[1] Ivan Illich, that severe social critic, writes: "For the primitive, the elimination of slavery and drudgery depends on the introduction of appropriate modern technology...."[2]

However, the mood of the times is no longer one of uncritical joy over our achievements. Doubts and worries beset technical and scientific specialists, as well as the public at large. I do not consider such worries unfounded, though they are often based on intuitive feelings rather than on strictly logical arguments. Let me try to sketch some of such attitudes.

A first group of critics do not attack scientific technology as such, but they point out that the rapid expansion of technology will, in a not too distant future, lead to exhaustion of the resources of the earth: we shall have to face shortages of food, of energy, of minerals. Only people will be superabundant. The reports of the Club of Rome have made many people aware of the simple fact that exponential growth, that is growth with a fixed percentage per year, cannot go on for a very long time. I apologize to the Club of Rome for reducing their elaborate studies to this obvious platitude, but I believe that the man in the street will not retain much more than that. "Exponential jokes"—one grain of wheat on the first square of a chessboard, two on the second, four on the third, and so on; a penny put out at three percent interest by Adam—are getting stale. Let me all the same give a further example. A healthy baby doubles its birthweight in half a year. Suppose it went on doing this, then by the time it went to school at the age of six it would have reached a weight of twelve tonnes,* the weight of a full-grown elephant. Now parents worry over many aspects of the development of their offspring but they are usually not afraid that their child will develop into an elephant. They know that there is a reliable internal control mechanism that will keep growth within acceptable limits. A niece of mine once adopted a stray puppy of indeterminate race and was rather worried when the animal kept growing and growing; yet she knew at least that it could not get larger than a Great Dane or an Irish Wolfhound. (In reality it stayed just below an Alsatian.) With our technology it is not clear that a reasonable saturation will be reached without crises and catastrophes.

A special resource is our natural environment, and many fear that irreparable damage will have been done by technology before the basic materials of technology run out. Vast quantities of oil can be obtained, are being obtained by offshore drilling. There is bound to be oil spillage now and then, and it is quite possible that beaches will

*For readers unfamiliar with this unit, by international agreement the French word tonne has been accepted to designate the metric ton of 1000 kilograms. It is a useful convention because it avoids confusion with the short ton of the United States (equal to 2000 pounds avoirdupois or 907.2 kg) and the long ton of Great Britain (equal to 20 hundredweight, a hundredweight being 8 stones or 112 pounds, hence to 2240 pounds or 1016.06 kg).

have become unsuitable as recreation areas before the oil supply runs out. Already today the water along many beaches is highly polluted by human and industrial sewage and is unfit for bathing. A short distance inland, plastic-lined pools then offer a poor substitute for a real swim. The same is true for several of the Swiss mountain lakes.

The disposal of waste from chemical industries poses increasing problems. This is even more true for the disposal of radioactive wastes from nuclear power plants. I think I am right in saying that few people are entirely happy about the proposed solutions. On the other hand, those that categorically reject every method proposed so far are probably also a small minority, albeit a vociferous one.

In modern feedlots the removal of cow dung—still considered a precious fertilizer in many parts of the world—presents difficulties. It is an amusing thought that this is an age-old problem. One of the labors of Hercules was to clean the stables of King Augeias, where manure had been piling up for many years. The method applied by Hercules would be unacceptable today. He led two rivers through the stables and let them carry the dung away. Such water pollution would certainly provoke an outcry of today's ecologists.

The carbon dioxide problem is less acute but possibly far more important. If we continue to burn more and more fossil fuel and to cut down forests indiscriminately, we may increase the carbon dioxide content of the atmosphere, which thereby becomes less transparent to infrared radiation. This would lead to an increase of the average temperature of the earth and to dramatic changes of climate.

Concern about irreparable damage to our environment is also part of the apprehensions of a second group of critics. They are convinced that, quite apart from possible limitations by scarcity of resources, an excessive use of mechanical energy, and of technological machinery in general, will be dehumanizing and degrading and that our industrial civilization has already passed the limits. I quoted just now the first half of a sentence of Illich. It goes on as follows: "and for the rich, the avoidance of an even more horrible degradation depends on the effective recognition of a threshold in energy consumption beyond which technical processes begin to dictate social relations." In this connection I want to point out that the appliances in our private life are partly time-saving, partly time-wasting; the time we save by devices of the first category can be idled away by devices of the second one. I should also like to try to reproduce a remark once made to me by F. Runge, engineer with Telefunken, son of the well-known Göttingen mathematician Carl Runge and brother-in-law of Richard Courant—intermarriage was a frequent feature of traditional German university life. "When the Devil, whose main objective is to keep man away from meditation and contemplation, noticed that thanks to improved techniques man needed less and less time and effort to provide for his needs, he became really worried. And so he invented—or had people invent —television. And a wonderful invention it was. It cut both ways. It gave plenty of people plenty to do during their working hours, and during their hours of leisure it kept them semi-dazed in front of a TV set. So for the time being the danger that mankind would start thinking about higher things was averted."

Those who think along such lines will, to a certain extent, welcome the conclusions of the Club of Rome: circumstances may force us to adopt another pattern of life, to do away with centralized large-scale operations, to use more human, more "convivial" (Illich's terminology) tools. In some ways such ideas remind one of the unrealistic pastoral illusions of former centuries. Yet I think it would be wrong to regard the whole idea of "Small is Beautiful" only as an extravagance of well-to-do people in a prosperous industrial state. E.F. Schumacher, who left Germany as a young man because he found Hitler's politics unacceptable, deserves to be taken more seriously. (Schumacher, by the way, was not of Jewish descent, which is relevant; he was also Heisenberg's brother-in-law, which is not.)

A feeling of dissatisfaction and distrust does not only relate to the material conditions in industrial societies. It goes deeper. Traditional religions contained a cosmogony, an explanation of the origin of our world and of life on earth; they also contained an ethical code. Modern science has replaced the cosmogony. Darwin's origin of species and Weinberg's first three minutes have ousted Genesis. Rejecting the biblical ideas about creation led many to question the Christian code of ethics as well. But here science does not offer any help. The feeling of incertitude that is characteristic of many young people in our industrial society is not primarily caused by the rapid change of our material surroundings: to that young people adapt with surprising ease. Far more important is the lack of an accepted ethical guideline. That explains the attraction of many more or less mystical cults —some of them no doubt started by well-intentioned and devoted people, some driven by mad fanaticism, some with a definite commercial flavor. It may also explain escape by suicide, or by drugs, a kind of slow or chronic suicide. It may even explain the grip of Marxism, which claims to contain a scientific base for a code of behavior, although most natural scientists will have difficulty in acknowledging its scientific character.

The God of the Old Testament was a god of wrath. He destroyed Sodom and Gomorrah by fire from heaven, punishing them for their sins. For most of us such notions have lost their meaning: we have turned science into our God but it still threatens to destroy our cities with fire from heaven if not for our sins then for our stupidity. Nowhere is the action of the science-technology spiral more manifest, nowhere are its results more alarming, nowhere are apprehensions more justified than in the case of nuclear armament. I shall not try to retell the story of the nuclear bomb: the main features are well known and I cannot add any first-hand information. Could and should England and the United States have refrained from developing the bomb? I do not think so. The danger that it would be developed in Germany was too great, and Hitler would have had no scruples in employing it against England and Russia. Could one have made some underhand arrangement between scientists on both sides not to develop such a weapon? If that was really what Heisenberg had in mind when he went to see Bohr during the Second World War, one can only wonder at his naïveté. Anyway, "with the Devil you cannot negotiate."

Could one have used the threat of the bomb to enforce Japanese surrender without actually destroying two major cities? I am not convinced that would have been impossible but I have to admit that I know little about warfare and international politics. If the memory of the horrors of Hiroshima and Nagasaki has helped so far to prevent both the further use of nuclear weapons and the outbreak of another major war then I am willing to agree that this destruction was not entirely unjustified.

The end of the war was for many of us the end of a nightmare, but since then another long-lasting nightmarish situation has been building up. Ever greater stockpiles of nuclear weapons with ever greater destructive power have been accumulating, in the United States and in Soviet Russia, on a smaller scale in England and France, on a less accurately known scale in China, and on a small scale in some other countries. Means for bringing these weapons accurately to any spot on earth have been perfected.

Niels Bohr was one of those who clearly foresaw this development. His open letter to the United Nations, dated 1950, shows that already in 1944 he was fully aware of the potential threats.[3] It cites the text of a memorandum he submitted early in 1944 to—and later discussed with—President Roosevelt. In that memorandum he speaks about "the terrifying prospect of future competition between nations about a weapon of such formidable character." He points out that "the enterprise . . . has proved far smaller than might have been anticipated" and emphasizes the need for "prevention of a competition prepared in secrecy." It is clear that, right from the start, he was convinced that Russia would have no difficulty in making nuclear weapons once it knew that they had been made elsewhere. Bohr believed it should be possible to use America's head start to force the Soviet Union into accepting international control if the appropriate steps were taken at an early moment. In a further memorandum of March 1945 he adds:

All such opportunities may, however, be forfeited if an initiative is not taken while the matter can be raised in a spirit of friendly advice. In fact a postponement to await further developments might, especially if preparations for competitive effort in the meantime have reached an advanced stage, give the approach the appearance of an attempt at coercion in which no great nation can be expected to acquiesce.

Bohr thought that Roosevelt paid some attention to his ideas but a meeting with Churchill was a complete catastrophe.[4] Churchill, whose tremendous force may have partly resided in a complete disregard of subtle distinctions like those of Bohr's complementary philosophy, did not realize that Bohr, far from being a muddle-headed thinker, saw the future more clearly than he did himself. Whether a policy along the lines Bohr advocated would have been feasible is doubtful; now that such a policy was not followed what happened was entirely in agreement with Bohr's predictions. Russia rapidly made its first nuclear bombs and from then on "competition prepared in secrecy" has been the order of the day. And the USSR has certainly not acquiesced to any kind of coercion.

Einstein was far less knowledgeable about the details of nuclear physics than Bohr, who, in the late thirties, made important contributions to the theory of nuclear reactions and nuclear fission and had many more contacts with experimental physics than Einstein, who was pursuing his solitary quest for a unified field theory. Einstein was involved only because he signed the fateful letter to Roosevelt that did so much to get the nuclear bomb project started. But in this case his views did not diverge in any essential way from those of Bohr. C.P. Snow summarizes Einstein's attitude as follows:

> The bomb was made. What should a man do? He couldn't find an answer people would listen to. He campaigned for a world state: that only made him distrusted both in the Soviet Union and in the United States. He gave an eschatological warning to a mass television audience in 1950:
>
>> And now the public has been advised that the production of the hydrogen bomb is the new goal which will probably be accomplished. An accelerated development towards this end has been solemnly proclaimed by the President. If these efforts should prove successful, radioactive poisoning of the atmosphere, and hence, annihilation of all life on earth will have been brought within the range of what is technically possible. *A weird aspect of this development lies in its apparently inexorable character. Each step appears as the inevitable consequence of the one that went before.* And at the end, looming ever clearer lies general annihilation.*
>
> That speech made him more distrusted in America. As for practical results, no one listened. Incidentally, in the view of most contemporary military scientists, it would be more difficult totally to eliminate the human species than Einstein then believed. But the most interesting sentences were the ones I have italicized. They are utterly true. The more one has mixed in these horrors, the truer they seem.[5]

The situation in which our scientific and technical proficiency has landed us is indeed a grim one. Like Gulliver, who had some difficulty in appreciating the fundamental differences between broad-endians and narrow-endians, a visitor from another world would find it hard to see much difference between ideologies in different parts of the world. He would probably conclude that the earth has entered

---

*The text of Einstein's TV broadcast as given in Einstein, *Ideas and Opinions* (Laurel edition), is somewhat different. For instance, the italicized sentences read there "The ghostlike character of this development lies in its apparently compulsory trend. Every step appears as the unavoidable consequence of the preceding one." That is closer to the original German version in *Mein Weltbild* (Einstein always wrote a text in German first) and is probably what Einstein said, but Snow's translation is the better one.

a suicide pact and is now carefully preparing it on a global scale. Or would he regard what is being prepared as a brutal and belated form of family planning rendered necessary because more humane methods have long been neglected and in some quarters even obstructed?

It is in this context that the question of the responsibility of scientists imposes itself. There are many that hold that programs of scientists should be judged, approved, or stopped on the basis of social relevance. The notion strikes me as singularly useless. First of all scientific beauty and philosophical depth are at least as important as practical applicability and that is rarely taken into account. Second, the time lag between scientific progress and practical application makes it impossible to determine practical social relevance at the time fundamental results are being obtained, and even more impossible to do it before.

The ethical and social implications of fundamental research can be assessed only when applications are being made or at least being considered. But at that moment it is usually too late for the scientist to exert much influence. In general, fundamental knowledge is in the public domain long before it is going to be applied, and it will be applied even if the original pioneers do not cooperate, whether for reasons of principle or because they are just not interested. Whatever applications are possible, however gruesome they may be, there will always be technicians willing to serve, whether for patriotic or for financial reasons: they will forget or defend the possible use and they will be fascinated by the problems as such. Oppenheimer once said that the hydrogen bomb is technologically sweet and the jarring incongruity of that term epitomizes the whole problem. I mentioned this aspect already when, in Chapter 4, I spoke of Daedalus. The case of the first nuclear bomb is exceptional: there the fundamental scientists cooperated wholeheartedly. I have already pointed out that this led to a considerable shortening of the usual time lag. But the bomb would have been made anyway, albeit later.

Again we must ask: What should scientists do? We cannot undiscover what has been discovered, not unexplain what has been explained. And we cannot prevent the application of what has been discovered and explained. Even if we could stop all basic research now, that would not have a *direct* influence on world affairs for the first nine years or so. A vow of chastity made today will not influence the birthrate during the next nine months. Stopping basic research would not reduce the threat of a nuclear war, for instance: for waging war, existing knowledge is quite adequate. We can only try to behave as mature citizens, even as knowledgeable citizens, and not only as citizens of our own country but even, because of our many international contacts, as citizens of the world. We can help fight fallacious ideas and we can try to expose hypocrisy and stupidity. But we alone cannot control or stop the ominous, the inexorable spiral.

All the same, I have some hope for the future. The science-technology spiral means an alliance of the philosopher and the engineer, of the *homo sapiens* and the *homo faber*, but it is a one-sided philosopher that has joined forces with a narrow-minded technician. They make an admirably efficient and productive duo but they have outrun control by wisdom and by charity. The reactions I have tried to sketch show that man begins to understand that this is so. Even if such reactions take the form of inept anti-intellectualism or lead to a fatuous wish to control the essence of knowledge by democratic despotism, even then they bear witness to a growing awareness that we should not let science and technology become our masters instead of our tools. Perhaps this final chapter of my book can contribute something to a better understanding of the nature of science and technology and of their mutual interaction and so, in a very small way, assist in gaining control of our destiny before it is too late.

The Discussions Between Bohr and Einstein on the Interpretation of Quantum Mechanics

In a way, it was unexpected that Einstein, who had introduced the notion of a particle-wave duality when he proposed the light quantum, who had in many of his papers been dealing with various aspects of statistical theory, and who even could be said to have been "godfather" of De Broglie's wave theory, did not take kindly to the statistical interpretation of the new formalism.[1] In his discussions with Bohr we can distinguish three distinct phases:

First Phase: Solvay Conference 1927. Einstein tries to find inconsistencies in the Heisenberg-Bohr theory, by means of shrewd "Gedankenexperimente" (fictional experiments). Of course, that was a method that had by no means been invented by Heisenberg or Bohr. Einstein had already used it to great advantage in discussing the implications of his theory of relativity. Bohr was able to refute all such experiments in detail (I vaguely recall Ehrenfest reporting at the colloquium about these discussions but have mainly to rely on Pais's article).

Second Phase: Solvay Conference 1930. This time Einstein introduced a new element: the relation $E = Mc^2$ makes it possible in principle to determine energy by weighing. Using this, Einstein tried to show that one can beat the equation $dE \cdot dt = h$. This gave Bohr considerable trouble, but, possibly after a sleepless night, he was able to refute Einstein's claim. The essential point in his argument was the red shift of the clock activating a shutter, which had a spread because the vertical position of the balance had to have a certain spread. I spent most of the autumn of 1930 in Copenhagen, and we were quite excited about this story. We did not exactly welcome the conquering hero by sounding trumpets and beating drums, but we—I think Gamow, Landau(?), and that versatile artist Piet Hein—had a nicely designed contraption made in the workshop. It may well have stood model for the elaborate drawing reproduced by Pais.[2]

Third Phase: Einstein conceded that the theory was free of contradictions, that it was consistent. He also admitted that it was extremely powerful and fertile. But he could not regard it as a complete description of nature and therefore he could not regard it as final. And here I can draw on my own personal experience. The scene is a colloquium at Leiden, almost certainly during the winter of 1931–32 when I was assistant to Ehrenfest. Ehrenfest, of course, was in the chair, but more silent than usual, and Einstein was giving a lecture. What exactly did he discuss? I think—but I have to admit that I am not completely certain—that he started with a box, containing a clockwork and a shutter and placed on a sensitive balance, as in his 1930 paradox. But this time he no longer tried to beat Heisenberg's principle. He stressed that after the light quantum has escaped we can still choose whether we want to read the clock or alternatively determine the weight of the box. So, without touching the light quantum in any way, we can either determine its energy or the time at which it will return from a distant mirror. As I have said, I am not completely certain this was the experiment he discussed. I am certain that it was a case where we can still choose whether we want to know one quantity or the other (energy or time, place or momentum), although the particle itself is already out of reach; and that was also the gist of the later paper by Einstein, Rosen, and Podolsky. Ehrenfest had entrusted me with the task of opening the discussion and I explained to the best of my ability the Copenhagen views on such questions. Einstein listened, perhaps slightly impatiently, and then he said—and I believe now I can vouch for the exact words: "Ich weiss es, widerspruchsfrei ist die Sache schon, aber sie enthält meines Erachtens doch eine gewisse Härte." Pais, who also mentions this episode, translates this as: "I know, this business is free of contradictions, yet in my view it contains a certain unreasonableness." I think "a certain unpalatability" comes closer to what Einstein had in mind.

This was the only time I faced Einstein in a discussion. I had heard him give a lecture once or twice but that was all. When I came to Princeton after the war I did not try to approach him. The tragic death of Ehrenfest had broken a link and had created a distance I was reluctant to cross.

There is, of course, a fundamental difference between the statistical theories worked out by Einstein early in the century and the statistical interpretation of quantum mechanics, although there is also a formal analogy. Einstein could derive the essential properties of Brownian motion without introducing a molecular model. He had only to postulate that the probability of a state is given by

$$W = \exp(S/k),$$

where S, the entropy, can be calculated from macroscopic data. The atomic world only enters into the picture by Boltzmann's constant, k, written by Einstein as R/N (R the "gasconstant" in the equation pV = RT, and N Avogadro's constant). If k were infinitely small, or, which is the same thing, if the number of atoms were infinitely large, there would be no fluctuations at all and no Brownian motion. Similarly all quantum effects are connected with Planck's constant. If h goes to zero, quantum effects disappear.

But in the case of Brownian motion we do know that what we observe is a manifestation of the atomic world, a manifestation the main characteristics of which can be calculated without a detailed model, but which can be regarded—and was regarded—as a clear proof of the reality of the atoms.

Should we not similarly look in quantum mechanics for a deeper underlying structure? And anyway, isn't the "Härte" mentioned by Einstein an indication that the theory is incomplete?

Yes, we must look for some deeper structure, said Einstein, and he spent the last twenty-five years of his life in a fruitless search for a field theory that would encompass quantum theory and restore complete causality.

No, said Bohr. The quantum-mechanical description is as complete as it can possibly be. That which Einstein calls a "Härte" is an essential feature of our human condition. The images we form, the language we use are only adequate to describe atomic phenomena in a "complementary" way. We have sometimes to use a wave-picture, sometimes a particle-picture. These pictures are mutually exclusive, but quantum mechanics has created a consistent mathematical formulation governing their use. And this philosophy of Bohr prompted a whole generation of physicists to accept wholeheartedly the methods of quantum mechanics and to apply them without qualms to an ever expanding range of problems.

So successful have these applications been that—using the terminology of Chapter 2 we may say that the quantum mechanics of atoms and molecules (I am not speaking about the nucleus and still less about elementary particles and high-energy physics) has reached the technical stage.

But where is the ultimate truth? The spectacular success of Bohr's school of thought and the sterility of Einstein's later attempts are no real measure. Personally, I think Bohr came closer to the heart of the matter. I cannot help feeling that there is a curious vicious circle in Einstein's way of thinking. As a philosopher he believed in causality; that he had learned from his work in physics. In physics he looked for causality because that corresponded to his philosophical convictions. But I know I am prejudiced. As I explained, I knew Einstein only slightly whereas my contacts with Bohr have had a profound influence on my whole outlook.

And anyway, who am I to pronounce judgment in this debate between the two greatest theoretical physicists of our century? I know that the account of their differences I have presented here is very superficial, and experts may well be of the opinion that I have misunderstood some of the essential points. But it is a fair account of the way I looked at these questions, the way I still look at these questions. And I think it is not very different from the way many average physicists of my generation looked at them.

Physics Around 1930

Although this section is more technical than my main text, it is not different in spirit: it does not purport to give a comprehensive and well-documented survey; it is a narrative of personal impressions. I trust that what I write here is also factually correct, but choice and emphasis are determined by my personal preferences and experiences.

Around 1930 non-relativistic quantum mechanics was gradually entering the "technical stage"—in the sense explained in Chapter 2, that is; no industrial or commercial applications were yet in sight. Let us first look at the study of atomic spectra. It was no longer expected to lead to new laws of nature. Neither was it felt that it should be further pursued in order to confirm the basic notions of quantum mechanics. A remark of O. Klein is characteristic of the situation. I once said to him that it would be nice to have large automatic calculating machines that would accurately determine the eigen-values of the Schrödinger equation for many interacting particles, and hence the energy-levels of complicated atoms. (By the way, even the largest modern computers, hardly dreamt of in those days, are not yet very good at this.) Klein said: "But such machines exist; they consist of a gas discharge tube and a spectroscope." All the same, details of complicated spectra remained interesting for their own sake and there was plenty of room for refining mathematical methods.

During my first months at Copenhagen, B. Trumpy from Norway was a guest in Bohr's Institute. He had measured intensities of some spectral lines and was comparing them with values he calculated. For that calculation he needed (approximate) wave functions, which he determined by numerical integration, using a hand-operated mechanical calculator. He was a hard worker and could calculate for hours and hours, but about once an hour he would get up, quietly walk once or twice around the table on his hands(!), and then, refreshed, resume his calculations.

Of course, in recent years the whole subject has been revived and considerably refined by the coming of lasers. By the way, it is amusing to consider here how lasers might have been invented already in those days. The notion of stimulated emission was well-established since Einstein's 1917 paper. Inverted populations of levels must not have been unknown to experts on gas discharges, and the optical technology that was used in making interferometers would have been quite adequate for getting a laser going.

Another new line in spectroscopy *was* however developed in the late twenties and thirties: the study of hyperfine structure. In the theory of Rutherford-Bohr it is assumed that the nucleus is a point-charge (with charge Ze and with a mass M). That was completely adequate to explain the structure of spectra. If, however, one looks at these spectra in great detail—with spectrometers of very high resolving power—then one discovers an even finer structure, the so-called hyperfine structure. Pauli had already observed in 1924 that this must be due to a property of the nucleus other than charge and mass and he suggested the existence of an angular momentum of the nucleus.[1] Somewhat later Goudsmit and Back carefully analyzed the hyperstructure in the bismuth spectrum and were able to derive both the angular momentum and the magnetic moment. From then on the analysis of hyperfine structure made useful, albeit modest, contributions to our knowledge of the nucleus. At Copenhagen, Ebbe Rasmussen headed a group that did work in this field, and the German physicist Kopfermann was a frequent visitor.

On one of the first days of the Copenhagen conference Goudsmit asked me if I could calculate the interaction between a magnetic moment of a nucleus and an electron in a so-called s-state, for instance in the ground state of the hydrogen atom. That was at that moment an unsolved problem. In such states the wave function is spherically symmetric around the nucleus and does not vanish at the nucleus. If we now assume that a magnetic moment μ is placed in the nucleus and that the total magnetic moment of the electron is distributed in space according to the probability density, and then try to calculate the interaction using the normal formula for the interaction of two magnetic dipoles, then we find that the integral diverges. I really had some beginner's luck that time.[2] I saw at once how the problem might be tackled and in a few days I had worked out the solution.

The question is that in Dirac's (relativistic) theory of the electron the magnetic moment of the electron is the result of a current density. I had been deeply impressed by that fact and also by the beauty of the current distribution in the ground state of hydrogen, which was explicitly written down in Weyl's book. It can even be described in words. The charge density (or to be accurate the probability density that a charge will be found) falls off exponentially with increasing radius. The current density can be described by saying that every spherical shell rotates around the z-axis at such a speed that the linear velocity at the equator is always αc, where c is the velocity of light and α the so-called fine structure constant, $\alpha = e^2/\hbar c = 1/137$. In retrospect it is curious that as far as I have been able to ascertain no one had thought of associating a current density with the electron spin before Dirac formulated his relativistic theory. Yet it would have been quite simple and straightforward to do this within the framework of Pauli's nonrelativistic theory. One has only to apply the formula of macroscopic electrodynamics

$$i = c \text{ curl } M$$

For an s-state with spin in the $-z$-direction this gives

$$i = c \frac{e\hbar}{2mc} \text{ curl } e_z \, |\Psi|^2$$

The formula for the magnetic field at the origin becomes then almost a tautology.

A slight historical—and didactical—digression. Soon after Ørsted had found that an electric current acts on a magnetic needle (he himself, by the way, did not speak about currents* but about the electric conflict), Ampère formulated the hypothesis that all magnetism is due to currents. This idea was taken over by Maxwell; it is also an essential feature of Lorentz's theory of electrons. Now it might seem for a moment that Goudsmit's and Uhlenbeck's idea of the spinning electron might lead to the existence of true magnetic dipoles. Dirac's theory did away with that notion, to my great relief. As a matter of fact, I found—and still find—magnetism a muddled subject unless you base it entirely on currents. Magnetic moments can then be introduced much later in the game as auxiliaries, useful for performing certain calculations but without real physical meaning.

Knowing the current distribution in an s-state I had no difficulty in calculating the magnetic field arising from that current distribution at the origin. As a matter of fact I needed little more than the formula for the tangent galvanometer—the "practicum" did have its uses after all—and some elementary integrations. One finds for this field (neglecting higher powers in α)

$$H(o) = \frac{8\pi}{3} \cdot \frac{e\hbar}{2mc} \, |\Psi(o)|^2$$

where $\Psi(o)$ is the normalized Schrödinger wave function at the center. It is then easy to find the interaction energy. Goudsmit provided some ingenious

*Compare Maxwell's comment on Ohm, in Chapter 2.

estimates for $|\Psi(o)|^2$ and found that my formula could be successfully applied to the Na-spectrum and a few other cases. But he insisted I should publish the thing alone, so I spent some of my time at Copenhagen in writing a paper and I then sent it to Goudsmit. I still have a copy of the text; it was a pretty clumsy paper and I can well understand why Goudsmit did not quite know what to do with it. Finally, he sent it back with some suggestions for improvements, but before I could take care of them, Fermi's well-known paper had appeared. I must admit it was a disappointment. And the result turned out to be even more important than could be foreseen at the time. It predicted the existence of an absorption and emission line of atomic hydrogen at a wavelength of 23 cm. Later, radio astronomers studied the emission of this radiation by various regions in space and this has contributed much to our knowledge of the structure of our galaxy. It would have been nice to have been the first to publish the underlying theory.*

However, it may have been just as well. I was sufficiently conceited anyway, and this early introduction to this field stood me in good stead later on.

Then there were the problems of atomic collisions. Mott,† who was just finishing his stay at Copenhagen when I arrived, did beautiful work on that.

The further applications of quantum mechanics went in two opposite directions. On the one hand was the direction of molecules and of condensed matter, especially solids. We have already seen that Pauli was dissatisfied with the unexact character of the further development of the Heitler-London theory of the chemical bond. And it must be admitted that even today questions can be raised as to the validity of the approximations involved in the calculations—now with big computers—of binding energies, etc. We shall see later that Pauli also had some objections to solid-state physics. Yet it cannot be denied that there the impact of quantum theory has been of decisive importance. It is amusing to ponder the flow of influences. Earlier in the century the specific heat of solids was a phenomenon the analysis of which helped to point the way towards quantum theory. Next, the center of the battle was in atomic spectra, and, finally, new quantum mechanics began to pay back its old indebtedness to solid-state phenomena.

Quantum mechanics also moved in the opposite direction: towards smaller particles and systems, towards the nucleus. There things were, for the time being, far more obscure. Here not only was it a question of equations too difficult to deal with but fundamental points were not understood, and one could certainly *not* speak about the theory being in the technical state.

---

*During a visit to Leiden in the early thirties, Fermi more or less apologized for having interfered with my work. That was very generous: he could not possibly have known I had been working on that problem!

†N.F. (Sir Neville) Mott. 1905– . Professor at Bristol 1933–1954. Cavendish Professor at Cambridge 1954–1971. Nobel Prize for Physics 1978.

Now where did Bohr himself stand? He was only mildly interested in the further development of mathematical techniques. I later became interested —following up a brilliant idea of Oscar Klein—in the theory of the assymetric top and that led me to some new results in the theory of the representation of the group of three-dimensional rotations that I could generalize to more general groups. That work was written down in the thesis with which I obtained my doctorate at Leiden on 3 November 1931.

This thesis—as well as a preliminary publication—contains what is often referred to as a Casimir operator.[3] To get this into the right perspective, however, one should consult a facetious English book, *How to be Famous*;[4] under the heading "Plato" it says: "His own inaccurate conception of Platonic Friendship has been superseded. However, in his day it was an improvement on the existing system of having no Platonic Friendships at all." Mutatis mutandis this might also be applied to Casimir operators.

From time to time Bohr would ask me "Hvordan staar det med det her Rotationsvaesen?" (How is it going with this rotation business?)—but more out of human kindness than because of any real interest.

In solid-state physics, Bohr was in principle interested; after all, he had written a magnificent thesis on the theory of electrons in metals,[5] in which he had among other things, made it abundantly clear that many fundamental properties of electrons in metals cannot be understood on the basis of classical theory. But his interest was not really active and in his institute little in this field was going on. Yet he must have voiced expectations about what quantum mechanics would be able to do: in Bohr's collection of reprints there was a paper of Felix Bloch on electrons in metals with an inscription by Pauli—who did not like solid-state physics but who knew an outstanding paper when he saw one—"Aus dem beiliegenden Separatum von Bloch wirst Du sehen, dass Dein Triumf über die Gelehrten vollständig ist" (From the enclosed reprint of Bloch you will see that your triumph over the learned is complete).

Neither was Bohr himself as yet very active in nuclear physics: that would come a few years later, almost as a second youth. But he did encourage Gamow and this must certainly have made Gamow more credible to Rutherford and thus have furthered the first Cockcroft-Walton experiments.

Bohr's own work consisted mainly in refining and deepening his ideas on the meaning of quantum mechanics, and trying to define the limits of its validity. The first problem he discussed with me was a further analysis of a presentation he had already made at the 1929 conference in which he had argued that the magnetic moment of an electron cannot be determined by a classical Stern-Gerlach experiment. We went through a number of possible experimental arrangements and always found that the uncertainty in the Lorentz force was larger than the effect to be measured. However, it is probably my fault, but I have never been able to see that this was very meaningful. And Mott showed, applying Dirac's relativistic theory, that electrons scattered by a nucleus are polarized to an increasing degree when their speed approaches that of light.[6] This polarization can in principle be detected—and many years later was experimentally detected—by double-scattering experiments.

There were, however, serious difficulties associated with Dirac's equation: the states of negative energy. Bohr may have hoped that they could be circumvented by a careful analysis of what was measurable and what not, but things did not work out that way. The trouble with Dirac's equation was that besides having solutions with positive energy it has also solutions with negative energy. For a free particle there is even complete symmetry between states with positive and states with negative energy. Now, at first, one might think that this was no real difficulty and that one might just discard the solutions with negative energy. After all, the well-known relativistic equation

$$\left(\frac{E}{c} \right)^2 - p^2 = m_0{}^2 c^2$$

does also admit negative values for the energy.

But things are not as simple as that. The point is that transitions from states with positive energy to states with negative energy cannot be excluded. Application of the usual theory of radiation would lead to the conclusion that all electrons should in the long run get more and more negative energy, the energy balance being made up by radiation. Klein had found even a more striking example: an electron running up against a static potential of more than $2mc^2/e$ would have a finite chance of getting through as a negative energy electron. (If one might still have some hope of getting out of the radiation difficulty by some modification of electrodynamics, one could hardly hope to modify the theory for the action of a static field.)

Then came Dirac's bold proposition: assume that all states of negative energy are occupied. (Since electrons obey Fermi-Dirac statistics, which is the quantum-mechanical expression of the Pauli principle, you need only one electron per state.) One often speaks—or rather, spoke—about the Fermi-sea. Dirac further proposed that sometimes there might be an unoccupied place in this Fermi-sea and that would act as a positively charged particle with positive energy; that might be the proton, so Dirac thought at first. In those days the number of elementary particles physicists believed to exist passed through a minimum: it was believed there existed only protons and electrons. The atomic weight of separated isotopes was always close to an integer times the atomic weight of hydrogen; that integer, the mass-number, was believed to be the number of protons in the nucleus. A number of electrons then had to be added to bring the charge number Z to the correct value. Dirac's attempt to reduce the number still further, namely to one, would in reality soon turn out to be the beginning of an increase in the number of elementary particles, an increase that is still continuing today.

Bohr was, at first, not at all charmed by this idea of a Fermi-sea. Dirac claimed that you would not notice the electric fields of all these negative energy particles because they were all around so that their electric fields would compensate one another, and I remember that Bohr explained to me that that was nonsense because the divergence of the field would have to be infinite anyway. I think Dirac later dropped that argument.

The rest of the story is well known. It was soon shown by several theoreticians that in such a Dirac theory there would always be symmetry between electrons and holes. Therefore, one could never get a proton that way. So Dirac suggested that the holes might be positive electrons, but why then did positive electrons never occur? In 1932 Anderson, and somewhat later, Blackett and Occhialini, showed that positive electrons do occur, though only temporarily.

The model of holes in the Fermi-sea of negative-energy electrons is a useful one to describe the positive electron and to account for its main properties, but it is no longer the way in which theory looks at things. Although in the theory of atomic spectra and similar problems, the Dirac equation can be considered just to provide relativistic corrections to the Schrödinger equation without changing the essence of quantum mechanics, today most theoreticians would rather regard the Dirac theory as something beyond the original quantum mechanics, namely as an early example of a field theory in which the number of particles is not conserved.

It is not always realized that almost simultaneously with Dirac's theory of electrons and (at first) protons, but slightly earlier, the idea of positive holes had also arisen in solid-state physics. It was at the base of Peierls's explanation of the positive Hall effect. In the solid-state case the Fermi-sea is very real: it consists of the electrons in the valency band. The charge of the Fermi-sea is here no problem: it is compensated by the positive charge of the atom cores. At the top of the band the electrons have a negative effective mass; in that way they are analogous to the negative-energy electrons (which were sometimes facetiously called donkey electrons because they moved against the direction of applied force). I do not think that at the time much attention was paid to that analogy, in any case not at Copenhagen.

In today's terminology we might say: the Dirac vacuum is a semiconductor with a band gap of $2mc^2$, that is, roughly a million electron volts. Since this band gap is so large there is practically no intrinsic conduction. However, the vacuum is n-type: it is doped with positive nuclei. Like holes in n-type germanium or silicon, positrons can survive for some time in the n-world. Positronium corresponds exactly to an exciton, and so on.

What about quantum electrodynamics? Dirac had shown how to quantize the radiation field and his theory was entirely adequate for calculating emission and absorption of radiation, radiation damping, line width, and so on. Heisenberg and Pauli gave a relativistic generalization (and so did Fermi in a slightly different form). A few years later Bohr and Rosenfeld would publish a careful analysis of uncertainty relations and possibilities of measurements prevailing in electromagnetic fields.

There remained, however, as Ehrenfest emphasized, a fundamental difficulty that was already present in Lorentz's theory of electrons. If one assumes that an electron is a point, has a radius zero, then the electromagnetic field will have an infinite energy. If in classical theory one assigns a finite radius to an electron then the question arises how this charge density is held together: one has to assume some kind of tensions of non-electromagnetic nature—so-called Poincaré tensions—and relativistic invariance poses further problems. Fortunately, many of the results of the classical theory of electrons are "structure-independent," but as soon as one approaches the problem of electromagnetic mass this is no longer the case. Now in Dirac's theory of radiation, and also in the more sophisticated formulation of Heisenberg and Pauli, the terms in the equation expressing the interaction of an electron and the electromagnetic field definitely describe a point charge. And certainly, already in a second approximation, there appear divergent terms in the expression for the energy. In those days there was a feeling that someone would soon find an elegant solution for this difficulty. Coupled with that was a hope this would at the same time lead to a theoretical value of the fine-structure constant. Long before the coming of new quantum mechanics Einstein had been impressed by the fact that h and e^2/c have the same dimension and are not orders of magnitude different: he felt this shows a relation between the quantization of electric charge and the quantization of energy, but he was unable to get beyond that.

"Who is going to create quantum electrodynamics?" (Wer wird die Quantenelektrodynamik machen?) was a question Landau frequently asked.

To what extent have the problems of quantum electrodynamics been solved? Great progress was made in the late forties. It seems to me that the problem of electromagnetic mass has not really been solved, but that a systematic method has been found to circumvent it. To put it in a very crude way: one cannot calculate electromagnetic mass but one can determine the very tiny differences in electromagnetic mass in different situations. The results of this "re-normalization theory" have been confirmed experimentally with almost unbelievable precision, but again any hopes that the theory would lead to a determination of the fine structure constant proved vain: the theory works for any value of α provided it is small.

In 1929 the great astronomer Sir Arthur Eddington made a courageous attempt to explain the fine structure constant, and many other constants of nature.[7] In his theory α should be equal to the number of components of a symmetrical tensor in sixteen dimensions, plus (added as an afterthought) one. That gives ½ × 16 × 17 + 1 = 137. Unfortunately I have been utterly unable to follow his reasoning. What is more important, none of the theoretical physicists whose opinions I respect have been able to do so either. Yet one cannot help wondering whether someday a grain of truth will be discovered in these weird speculations.

A problem discussed at several successive conferences was posed by the continuous beta spectrum. Whereas alpha particles leave a nucleus with well-defined energies, the energy of beta particles, that is of electrons leaving the nucleus, is distributed over quite a wide range. For radium E, for instance, the average energy is roughly 340 keV, the maximum energy 1 MeV. Yet it became increasingly clear that the nuclei before and after radio-active disintegration can have no spread in energy. It was at first thought that the energy balance might be established by gamma rays, but careful measurements by Lise Meitner ruled out that possibility. So there were two possibilities: either in this case the conservation of energy does not hold for the elementary radioactive processes, or the missing energy escapes in the form of an as yet undetectable radiation. Bohr inclined to the first point of view; at least he felt one should be prepared to give up the idea of rigorous conservation of energy. (An analogous proposal relating to the emission of light by atoms had been made by Bohr, Kramers, and Slater in 1924, but it had been refuted by experiments of Geiger and Bothe.) Pauli, on the other hand, was not willing to renounce rigorous conservation of energy and he postulated a new particle, subsequently called neutrino, that would carry away a varying proportion of energy and that had so little interaction with matter that it could escape unnoticed. It therefore could have no charge and no magnetic moment. Pauli did not dare to publish his idea but on 4 December 1930 he sent an open letter to the "group of radioactive ones" at a meeting in Tübingen. "Dear radioactive Ladies and Gentlemen" it begins, and it explains his idea. The essential point is, that not only does this scapegoat carry away the missing energy, but it can also account for discrepancies in angular momenta and in statistics. Of course, we all know that Pauli's idea has later been fully vindicated both theoretically and by direct experiments. Pauli himself has related the history of the topic in a lecture published posthumously.[8]

One day—it must have been in 1930, I think—Bohr received a letter from Pauli which he found rather difficult to answer. So he asked his wife to write a kind letter to Pauli and to tell him that Bohr himself would write on Monday.* Two weeks later there came a reply from Pauli to Mrs. Bohr saying that Niels had been very wise to promise to write on Monday without specifying which Monday. But he should not feel in any way tied to Monday (Er soll sich keineswegs an Montag gebunden fühlen); a letter written on any other day would be just as welcome. I have so far been unable to trace these letters of Pauli. It would be nice if the letter to Bohr contained the first intimation of the neutrino idea, but I am afraid this is not possible. Pauli's letter to the "radioactive ones" is dated 4 December 1930 and the idea is also mentioned in a letter to O. Klein dated 12 December 1930, in which he writes "I should like to know Bohr's opinion," which means that he had not yet written to Bohr. I left Copenhagen before Christmas and although I came back for a month in March 1931, I am almost certain the episode did not take place during that time. So the remark I made elsewhere[9] must be relegated to the domain of embellished legend.

*Frisch, loc. cit. p. 48 tells the same story but makes it Thursday. My recollection —I was present when Pauli's answer came—is Monday and this seems more logical: one likes to postpone things to the next week. Not that it makes much difference.

Low-temperature Physics in the Thirties

Low-temperature physics was then—and still is today—a fascinating subject, but its fascination lies in a multiplicity of elegant and surprising details rather than in one major contribution to the basic principles of natural philosophy, and since this is not a textbook where such details can be discussed at length I am afraid it may be difficult to convey an impression of that characteristic. The best I can do is to present a kind of "catalogue raisonné" in which I shall pay special attention to the Leiden work. Even so, this appendix has become rather long. There is a good reason for that. I do not want to say that I did particularly valuable work during my years at the Kamerlingh Onnes Laboratory, but I came closer to having a regular program than at any time before or after, and I was something of a specialist in a well-defined field.

Heat is a mode of motion. One might therefore expect that when the temperature goes down to zero all internal motion will gradually cease and the matter will settle down to its true equilibrium state; why should anything very remarkable happen en route? De Haas once told me that even the great Lorentz originally had some misgivings about the program of Kamerlingh Onnes. Indeed, if the atomic motion would follow the laws of classical mechanics, low-temperature physics might well be a dull subject. But low-temperature physics is quantum physics; that is what makes it interesting. It means that there are discrete quantum states and when the temperature goes down, states of higher energy will no longer be excited: fewer and fewer states will be involved in the thermal motion and will reveal their true nature. Sometimes this leads to simplification; behavior at low temperatures is then easier to explain than high-temperature behavior. Sometimes effects that are barely observable at room temperature become very pronounced at low temperatures. And in two instances the low quantum states have most remarkable properties: superconductivity and superfluidity.

Simplification prevails in the case of gases and liquids: they cease to exist. At liquid-helium temperatures the vapor pressure of all gases—helium itself excepted—is exceedingly small. That is convenient for vacuum technique, but the very, very dilute gases have otherwise no interesting properties I know of. At helium temperatures there exist no liquids, of course again except liquid helium itself, about which more later.

Non-magnetic, non-conducting substances also show simplification in their thermal behavior. The thermal motion can be thought of as consisting of a superposition of elastic waves. However, unless the wavelength is much longer than the distance between neighboring atoms these waves will show dispersion: their velocity depends on their frequency. This relation has been found to be a rather complicated one. Today we know a lot about the so-called phonons—quantized waves—both from theoretical calculations and from refined experimental procedures. But at sufficiently low temperatures—that is, at temperatures of liquid helium (4°K and less)—only the very long waves are excited. Their velocity does not depend on frequency and can be calculated from the macroscopic elastic constants of the crystal. The specific heat is then proportional to the cube of the temperature.

New quantum mechanics added one important feature. It established definitely the existence of a zero point energy. The lowest value of the energy of a harmonic oscillator is not zero but $\frac{1}{2}$ $h\nu$. This can be qualitatively understood on the basis of Heisenberg's principle of indeterminacy. A particle cannot be exactly in its equilibrium position, for then the velocity would have an infinite spread. The best possible compromise gives the value $\frac{1}{2}$ $h\nu$. This means that even at absolute zero, crystal vibrations contribute to the total energy of a crystal and the value of this contribution can be estimated on the basis of Debye's simple theory. However it is difficult to measure this energy directly. Its value is so small compared with the total binding energy of a crystal and the theoretical calculations of such binding energies are so inexact, that it is completely illusory to think that comparison of theoretical and empirical values for the binding energy would provide a proof. But there is another possibility. If we compare two isotopes, for instance the isotopes neon 20 and neon 22, all frequencies will be in the proportion $(22/20)^{1/2}$ and therefore the zero-point energies will be different by about 5%. On the other hand the leading terms in the formula for the vapor pressure of a solid at low temperature are $\log p = C - \lambda_0/RT$ where λ_0 is the heat of vaporation per mole at the absolute zero and R is the so-called gas constant. It follows:

$$\log\frac{p_{22}}{p_{20}} = -\frac{(\lambda_0)_{22} - (\lambda_0)_{20}}{T} = -\frac{(\lambda_0)}{T} \cdot \left\{\sqrt{\frac{22}{20}} - 1\right\}$$

Measurements by Haantjes and Keesom, who had separated—or, better, enriched—neon isotopes by fractionated distillation, certainly confirm this idea. There remain some discrepancies, but the agreement between these measurements and the simple theory is close enough to be regarded as convincing proof of the existence of zero-point energy.

Not only can one study the specific heat—as was done by Keesom and collaborators—one can also investigate the conduction of heat. That was done by De Haas and Biermass for thin crystal rods of quartz and of potassium chloride. They found the somewhat surprising result that the heat resistance is *not* proportional to $1/d^2$ (d = diameter of rod) as one would expect in analogy to Ohm's law, but to $1/d^3$. This can be understood on the basis of a picture already used by Debye: assume that the elastic waves are diffusely scattered when they hit the boundary of the rod but are otherwise unperturbed. Then a simple argument shows that the heat flow W through a rod with diameter d and length l the ends of which are maintained at temperatures T_1 and $T_1 + \delta T$ is given by $W = C \cdot d^3 / l \cdot T_1^3 \cdot \delta T$ and I showed that it is even possible to calculate the constant from the elastic constants of the material.

It is a rather amusing situation: because of quantization only the long waves are excited, but their properties can be derived from macroscopic theory of elasticity without further reference to quanta and atomic models. (Of course, if we should wish to derive the value of the elastic constants from first principles we would have to go deeply into atomic theory.)

Paramagnetism is a weak phenomenon at room temperature; it becomes far more pronounced at helium temperatures. It has been studied especially in ionic crystals, viz. in salts of elements of the iron group and of rare earths. Ions of the iron group and of the rare earths have magnetic moments. These moments can orient themselves in an external magnetic field and theory can make rather precise predictions about how and to what extent this happens. Measurements on gadolinium sulphate were already made in the twenties by Kamerlingh Onnes and Woltjer. They clearly showed the existence of paramagnetic saturation. Such measurements were continued during the thirties, mainly by Wiersma and Gorter. Van Vleck's classic book appeared in 1932; it refers frequently to the Leiden measurements, and it was certainly a source of inspiration to later workers there. Out of this collaboration there arose a warm friendship between Van Vleck and many of the Dutch physicists.

On 28 September 1974, Van Vleck received the Lorentz Medal of the Royal Netherlands Academy of Arts and Sciences. I had the privilege of making the presentation, and in my speech I tried to express our admiration for his work and our gratitude for his friendship. Let me quote my final sentence:

Dear Van, this has been a very incomplete and very superficial summary, but how could one do justice in half an hour to half a century of work by a great physicist, and a hard working man at that? But I can sum up my summary in a very few words: I know you are proud of the Dutch ancestry of the Van Vlecks. So are we.

The study of paramagnetism had many interesting developments and sidelines. One of these was the investigation of the Faraday effect, that is the rotation of the plane of polarization by a substance in a magnetic field. Such work was done by Jean Becquerel (1878–1953) who for many years was a regular visitor to the Leiden Laboratories. He was, like his great-grandfather Antoine César (1788–1878), his grandfather Alexandre Edmond (1820–1891) and his father Antoine Henri (1852–1908) had been, associated with the Musée d'Histoire Naturelle in Paris, where he had a heavy teaching load and little opportunity for experimentation, but which provided him with a beautiful collection of mineral crystals containing rare earths. In some of them, especially in tysonite (a fluoride of a mixture of cerium, lanthanum, neodymium, and praseodymium), the Faraday effect is very large at low temperatures: at 1.7°K and in a field of 27,000 Gauss (2.7 Tesla), the rotation per millimeter for the green line of mercury is 1398°. That means that under those conditions even in the thinnest crystal investigated by Becquerel—0.675 mm—the polarization turned round more than two and a half times! Becquerel pursued his special line of investigation with great assiduity, found many interesting phenomena, and inspired several theoretical papers by H.A. Kramers. It seems to me that this work is not as well known as it deserves to be.

Becquerel was a lively little fellow, quick of speech and quick in his movements. On helium days he would always get one or two refills and while waiting for the cryostat to come back from the liquefier he would walk up and down the corridor explaining with eloquent gestures and in rapid French—notwithstanding his frequent visits, he never learned a word of Dutch—to anyone who would listen how things were going.

Van Heel puts a very diplomatic acknowledgment in the preface to his thesis (I translate from the French):

Monsieur Becquerel, I enjoyed the privilege of working under your inspiring guidance; so I had the opportunity to witness and to admire your skill in subtle experiments and your tenacity on helium days, days that were extremely exhausting but all the same most interesting.

The most vulnerable part of his equipment was the Dewar vessels. As usual at Leiden in those days there were three coaxial Dewar vessels: the inner one contained the liquid helium, the middle one liquid hydrogen, and the outer one liquid nitrogen. Since the crystal had to be put in a magnetic field between pole pieces of an electromagnet that were only 15 mm apart, the lower end of the Dewars, which were some 10 cm in diameter at the top, had to be very narrow: the interior diameter of the helium vessel was 5 mm, the outer diameter of the nitrogen vessel only 14 mm. To make such an array of Dewars was quite a feat, but the Leiden glass blowers were able to cope with the problem. Once a young Dutch physicist, whom De Haas had assigned the task of assisting Becquerel, wanted to check the helium level and used a flashlight for doing so. However, he had forgotten the magnetic field was still on, the flashlight was pulled from his hand, and the whole arrangement was smashed. A new set of Dewars was rapidly made and the assistant was replaced. Somewhat later fate caught up with him. He had built a complicated vacuum system—entirely of glass, as was usual in those days. A younger helpmate let a cathetometer topple over and it thoroughly broke up the whole installation. Again, the damage was soon repaired, but one wonders whether these two incidents did influence the subsequent career of the man in question: he later distinguished himself in mathematical physics; his name: H. Bremmer.

A new chapter in paramagnetism began in 1933: adiabatic demagnetization. With liquid helium boiling under reduced pressure one could not get much below 1°K (The vapor pressure there is about one tenth of a millimeter of mercury and it decreases rapidly with decreasing temperature: at 0.6°K it is about two one-thousandths of a millimeter of mercury.) Now in 1926 Debye and Giauque had suggested independently that adiabatic demagnetization of a paramagnetic salt might be a suitable way to obtain much lower temperatures. The idea was simple: magnetize a paramagnetic salt at a temperature of about 1°K. Heat is developed and carried away to the helium bath. Then break the thermal contact with the helium bath (that can easily be achieved if the salt is contained in a capsule suspended in a tube that can be evacuated and that is immersed in liquid helium.) The thermodynamic argument was incontestable, but neither Giauque nor Debye had liquid helium.[1] Giauque at Berkeley, California, started building low-temperature equipment, but I have been told that the economic depression—which hit California badly—retarded his progress, and, in any case, he was not inclined to use makeshift solutions. As he wrote himself,

> it was evident that the method rested on such a firm thermodynamic basis that all the equipment was designed not merely with the idea of producing low temperatures but to enable other investigations to be carried on in a considerable volume cooled to low temperatures.

It was 1933 before he could announce his first results. By that time De Haas, Wiersma, and—as a theoretical advisor—Kramers had also taken up Debye's suggestion. Of course they could use or adapt existing Leiden equipment. Their first announcement was almost simultaneous with that of Giauque. Soon afterwards in 1934 Simon and Kurti at Oxford published their first results on magnetic cooling. Today we have a more convenient method for obtaining temperatures well below 0.01°K (helium 3–helium 4 dilution refrigerators), but for several decades adiabatic demagnetization was the only method available for entering that new range of temperatures. For the very lowest temperatures it still holds sway and the idea can then also be applied to magnetization and demagnetization of nuclear spins.

My own experimental work dealt mainly with adiabatic demagnetization; I could take over the equipment that had been designed by Wiersma, which I left almost unchanged as far as the vacuum system was concerned, but I changed the measuring technique. The work I did together with some younger students was mainly concerned with the magnetic properties of the paramagnetic salts themselves. Only once did I apply the low temperature obtained to the study of properties of another substance imbedded in the salt.

Yet another aspect of paramagnetism was the study of paramagnetic relaxation: in an alternating field the magnetization cannot always keep step with the magnetic field. It lags behind in phase and its value remains below the value in a static field. This phenomenon was discovered and studied by Gorter and collaborators. At nitrogen temperatures, frequencies have to be of the order of megacycles before the time lag is noticeable. De Haas and Dupré found that at helium temperatures relaxation effects occur already at frequencies as low as 100 cycles and less. Together with Dupré I formulated a simple theory, valid at both temperatures. The main idea was that there exists a spin temperature, a temperature of the system of magnetic moments, that can be different from the temperature of the crystal lattice. This presupposes that the temperature equilibrium among the magnetic moments is established far more rapidly than the equilibrium between spin and lattice. This simple idea led to formulae that were in many cases in excellent agreement with experiment. The idea of a spin temperature turned out to be a fruitful one. It can even be applied to the temperature of a nuclear spin system at temperatures below 0.001°K.[2]

Let us now have a look at the work on metals. There was the work of Keesom and his co-workers on the specific heat. In the theory of Lorentz, based on the idea of free electrons following the laws of classical statistical mechanics, this presented a puzzle. If there were as many free electrons as atoms, the law of Dulong and Petit could not possibly hold: the specific heat should be much larger. Sommerfeld had shown that in the Fermi-Dirac statistics the specific heat is much smaller and proportional to the absolute temperature. We may write

$$C_v = \frac{3}{2} \, NkT/T_F$$

where the Fermi temperature T_F happens to be of the order of 10,000°K. It is true that this specific heat decreases with decreasing temperature, but as we have seen, the specific heat of the crystal lattice decreases even more rapidly. At very low temperatures the specific heat becomes of the form $C_v = AT + BT^3$, and it so happens that for many metals the two terms become equal at a helium temperature. Finding the linear term was a very direct confirmation of the Fermi-Dirac statistics. Moreover a fairly good quantitative agreement with the calculations of Sommerfeld and later authors was obtained.

As for the electric resistance, the theory of Bloch led to the following conclusions. At high temperatures (that is at temperatures where the specific heat of the lattice assumes the Dulong and Petit value) the resistance should be proportional to T. At absolute zero it should be zero for a perfect crystal, a crystal without chemical impurities and without any flaws in its structure ("physical impurities"). In the range where the T^3 law holds for the specific heat, the resistance should be proportional to T^5. For an imperfect crystal there should be a residual resistance R_0 and the total resistance should be of the form $R = R_0 + \alpha T^5$. Broadly speaking this was in agreement with experiment. It is said that Pauli was disgusted by the idea of making this residual resistance the subject of a theoretical investigation. "It is a muck-effect" he said, "and one should not wallow in muck."

De Haas was a man who was always happy when experiments did not agree with theory. So it must have given him great pleasure when G.J. van den Berg found that the electric resistance of gold decreased with decreasing temperature to what at first appeared to be a constant value but increased again when the temperature was lowered still further: the resistance passed through a minimum. Van den Berg and I even measured the resistance of a gold wire at temperatures below 1°K obtained by adiabatic demagnatization and found that it continued to rise. This effect, which was later studied in great detail by Van den Berg and co-workers, disagreed with very general and clearcut predictions of theory, and it would be several decades before Kondo was able to show by what refinements of the theory of electrons in metals it could be explained. Today the phenomenon is generally referred to as the Kondo effect. There is not much to be said against this, as long as one remembers that it was not Kondo who discovered this effect: he explained it.

Around 1930 the Russian physicist Shubnikov, together with his charming wife A. Trapeznikova, spent quite some time at Leiden and did remarkable work on the magneto-resistance of bismuth. It was known since the end of the nineteenth century that the electrical resistance of bismuth increases considerably in a magnetic field and it was also known that this effect becomes more pronounced at low temperatures.*

Du Bois and Wills for instance found in 1899 that at 93°K the resistance of a bismuth wire increased by a factor 230 in a magnetic field of 37,500 Gauss (3.75 Tesla), whereas this factor is less than 2 at room temperature. It was therefore perhaps to be expected that the effect would be even larger at hydrogen temperatures, but I do not think anybody had expected an increase by a factor of several hundred thousand in a field of 30,000 Gauss at a temperature of 14°K. Such spectacular results were obtained with extremely pure single crystals: Shubnikov started from the purest metal he could get and recrystallized it numerous times.

Post–World War II semiconductor technology has familiarized us with the fact that minute quantities of "dope" may drastically modify physical properties, and the art of growing pure crystals has reached a high level of perfection. In a way, Shubnikov's results anticipated these developments. Incidentally, we may describe the properties of bismuth by saying that it is a semiconductor with a negative bandgap.

Shubnikov went back to Russia, but during the Stalin purges preceding the Second World War he disappeared; in his case there was no Kapitza to save him.

As early as 1914 De Haas had pointed out that there might be a correlation between diamagnetism and change of resistance in a magnetic field, so it is not surprising that after the spectacular results of Shubnikov he suggested to one of his students, P.M. van Alphen,† that he should investigate the diamagnetism of bismuth at low temperatures. That hunch yielded rich rewards. It was found that the diamagnetic susceptibility of bismuth at low temperatures is not constant but shows periodic variations as a function of the magnetic field. That is the famous De Haas–Van Alphen effect that later in the hands of Shoenberg and others became an important tool for finding details of the states of electrons in metals.

The story of liquid helium itself shows both the virtues and the drawbacks of the "door meten tot weten" doctrine prevailing at Leiden. Systematic quantitative measurements did yield important results, but a number of very surprising properties that can be discovered by simple qualitative observations went unnoticed or received little attention because they did not fit into

---

*This property of bismuth was often used for measuring magnetic fields, especially when the field had to be measured in a small and rather inaccessible region of space. In a 1901 edition of Kohlrausch, *Lehrbuch der Praktische Physik*, the bible of experimental physicists of those days, this method is described in some detail. In the twenties it was used by Gerlach for measuring the fields in the Stern-Gerlach experiment and in the thirties at Leiden for measuring magnetic fields inside and in the neighborhood of superconductors.

†Pieter M. van Alphen got his doctor's degree on 6 July 1933. Later he joined the Philips Laboratory, where he worked mainly on applied optics.

the program of measurements. It was in this area that the newer low-temperature groups, at Oxford, Cambridge, and Moscow, were able to take the lead, although Leiden remained the incontestable leader so far as the diagram of the state of helium was concerned.

Let us first look at the various states of helium. Whereas other liquefied gases become solid under their own vapor pressure at some temperature below their boiling point, this is not the case for helium. In 1926 Keesom found that helium does solidify under pressure. At 4°K this happens at 130 atmospheres, at 1°K at 25 atmospheres, and at absolute zero the melting pressure will be not much lower. Liquid helium occurs in two different phases known as helium I and II. The transition occurs at 2.19°K when the helium is under its own vapor pressure (39 mm of mercury) and at 1.78°K at 30 atmospheres, the melting pressure at that temperature. All these facts were ascertained by Keesom and his daughter. They also carefully measured densities and specific heats. By 1936 the thermodynamic properties of liquid helium I and II were rather accurately known. But it was not yet clear that helium II is a most peculiar liquid. However, it was noticed during these caloric measurements that temperature equilibrium was established more rapidly in helium II than in helium I. It was only then that special measurements were made to determine the heat conductivity: in 1936 Keesom and Miss A.P. Keesom concluded a preliminary communication with the sentences:

> It appears that the heat conductivity of liquid helium II at the temperatures mentioned is about 3×10^6 times that of liquid helium I. Connecting this with the abrupt change of the heat conductivity when passing the lambda-point [the name given to the transition point because of the shape of the specific heat curve] we may perhaps be justified in calling liquid helium II supra-heat-conducting.

Keesom and his daughter were just in time: from 1936 onwards one remarkable effect after the other was discovered. Rollin at Oxford established the existence of rapidly moving surface films (which also explained some observations made already in 1922 by Kamerlingh Onnes but not pursued further). Superflow—helium II flows almost unobstructed through narrow capillaries or between optically polished plates—was reported by Allen and Misener at Cambridge and by Kapitza, who, moreover, made the very important discovery that the helium that flows through is lower in temperature; the so-called fountain effect discovered by Allen and Misener is in a way the reverse of this. The upshot of all this is that helium II does not behave like any normal liquid; it does not follow the normal macroscopic equations of hydrodynamics. It is often referred to as a quantum fluid. Today we can also use (small quantities of) the helium isotope with atomic mass 3 and mixtures of helium 4 (the normal helium) and helium 3, so the range of "quantum fluids" has been somewhat extended. Experiments of ever greater sophistication are still going on and on the other hand understanding these curious phenomena has been one of the major challenges for statistical mechanics. On the whole, I think theory acquitted itself fairly well of its task, although I wonder whether it would have predicted these phenomena if they had not been first discovered by experimenters. But there I am rather outside the limits of my field of competence.

There exists a profound analogy between superfluidity and superconductivity, both from an experimental and from a theoretical point of view. However, the development took place in the reverse order. The first observation, published by Kamerlingh Onnes in 1911, was that the electric resistance of mercury disappears below 4°K. Soon other superconductors were found, tin and lead among them. The next discovery was at the same time a disappointment: superconductivity disappears in fairly low magnetic fields. These "threshold fields" depend on temperature: they are zero at the transition point and increase with decreasing temperature; for the metals mentioned they remain below 1,000 Gauss when the temperature approaches zero. That meant that to begin with, Kamerlingh Onnes's hopes of producing high magnetic fields with superconducting coils came to nothing.*

Soon one began to look for other changes of properties that might be associated with superconductivity and one did not find anything. There was no latent heat at the transition point, no change of crystal structure, no discontinuity in thermal conductivity.

Whereas the transport phenomena in helium II are definitely unclassical, the electromagnetic behavior of superconductors seemed to obey the normal laws of electromagnetic theory, albeit with the remarkable stipulation that the conductivity is infinite. As a matter of fact, long before the discovery of superconductivity, Maxwell in his famous treaty on electricity and magnetism had already worked out in which way perfect conductors would react when a magnetic field is applied. One very simple case is that of persisting currents in a ring: cool the ring in a magnetic field, then switch off the field; because of the infinite conductivity the magnetic flux cannot change. A current is induced so that this flux remains constant and that current keeps circulating ad infinitum, or rather, as long as the liquid helium lasts. My wife and I, in youthful enthusiasm, once observed a whole night long, by means of a sensitive magnetometer,† the current circulating in a circuit consisting

*In the thirties some alloys with considerably higher threshold fields were discovered and some of their properties were studied. Yet about half a century passed between Kamerlingh Onnes's discovery and the first practical applications of superconducting magnets.

†Although the apparatus was, in principle, astatic it did react somewhat to magnetic stray fields, hence the nightwork; and there was no cavalry and in any case no suitable park surrounding the laboratory at Leiden.

of a several-meters-long bifilar section and one circular loop. We did not find any change in the current, and this was already a quite sensitive resistance measurement. More recently such measurements have been repeated with much greater precision, with identical results. I like to say that the resistance of superconductors is about the zeroest quantity we know.

Further progress was retarded by a curious erroneous idea, by the myth of the frozen-in field. It was taken for granted that when a sphere was cooled in a magnetic field and became superconductive the magnetic field distribution inside the sphere would not change. If then the magnetic field was switched off, that internal field should still be unchanged, and the sphere should behave as a magnetic dipole. In other words, one applied what was true for a material ring to any closed ring that could be drawn inside the sphere. The experiment was tried with a hollow sphere and indeed it became a magnet. But here Kamerlingh Onnes's advice was not followed: had a precise measurement of the remaining magnetic moment been carried out, one would almost certainly have found that it was smaller than the supposed value. Then one might have repeated the experiment for a number of spheres and one might have arrived at the conclusion that in the ideal case of pure metals and perfectly shaped solid spheres no frozen-in field exists. And one might have arrived at the conclusion that the superconducting state is characterized not only by infinite conductivity but also by absence of an internal magnetic field.

As things were, this solution was approached only step by step. Keesom and co-workers found that though there is no latent heat at the transition point there is a sudden jump in the specific heat. De Haas and Bremmer found that though there is no discontinuity in the heat conductivity at the transition point there is a pronounced change in heat conductivity below the transition point when a magnetic field above the threshold field is switched on.*

De Haas, Voogd, and my wife performed a number of experiments on threshold fields with monocrystalline tin wires and found that contrary to the frozen-in field idea it did not make any difference whether one cooled first and then switched on a field or the other way round. Rutgers applied in an intuitive way a formula Ehrenfest had derived for transitions of the second kind and found a relation between the slope of the threshold curve and the discontinuity in the specific heat that was in excellent agreement with the experimental data, and Gorter worked out this idea in greater detail. By that time—it was 1933 and I had recently returned to Leiden from Zürich—I felt

---

*At temperatures well below 1°K the thermal resistance in the superconducting state becomes very large, whereas it remains small in the normal state. Lead wires can therefore be used as convenient thermal switches in experiments on ultra-low temperatures obtained by demagnetization of nuclear spins.

inclined to doubt the idea of the frozen-in field. My wife was going to see whether the magnetic-field distribution remained really constant when a cylinder of tin was cooled in a magnetic field. But before this was done Meissner and Ochsenfeld published their result: in the superconductive state B is always zero. What could have happened much earlier happened now. The whole picture changed: the superconducting state was a particular phase of the electrons in the metal characterized not only by infinite conductivity but also by the absence of an internal magnetic field. Gorter and I then formulated the thermodynamic theory I mentioned before. I should like to add that the thermodynamics of matter in a magnetic field, though elementary, is not without pitfalls. Even the remarkably accurate textbook of thermodynamics of Guggenheim contained erroneous formulae in its first two editions.[3] They were corrected—at my instigation—in the third and later editions. The great Heisenberg once thought he could derive the existence of the Meissner effect from thermodynamics because he confused free energy at constant field and free energy at constant magnetic moment. In short, this is a field where it pays to be a bit of a pedantic schoolmaster.

And then Keesom got into the action. Together with Van Laer he made accurate measurements on the specific heat and latent heat with and without magnetic field. Their results confirmed the thermodynamic theory. This was not entirely trivial: it was not certain that the transitions would be reversible. They were, at least to a very high degree of approximation.

The problem of finding a theoretical explanation of superconductivity assumed a different aspect. One had so far been looking for mechanisms that might lead to infinite conductivity. Now it became clear that one had first of all to look at a phase transition of the Fermi-Dirac gas. But phase transitions cannot be explained in terms of a model with independent particles. Kronig once introduced the idea of an electron crystal but was unable to get any further. Heisenberg tried to make the Coulomb repulsion responsible for superconductivity; that did not work at all. The famous maxim of Felix Bloch that the common feature of all theories of superconductivity was that they could be refuted remained valid until well after World War II when Bardeen, Cooper, and Schrieffer formulated their now famous theory.

It is interesting to compare the main lines of thought of this theory with the development of experimental findings.

As a first step one has to get rid of the Coulomb repulsion, that is, one has to understand why a picture with non-interacting Fermi-Dirac particles gives such surprisingly good results. Next one has to show that because of interaction with the lattice vibrations these particles, which are really pseudoparticles, have even a small mutual attraction. This leads to a ground state fundamentally different from a filled Fermi surface: electrons with opposite momentum combine in so-called Cooper pairs. This particular ground state can be shown to be perfectly diamagnetic. And finally one has to understand infinite conductivity. That is more difficult because a ring with a circulating persistent current is not in a state of true equilibrium: a state with zero current has lower free energy. Why then do transitions to that state not occur? The reason is to be found in flux quantization: the magnetic flux passing through a closed loop has to be an integer multiple of a fundamental flux quantum, $hc/2e$. The existence of such a flux quantum is tied to the existence of a macroscopically coherent many-particle wave function.* Now one flux quantum corresponds to some 10^4 electrons circulating. If we want to change the flux by one quantum we have either to have a simultaneous many electron transition, which is unlikely, or we have to pass through states where flux is not quantized, which means through states of higher free energy, which is very improbable too.

In any case: the theoretical order of things is:

1. New fundamental state.
2. Perfect diamagnetism.
3. Infinite conductivity.

That is just the reverse of the story of experimental discovery, but roughly what happened in the case of liquid helium.

These theoretical developments happened much later. During the thirties there was one notable step forward: the theory of London, which was halfway between a thermodynamic description and a true atomistic theory, and which has had a strong influence both on subsequent experiments and on the development of theory.

In a way, this step forward began with a paper of Becker, Sauter, and Heller[4] published before the discovery of the Meissner effect, in which they extended Maxwell's calculations for infinitely good conductors by taking into account the inertia of electrons;† incidentally, during a colloquium at Zürich, Pauli had already asked, why does one not write $di/dt = C\cdot E$? This resulted in a finite penetration depth depending on the number of electrons.

Now the London equations are essentially the same: London assumes that the state of a superconductor in a field is always the same as that described by the Becker-Sauter-Heller equations when one starts from field zero, but his "philosophy" is different.

It already contains, though in embryonic form, the notion of a macroscopically coherent wave function that is "not changed much" by an applied external magnetic field. Bardeen once told me that that idea had had a

---

*This macroscopic wave function appears in an even more spectacular way in the Josephson effect.

†The idea of doing this for materials with *finite* conductivity occurs already in H.A. Lorentz's doctoral thesis and is referred to by Maxwell.

profound influence on his theoretical approach. For the time being London's work encouraged research aiming at measuring the penetration depth. The work of Gorter and myself suggested that the number of superconducting electrons would increase with decreasing temperature and I devised a method for measuring the change of penetration depth with temperature. The method worked extremely well—in the hands of others. I am sorry to say that I bungled the experiment. That was in 1940, and it was almost the end of my activities as an experimenter. The Germans entered the country and in the spring of 1942 I left Leiden.

Bibliographical Notes

1. Family Background and Schooldays

1. For the early history of the Society of the Common Good, see Simon Schama, *Patriots and Liberators* (New York: Knopf, 1977), pp. 533–536.
2. G.L. de Haas-Lorentz, ed., *H.A. Lorentz, Impressions of His Life and Work* (Amsterdam: North Holland Publishing Co., 1957). Unfortunately, there are a number of rather serious slips in some of the translations.
3. Ibid., pp. 150–152.
4. The text is taken from A. Einstein, *Ideas and Opinions* (New York: Dell, Laurel edition, 3rd printing, 1978). This volume also contains the essays "H.A. Lorentz's Work in the Cause of International Cooperation" and "H.A. Lorentz, Creator and Personality." The original German versions can be found in *Mein Weltbild*, (Zürich: Europa Verlag, 1953).

2. Development of Physics

1. Cf. Maxwell, *Treatise*, 3rd ed., p. 381.
2. *Versuch einer Theorie der elektrischen und optischen Erscheinungen in bewegten Körpen.* In *Collected Papers* V, 1 (Leiden, 1895).
3. I follow Oesper's translation (New York: Academic Press, 1950), pp. 120–121.
4. Cf. Otto Frisch, *What Little I Remember* (Cambridge: Cambridge University Press, 1979), p. 45.
5. M. Born and P. Jordan, *Elementare Quantenmechanik* (Berlin: Julius Springer, 1930). W. Pauli, *Naturwiss.* 18 (1930): 602.
6. R. Courant and D. Hilbert, *Methoden der Mathematischen Physik* (Berlin: Julius Springer, 1924).

7. P. Jordan and R. de L. Kronig, *Movements of the Lower Jaw of Cattle During Mastication, Nature* 120 (1927): 807.
8. Tennyson, "Morte d' Arthur" from *Poems* (1842).

3. Early Years at Leiden

1. J.H. van Vleck, *The Theory of Electric and Magnetic Susceptibilities* (Oxford: Oxford University Press, 1932).
2. Martin Klein, *Paul Ehrenfest*, Vol. I (Amsterdam: North Holland Publishing Co., 1970).
3. M. Klein, ed., *P. Ehrenfest. Collected Scientific Papers* (Amsterdam: North Holland Publishing Co., 1959).
4. Ibid., p. 185.
5. P. Ehrenfest, *Golfmechanika* (The Hague: Van Stockum, 1932). The comb experiment is described in Appendix A of that booklet.
6. Henderson's literal translation, from Harold G. Henderson, *An Introduction to Haiku* (Garden City, N.Y.: Doubleday, Anchor Books, 1958).
7. Cf., for instance, *The Physicist's Conception of Nature*, edited by J. Mehra. (Dordrecht: Reidel, 1973), pp. 815–819.
8. Ibid., p. 819.
9. On 28 May 1927; see H.A. Lorentz, *Collected Papers* IX (The Hague: Martinus Nyhoff, 1938), 343.
10. G.L. de Haas-Lorentz, op. cit., p. 104.
11. Walter M. Elsasser, *Memoirs of a Physicist in the Atomic Age* (New York: Science History Publications, 1978; Bristol: Adam Hilger, 1978).
12. Ibid., p. 85.
13. Wolfgang Pauli, *Wissenschaftlicher Briefwechsel* Vol. 1 (New York: Springer, 1979), p. 474.
14. A convenient reference is J. Mehra, *The Solvay Conferences on Physics* (Dordrecht: Reidel, 1975).
15. Pauli, op. cit., p. 307.
16. Cf. Peter Robertson, *The Early Years* (Copenhagen: Akademisk Forlag, 1979), p. 138.
17. *Collected Papers*, pp. 617–622.
18. P. Ehrenfest, *Z. Physik* 78 (1932): 555–559.
19. W. Pauli, *Z. Physik* 80 (1933): 573–586.
20. W. Pauli, *Naturwiss.* 21: 841–843.

4. Copenhagen

1. For details on Niels Bohr's life and work a valuable source is *Niels Bohr*, edited by S. Rozental (Amsterdam: North Holland Publishing Co., 1967). (The Danish edition of the same book appeared in 1964, published by J.H. Schultz Forlag, Copenhagen.) It will be quoted as *Life & Work* in the following. I have freely borrowed from my own contribution to that volume, essentially the text of a talk given at a commemorative meeting

in 1963. Bohr's collected works are still in the course of publication. The first volume contains an important introduction by L. Rosenfeld. (The same author wrote the article for the *DSB: Dictionary of Scientific Biography,* 14 vols. and a suppl. [New York: Scribners, 1970–1978.])

2. Institute for Theoretical Physics, The Niels Bohr Institute 1921–1971 (Copenhagen: Rhodos, International Science and Art Publisher). See also *Selected Papers of Léon Rosenfeld* (Dordrecht: Reidel, 1977).
3. Cf. Frisch, op. cit., pp. 48–49.
4. Ibid., p. 85.
5. *Sonatorrek* in the original Icelandic. An English translation (by Palsson and Edwards) can be found in the Penguin Classics.
6. *Life & Work,* p. 307.
7. Ibid, p. 24.
8. Freeman Dyson, *Disturbing the Universe* (New York: Harper & Row, 1979), Ch. 15.
9. Ibid., p. 172.
10. See R. Graves, *The Greek Myths* (Harmondsworth, Eng.: Penguin Books, 1955).
11. Alexander Dorozynski, *The Man They Wouldn't Let Die* (London: Secker & Warburg, 1966).
12. Anna Livanova, *Landau: A Great Physicist and Teacher* (London: Pergamon, 1980). I reviewed this book in *Nature* 288 (6 November 1980): 31.
13. See Dorozynski, op. cit., p. 62.
14. For instance, in Robertson, op. cit.
15. Dorozynski, op. cit., p. 54.
16. G. Gamow, *Constitution of Atomic Nuclei and Radioactivity* (Oxford: Oxford University Press, 1931). It is the first text in which nuclear physics is treated from a theoretical point of view.
17. *Life & Work,* p. 111.
18. G. Gamow, *Thirty Years That Shook Physics.* (Garden City, N.Y.: Doubleday, 1966), contains a detailed account.
19. *Scientific American,* 194 (March 1956): 96; *IEEE Student Journal,* September 1963, p. 36.
20. Frisch, op. cit., p. 85.

5. *Berlin, Zürich, and Back to Leiden*

1. A short biography of C.J. Gorter is given in Chapter 6.
2. For the history of nuclear physics during that period compare *Nuclear Physics in Retrospect,* edited by R.H. Stuewer (Minneapolis: University of Minnesota Press, 1979).
3. Fowler's *Modern English Usage,* 2nd ed. (Oxford: Oxford University Press, 1968).
4. K. Mendelssohn, *The World of Walter Nernst* (Pittsburgh: University of Pittsburgh Press, 1973).

5. H.B.G. Casimir, *"Walter Nernst und die Quantentheorie der Materie,"* *Berichte der Bunsengesellschaft* 68 (1964): 530.
6. Rainer Maria Rilke, from the *Buch der Bilder:* "Zum Einschlafen zu sagen."
7. For his role in "Aryan Physics" compare Alan D. Beyerchen, *Scientists Under Hitler* (New Haven: Yale University Press, 1977).
8. See *Phil. Trans. Royal Society London,* A 292 (1979): 137.
9. Elsasser, op. cit., p. 161.
10. C.G. Jung, W. Pauli: *Naturerklärung und Psyche* (Zürich: Rascher, 1952).
11. The underlying mathematics is described in a paper I published in 1960, the year of Pauli's sixtieth birthday: *Helvetica Physica Acta* 33, (1960): 849–854.
12. Cf., for instance, Broda, *Ludwig Boltzmann* (Vienna: Deuticke, 1955).
13. A. Einstein, *Out of My Later Years* (Secaucus, N.J.: Citadel, paperback, 1977).
14. Lorentz, *Collected Papers* IX, p. 181.
15. Cf. Heisenberg, *Der Teil und das Ganze,* p. 122.
16. Cf. my article in the *Nederlands Tijdschrift voor Natuurkunde* and also my article in *DSB* for a closer analysis.

6. *Low Temperatures*

1. Cf. R.V. Jones, *Most Secret War* (London: Hamish Hamilton, 1978) for a well-documented account of this role.
2. H.L. Curtis, *Electrical Measurements* (New York: McGraw-Hill, 1937).
3. Cf. Jagdish Mehra, op. cit. p. 100.
4. C.J. Gorter, *Paramagnetic Relaxation* (Amsterdam: Elsevier Publishing Co., 1947).
5. The relevant correspondence can be found in P.L. Kapitza, *Experiment, Theory, Practice.* (Dordrecht: Reidel, 1980), (Boston Studies in the Philosophy of Science, Vol. 46).

7. *War Times*

1. Beyerchen, op. cit.
2. The text is reproduced in full in *Berichte und Mitteilungen der Max-Planck-Gesellschaft,* 1979.
3. Cf., for instance, Elsasser, op. cit., p. 161.
4. Laura Fermi, *Illustrious Immigrants* (Chicago: University of Chicago Press, 1968).
5. Beyerchen, op. cit., pp. 15–39.
6. Cf. *Einstein/Sommerfeld Briefwechsel,* edited by Armin Hermann (Basel/Stuttgart: Schwabe & Co., 1968), pp. 114–115; *DSB* XII, 530.
7. Cf. Jones, op. cit.
8. M. Planck, *Ann d Phys.* 19 (1934): 759. Cf. also *Vorlesungen über Thermodynamik,* 11th ed. (Berlin: De Gruyter & Co., 1964).

9. P. Ehrenfest, *Z. phys. Chemie* 77 (1911): 227–244. Also *P. Ehrenfest. Collected Scientific Papers* 167.
10. F.J. Philips, *45 Jaar met Philips* (Rotterdam: Donker, 1976).
11. Armin Hermann, *Max Planck* (Reinbek: Rohwolt, 1973).

8. *Holst and the Philips Research Laboratory*

1. Robert S. Cohen in *DSB*.
2. For the early history of the Natuurkundig Laboratorium I have been able to draw upon an unpublished account written by A. U. Garrett, assisted by a committee consisting of former members of the research staff.
3. Horace, *Epistles* I, 1, line 19: "et mihi res, non me rebus, subiungere conor."
4. Lewis Carroll, *Through the Looking-Glass*, Ch. 2.
5. Elisabeth Hawes, *Fashion Is Spinach* (New York: Grosset & Dunlap, 1940).

9. *Industry and Science After the Second World War*

1. Cf. also *An Anthology of Philips Research*, edited by H.B.G. Casimir and S. Gradstein (Eindhoven: N. V. Philips' Gloeilampenfabrieken, on the occasion of their 75th anniversary, 1966).
2. Blaise Pascal, *Pensées: La place de l'homme dans la nature.*
3. Cf. Steven Weinberg, *The First Three Minutes* (London: Fontana/Collins, paperback, 1978).
4. Dyson's book gives an inside view of that development.
5. B.D. Josephson in *Quantumfluids*, edited by D.F. Brewer (Amsterdam: North Holland Publishing Co., 1966), p. 174.
6. *Notes and Records of the Royal Society* 25 (1969): p. 4.
7. E. Espe and M. Knoll, *Werkstoffkunde der Hochvakuumtechnik* (Berlin: Julius Springer, 1936).
8. H.J. Lipkin, *Lie Groups for Pedestrians* (Amsterdam: North Holland Publishing Co., 1965).
9. Translation by W.H.D. Rouse in the Loeb edition.
10. *The Apology and Treatise of Ambroise Paré*, edited by G. Keynes (Chicago: University of Chicago Press, 1952) p. 102.
11. A useful summary is J. Smit and H.P.J Wijn, *Ferrites* (Eindhoven: Philips Technical Library, 1959).
12. A. Bouwers, *Elektrische Höchstspannungen*, (Berlin: Springer, 1939).

10. *The Science-Technology Spiral*

1. H. Marcuse, *An Essay on Liberation*, 1969, Penguin edition, p. 90.
2. I. Illich, *Energy and Equity* (New York: Harper & Row, Perennial edition, 1974), p. 8.
3. It is reproduced in Niels Bohr, *Life & Work.*

4. Cf. Jones, op. cit., p. 477.
5. C.P. Snow, *Variety of Men*, "Einstein" (Harmondsworth, Eng.: Penguin Books, 1969).

Appendix A: *The Discussions Between Bohr and Einstein on the Interpretation of Quantum Mechanics*

1. For a thorough study of this topic, I refer to A. Pais, "Einstein and the Quantum Theory," *Rev. Mod. Physics*, vol. 51, no. 4 (October 1979) pp. 863–914, especially p. 899.
2. Ibid., p. 902, Fig. 4.

Appendix B: *Physics Around 1930*

1. W. Pauli, *Naturwissensch.* 12 (1924): 741. It should be noted that Pauli does not consider the possibility of a magnetic moment of the nucleus. He estimates the order of magnitude of the interaction by assuming an electric quadrupole-moment.
2. See also Rosenfeld, op. cit.
3. "Rotation of a rigid body in quantum mechanics." Thesis, Leiden, 1931. *Proc. Roy. Ac. Amsterdam*, 34 (1931): 144.
4. Theodora Benson and Betty Asquith, *How to Be Famous: or, The Great in a Nutshell* (London: Gollancz, 1937).
5. "Studier om Metallernes Elektronteori." Thesis, Copenhagen, 1911. The first volume of his collected works contains an English translation.
6. N.F. Mott, *Proc. Royal Soc., London* (A) 124 (1929): 425.
7. His theory is presented in its final form in the posthumously published *Fundamental Theory* (Cambridge: Cambridge University Press, 1946).
8. W. Pauli, *Aufsätze und Vorträge über Physik und Erkenntnistheorie* (Braunschweig: Vieweg, 1961). The original letter is also reproduced there.
9. *History of Twentieth Century Physics*, Proc. International School of Physics "Enrico Fermi" Course LVII. (New York: Academic Press, 1977), p. 184.

Appendix C: *Low-temperature Physics in the Thirties*

1. For a description of early work on adiabatic demagnetization, compare my monograph *Magnetism and Very Low Temperatures* (Cambridge, 1940; also Dover publications, 1961). For a more recent survey, R. P. Hudson, *Principles and Application of Magnetic Cooling* (Amsterdam: North Holland Publishing Co., 1972).
2. For an account of early work on paramagnetic relaxation, compare C. J. Gorter, op. cit.
3. E.A. Guggenheim, *Thermodynamics* (Amsterdam: North Holland Publishing Co.).
4. R. Becker, F. Sauter, and G. Heller, Z. *Physik* 85 (1933): 772.

Index